Signals and Communication Technology

This series is devoted to fundamentals and applications of modern methods of signal processing and cutting-edge communication technologies. The main topics are information and signal theory, acoustical signal processing, image processing and multimedia systems, mobile and wireless communications, and computer and communication networks. Volumes in the series address researchers in academia and industrial R&D departments. The series is application-oriented. The level of presentation of each individual volume, however, depends on the subject and can range from practical to scientific.

Indexing: All books in "Signals and Communication Technology" are indexed by Scopus and zbMATH

For general information about this book series, comments or suggestions, please contact Mary James at mary.james@springer.com or Ramesh Nath Premnath at ramesh.premnath@springer.com.

More information about this series at http://www.springer.com/series/4748

Joachim Speidel

Introduction to Digital Communications

Second Edition

 Springer

Joachim Speidel
Institute of Telecommunications
University of Stuttgart
Stuttgart, Baden-Württemberg, Germany

ISSN 1860-4862   ISSN 1860-4870 (electronic)
Signals and Communication Technology
ISBN 978-3-030-67359-8   ISBN 978-3-030-67357-4 (eBook)
https://doi.org/10.1007/978-3-030-67357-4

This Springer imprint is published by the registered company Springer Nature Switzerland AG
The registered company address is: Gewerbestrasse 11, 6330 Cham, Switzerland

Preface

Digital communication has found an increasing interest in the past 70 years starting with the telephone network on copper wires, the development of the optical transmission, and the emerging Internet based on wire-line and wireless transmission technologies. Today, the trend to serve an increasing number of mobile users and also machines with information through digital networks is unbroken.

The new book *Introduction to Digital Communications* is aiming at graduate students, scientists, and engineers, who are interested in getting an introduction to modern digital communications. The main focus is on the fundamentals of the physical layer from the perspective of the theory of linear time-invariant as well as time-variant systems. The book draws a bow from single input single output to multiple input multiple output systems with an emphasis on wireless transmission over time-variant channels. The main concern lies on an accurate mathematical description, wherein the findings and lemmas are proven in detail. Various chapters are enriched by numerical examples and also illustrated with results from computer simulations provided by the open platform "webdemo" of the Institute of Telecommunications at the University of Stuttgart, http://www.inue.uni-stuttgart.de.

Organization of the Book

The book covers three main parts and a fourth part with two Appendices.

Part I

Deals with the principles of digital transmission, which are important for wire-line as well as wireless communications. It describes the main building blocks for Single Input Single Output (SISO) systems. The concept of quadrature amplitude modulation is introduced. An important part is the design of the overall system for minimal intersymbol interference with Nyquist's first criterion. The introduction of the equivalent baseband system allows the concise definition of the link between the transmitter input and the receiver output as a "black box" without details of the modulation, the spectral signal shaping, and the channel. For the receive signal, several detection methods are described in detail, such as threshold decision, maximum likelihood, and maximum a posterior detection. Also the difference between symbol-by-symbol and sequence detection is addressed and the maximum likelihood sequence estimator is described as an example. With an adequate model of the noise at the receiver, the symbol error probability is calculated.

The following chapters in Part I are devoted to the wireless transmission. The main difference is the wireless channel, which changes its characteristic with time. Therefore, the theory of linear time-variant systems is introduced to describe the building blocks of the system with time-variant impulse responses and delay spread functions. As not all students and engineers are frequently involved with this topic, the book contains an own Part II devoted to the theory of linear time-variant systems. Selected points are briefly reported for Part I, hence the reader is not required to study Part II beforehand. However, for a deeper understanding, the reader should get involved in Part II. The introduction of the equivalent baseband system, which is then time-variant, follows. With this model the increase of the output signal bandwidth at the receiver compared to the transmit signal is shown as an example. The multipath channel model is described in detail. As the wireless transmission link is multifaceted, a statistical characterization of the channel is helpful. To this end, various channel models are reviewed, such as the Rayleigh and Nakagami-m fading as well as the model according to Clarke and Jakes.

Part II

Is devoted to the theory of linear time-variant systems. In many cases, this topic is just touched upon during the education of graduate students in Electrical Engineering and Computer Science. Therefore, this dedicated Part II is provided. The input-output relation given by the time-variant convolution is addressed in detail and the mathematical properties are derived. We outline the relation with the well-known (time-invariant) convolution used by engineers in most applications. The time-variant impulse response and the delay spread function turn out to be the proper system descriptions in the time domain. Also the system functions in the frequency domain are presented, such as the time-variant transfer function and the Doppler spread function. For the statistical description of randomly changing time-variant systems autocorrelation functions as well as power spectral densities of the system functions are studied.

Part III

Deals with Multiple Input Multiple Output (MIMO) systems. First, the input-output relation is derived using matrix notation. We discuss the principle MIMO channel models, such as the time-variant finite impulse response and the i.i.d. Gaussian model. Furthermore, spatial correlations at the transmitter and the receiver are incorporated leading to the Kronecker model. Linear and nonlinear MIMO receivers are investigated in detail, such as the zero-forcing, the minimum mean squared error, and the maximum likelihood receiver. An important question is how many bits per channel use can be transmitted over MIMO channels. This issue is studied together with the maximization of the channel capacity. Next, the principles of spatial prefiltering and space-time encoding are investigated to improve transmission quality and to increase the data rate. In the last chapter, we leave the single-user transmission and consider the MIMO principle for a multitude of users in a network. Various multi-user MIMO schemes for the uplink and downlink are discussed, which can reduce the interference when the users transmit their signals in the same time slots and frequency bands.

Appendix

In Appendix A a summary on the characterization of random variables and stochastic processes is given.

Appendix B provides an overview on the most important lemmas of linear algebra required for the understanding of some topics of this book.

Second Edition of the Book

In Part I of the Second Edition, a new chapter deals with block-wise digital signal transmission over channels with finite impulse response (FIR) and the corresponding matrix descriptions are derived. As an alternative to Nyquist impulses for reduction of intersymbol interference, block-wise transmission with and without cyclic prefixes is studied. An introduction of the Discrete Fourier Transform (DFT) enables equalization in the DFT domain. On top of that the Second Edition is enhanced by a new chapter on Multicarrier Modulation and Orthogonal Frequency Division Multiplexing (OFDM). Part II contains more examples and diagrams on time-variant systems. In several other parts of the book an increased number of examples, tables, graphs, and figures illustrates the material. Finally, a nomenclature list is provided and extended by a summary of formulas, transforms, and important definitions used throughout this book.

Acknowledgments

The author is indebted to Prof. Stephan ten Brink for providing the facilities of the Institute of Telecommunications at the University of Stuttgart and to Mrs. Agnes Schoen-Abiry for drawing a lot of figures.

Stuttgart, Germany

Prof. Dr.-Ing. Joachim Speidel

Actual notes can be found at: https://www.inue.uni-stuttgart.de/institute/team/Speidel-00003/

Contents

About the Author

Joachim Speidel earned his Dipl.-Ing. (M.Sc.) degree in Electrical Engineering with a major in Communications Engineering from the University of Stuttgart in 1974 followed by a Dr.-Ing. (Ph.D.) degree in 1980 in the field of signal processing and system theory. From 1980-1992, he worked in various R&D positions for Philips in the broad field of digital video and data transmission. Among others, he and his team contributed significantly to algorithms and standards for video data compression and transmission. In 1992, he became a Professor at the Faculty of Computer Science, Electrical Engineering and Information Technology at the University of Stuttgart and was appointed Director of the Institute of Telecommunications. His research area includes telecommunications in wireless, fixed, electrical, and optical networks with a focus on encoding, modulation, detection, and MIMO systems. Since 2017, he is a Professor Emeritus. Through his numerous publications and patents, Prof. Speidel has made extensive contributions in the advancement of the field of telecommunications, the success of its products, and international standards. He is a member of various national and international organizations and advisory and review boards.

Nomenclature

Transforms

$\longrightarrow$	Fourier transform
$\longleftarrow$	Inverse Fourier transform
$\overset{t,\tau}{\longrightarrow}$	Two-dimensional Fourier transform
$\overset{f_t,f_\tau}{\longleftarrow}$	Two-dimensional inverse Fourier transform
$x(t) \longrightarrow X(f) = \int_{-\infty}^{\infty} x(t)\mathrm{e}^{-\mathrm{j}2\pi ft}\,dt$	Fourier transform of $x(t)$
$X(f) \longleftarrow x(t) = \int_{-\infty}^{\infty} X(f)\mathrm{e}^{\mathrm{j}2\pi ft}\,df$	Inverse Fourier transform of $X(f)$
$x(t){*}g(t) \longrightarrow X(f)G(f)$	Fourier transform of $x(t){*}g(t)$
$x(t)\circledast g(t,\tau) \overset{t}{\longrightarrow} \int_{-\infty}^{\infty} X(\zeta)G(f_t - \zeta, \zeta)\,d\zeta$	Fourier transform of $x(t)\circledast g(t,\tau)$
$X\left(\mathrm{e}^{\mathrm{j}2\pi f/f_s}\right) = \sum_{n=-\infty}^{\infty} x(n)\mathrm{e}^{-\mathrm{j}2\pi nf/f_s}$	(Discrete-time) Fourier transform of $x(n)$
$X(z) = \sum_{n=-\infty}^{\infty} x(n)z^{-n}$	z-transform of $x(n)$
$x_n \longrightarrow X_\nu = \frac{1}{\sqrt{N}}\sum_{n=0}^{N-1} x_n w^{\nu n}, \; w = \mathrm{e}^{-\mathrm{j}2\pi/N}$	Discrete Fourier transform (DFT)
$X_\nu \longleftarrow x_n = \frac{1}{\sqrt{N}}\sum_{\nu=0}^{N-1} X_\nu w^{-\nu n}, \; w = \mathrm{e}^{-\mathrm{j}2\pi/N}$	Inverse Discrete Fourier Transform (IDFT)

Convolution

$u(t){*}g(t) = \int_{-\infty}^{\infty} u(\tau)g(t-\tau)\,d\tau$	Time-invariant convolution
$x(t)\circledast w(t,s) = \int_{-\infty}^{\infty} x(s)w(t,s)\,ds$	Time-variant convolution
$x(t)\circledast g(t,\tau) = \int_{-\infty}^{\infty} x(t-\tau)g(t,\tau)\,d\tau$	Time-variant convolution, $g(t,\tau) = w(t, t-\tau)$
$x(t,\tau){*}g(t,\tau) = \iint_{-\infty}^{\infty} x(t-u, \tau-v)g(u,v)\,du\,dv$	Two-dimensional convolution
$u(n){*}h(n) = \sum_{m=-\infty}^{\infty} u(m)h(n-m)$	Discrete-time, time-invariant convolution

$$u(n) \otimes h(n) = \sum_{m=0}^{N-1} u(m)h((n-m) \bmod N)$$

Discrete-time, time-invariant cyclic convolution

$$x_s(k) \circledast g_s(k, m) = \sum_{m=-\infty}^{\infty} x_s(k-m)g_s(k, m)$$

Discrete-time, time-variant convolution

Symbols

$a(k) = \mathrm{Re}[a(k)] + j\mathrm{Im}[a(k)]$	Complex QAM symbol
a^*	Complex conjugate of a
$\mathbf{a}(k) = \big(a_0(k)\, a_1(k) \cdots a_{N-1}(k) \big)^T$	QAM symbol vector (block)
$\mathbf{a}^T$	Transpose of $\mathbf{a}$
$\tilde{\mathbf{a}}(k)$	OFDM symbol vector (block)
$\mathbf{A} = \big(a_{ij}\big)_{N \times M}$	$N \times M$ matrix $\mathbf{A}$ with entries a_{ij}
$\mathbf{A}$	Prefilter matrix
a_m	MIMO transmit power coefficient
$\hat{a}$	Estimate of a
$\mathcal{A}$	Set of MIMO transmit vectors
$\arg\{x\}$	Argument of x
$0 < \alpha_\nu(t) \leq 1$	Path loss coefficient of multipath channel
$b(l)$	Sequence of bits
$\mathcal{B}$	Symbol alphabet
C	Channel (or system) capacity
C_{erg}	Ergodic channel (or system) capacity
$\mathbf{c}(k),\ \mathbf{c}(k\prime)$	QAM symbol vector
$\mathbf{D}$	Matrix for singular value decomposition of $\mathbf{H}$
$\det(\mathbf{X})$	Determinant of $\mathbf{X}$
d_k	Branch metric
D_K	Path metric
$\delta(t) \begin{cases} \to \infty &; t = 0 \\ = 0 &; t \neq 0 \end{cases} ;\ \int_{-\infty}^{\infty} \delta(t)dt = 1$	Dirac impulse (distribution)
$\delta_k = \begin{cases} 1 &; k = 0 \\ 0 &; k = \pm 1, \pm 2, \ldots \end{cases}$	Unit impulse

$\delta(t) \rightarrowtail 1$	Fourier transform of $\delta(t)$
$\delta(t - s)$	Dirac impulse at $t = s$
Δf	Roll-off frequency of Nyquist lowpass
$\Delta f = 1/T$	Carrier spacing for OFDM
Δt_{coh}	Coherence time
$\Delta f_{\tau,coh}$	Coherence bandwidth
$\Lambda = \mathrm{diag}(\lambda_0, \lambda_1, ..., \lambda_{N-1})$	Diagonal matrix
$\mathbf{E}[X]$	Expected value (mean) of X
$\mathbf{E}$	Equalizer matrix
$\mathbf{F}$	Matrix of discrete Fourier transform (DFT)
f	Frequency
f_D	Doppler frequency shift
f_I	Cut-off frequency of pulse shaper
$f_\mu = \mu \frac{f_s}{N}$	OFDM carrier frequency number μ
f_ν	Carrier frequency number ν
$f_N = \frac{1}{2T}$	Nyquist frequency
f_{RF}	Carrier frequency for radio modulator
f_s	Sampling frequency
f_t	Frequency corresponding to time t
f_τ	Frequency corresponding to delay τ
$G \geq L$	Length of guard interval
$g(t, \tau)$	Delay Spread Function of time-variant system
$G(f_t, f_\tau) \overset{t,\tau}{\leftharpoonup} g(t, \tau)$	Doppler Spread Function
$G_t(t, f_\tau) \overset{\tau}{\leftharpoonup} g(t, \tau)$	Time-Variant Transfer Function
$G_\tau(f_t, \tau) \overset{t}{\leftharpoonup} g(t, \tau)$	Delay Doppler Spread Function
$g_C(t, \tau)$	Delay Spread Function of time-variant bandpass channel

$g_{C,ij}(t, \tau)$ — Channel Delay Spread Function between transmitter j and receiver i

$G_C(f) \begin{cases} \neq 0 \; ; f_0 - f_I \leq |f| \leq f_0 + f_I \\ = 0 \; ; \; else \end{cases}$ — Bandpass channel transfer function

$g_e(t, \tau)$ — Delay Spread Function of equivalent time-variant baseband system

$G_e(f_t, f_\tau)$ — Doppler Spread Function of equivalent time-variant baseband system

$G_I(f) \begin{cases} \neq 0 \; ; \; |f| \leq f_I \\ = 0 \; ; \; else \end{cases}$ — Lowpass transfer function with cut-off frequency f_I

$g_s(k, m)$ — Discrete-time Delay Spread Function

$g_T(n) = \begin{cases} 1 \; ; \; n = 0, 1, ..., N - 1 \\ 0 \; ; \; else \end{cases}$ — Discrete-time rectangular impulse

$G_T(z) = \sum_{n=0}^{N-1} g_T(n) z^{-n} = \frac{z^{-N} - 1}{z^{-1} - 1}$ — z-transform of rectangular impulse $g_T(n)$

$G_T\left(e^{j2\pi f / f_s}\right) = \frac{\sin(\pi N f / f_s)}{\sin(\pi f / f_s)} e^{-j\pi(N-1)f / f_s}$ — Fourier transform of rectangular impulse $g_T(n)$

γ_y — Signal-to-noise ratio of $\mathbf{y}_s$

$\Gamma(m)$ — Gamma function

$\mathbf{G}(k)$ — MIMO inter-channel interference matrix

$\mathbf{G} = \text{diag}(\mathbf{G}_1 \mathbf{G}_2 \ldots \mathbf{G}_U)$ — Block diagonal matrix

$h(k)$ — Discrete-time impulse response of time-invariant system

$h(n)$ — Discrete-time equivalent baseband channel

$h_e(t)$ — Impulse response of time-invariant equivalent baseband system

$H_e(f)$ — Transfer function of equivalent baseband system

$h_{i,j}(k,m)$	Discrete-time Delay Spread Function of equivalent baseband system between j and i
$\mathbf{h}$	FIR channel coefficient vector
$\mathbf{h}_0$	FIR channel vector with zero padding
$\mathbf{H}(k,m), \mathbf{H}(k), \mathbf{H}$	Matrix of the Delay Spread Functions
$\mathbf{H}$	Matrix of FIR channel
$\mathbf{H}_c$	Matrix of FIR channel with prefix input
$\mathbf{H}_{circ}$	Circulant matrix of FIR channel model
$\mathbf{H}(k,m)$	discrete-time equivalent MIMO channel matrix
$\mathbf{H}^T$	Transpose of $\mathbf{H}$
$\mathbf{H}^H = (\mathbf{H}^*)^T$	Hermiteian matrix (conjugate transpose)
$\mathbf{H}^+ = (\mathbf{H}^H \mathbf{H})^{-1} \mathbf{H}^H$	Pseudo inverse of $\mathbf{H}$, $M \leq N$
$\mathbf{H}^{++} = \mathbf{H}^H (\mathbf{H}\mathbf{H}^H)^{-1}$	Pseudo inverse of $\mathbf{H}$, $M \geq N$
$\|\mathbf{H}\|_F^2 = \mathrm{tr}(\mathbf{H}\mathbf{H}^H)$	Squared norm of $\mathbf{H}$ (Frobenius norm)
$\mathbf{H}_N = \frac{1}{\sqrt{g_P}}\mathbf{H}, \; g_P = \frac{M}{\|\mathbf{H}\|_F^2}$	Normalized MIMO channel matrix
$\mathbf{H}^{(\nu)}$	MIMO channel matrix, iteration step ν
$\mathbf{H}_u$	MIMO sub-channel matrix for user u
$\mathbf{H}_w$	MIMO channel matrix with independent, identically distributed (i.i.d.) Gaussian entries
i	Index of receiver branch i
$I(k), I(k,k)$	Intersymbol interference
$I(X,Y)$	Mutual information between X and Y
$I_0(x)$	Modified Bessel function, first kind, zero order
$\mathbf{I}_M$	$M \times M$ identity matrix

j	Index of transmitter branch j
j	Imaginary unit $\mathrm{j} = \sqrt{-1}$
J	Mean squared error (MSE)
$J_0(x)$	Bessel function, first kind, zero order
k	Discrete time
$K_{ij} + 1$	Length of channel Delay Spread Function between transmitter j and receiver i
$\kappa = \log_2 L$	Number of bits per QAM symbol
l	Discrete time of bit sequence
L	Number of QAM symbols in alphabet
$L + 1$	Length of response of FIR channel
λ_i	Eigenvalue of a matrix
λ	Wavelength
m	Nakagami fading parameter
m	Discrete delay variable
M	Number of MIMO transmit antennas
μ_0	Mean value
$n(k)$	Discrete-time noise at receiver
$n_C(t)$	Continuous-time noise at receiver
n	Discrete-time variable
N	Number of multicarrier
N	Number of MIMO receive antennas
$\mathbf{n}(k)$	Noise vector at receiver
$\mathbf{n}_c(k) = \begin{pmatrix} \mathbf{n}_{pre}(k) \\ \mathbf{n}(k) \end{pmatrix}$	Compound noise vector
$O(\ldots)$	Order of complexity
$\mathbf{0}_m = \begin{pmatrix} 0 \cdots 0 \end{pmatrix}^T$	Column vector with m zeros
$p(x)$	Probability density function (pdf) of X

$p_0(\psi)$ — Probability density function of shadowing

$p(x) = \frac{1}{\sqrt{2\pi}\sigma_x}e^{-\frac{x^2}{2\sigma_x^2}}$ — Gaussian probability density function with zero mean

$p_{xy}(x, y)$ — Joint probability density function

$p_{\mathbf{n}}(n_1, n_2, ..., n_N)$ — Multivariate probability density function

$p_L(q|a)$ — Probability density function of q conditioned on a

$p_L(\mathbf{r}|\mathbf{Hs})$ — Conditional multivariate probability density function

$p_K(q(0), q(1), ..., q(K-1)|w(0), w(1), ..., w(K-1))$ — Conditional multivariate probability density function

$P[a \le X \le b] = \int_a^b p(x)dx$ — Probability of X ranging between a and b

P — Rank of a matrix

P_{out} — Outage probability

P_S — Symbol error probability

$P_g(t, t'; \tau)$ — Delay Cross Power Spectral Density

$q(k)$ — Discrete-time receive signal

$q_R(t)$ — Continuous-time receive signal

$Q_R(f)$ — Fourier spectrum of receive signal $q_R(t)$

$q_v(t)$ — Signal in branch v of multicarrier receiver

$Q_v(e^{j2\pi f/f_s})$ — Spectrum in branch v of multicarrier receiver

$Q(\alpha) = \frac{1}{\sqrt{2\pi}}\int_\alpha^\infty e^{-\frac{u^2}{2}}du = P[X > \alpha\sigma_x]$ — Gaussian Q-function

$\mathbf{Q} \otimes \mathbf{R}$ — Kronecker product of $\mathbf{Q}$ and $\mathbf{R}$

$\mathbf{Q}_N$ — NxN Hermiteian matrix

$\text{rank}(\mathbf{H})$ — Rank of matrix $\mathbf{H}$

$\text{rect}(\frac{t}{T}) = \begin{cases} 1 \; ; \; |t| \le T/2 \\ 0 \; ; \; else \end{cases}$ — Rectangular impulse

$\text{rect}(\frac{t}{T}) \longmapsto T\text{sinc}(fT)$ — Fourier transform of $\text{rect}(\frac{t}{T})$

$r_i(k)$ — Receive signal of branch i

r_s — Spatial code rate

$R_{gg}(t, t\prime; \tau, \tau\prime) = \mathbf{E}\big[g^*(t, \tau)g(t\prime, \tau\prime)\big]$ — Autocorrelation of Delay Spread Function

$R_{GG}(f_t, f_{t\prime}; f_\tau, f_{\tau\prime}) = \mathbf{E}\big[G^*(f_t, f_\tau)G(f_{t\prime}, f_{\tau\prime})\big]$ — Autocorrelation of Doppler Spread Function

$R_{G_t G_t}(t, t\prime; f_\tau, f_{\tau\prime}) = \mathbf{E}\big[G_t^*(t, f_\tau)G_t(t\prime, f_{\tau\prime})\big]$ — Autocorrelation of Time-Variant Transfer Function

$R_{G_\tau G_\tau}(f_t, f_{t\prime}; \tau, \tau\prime) = \mathbf{E}\big[G_\tau^*(f_t, \tau)G_\tau(f_{t\prime}, \tau\prime)\big]$ — Autocorrelation of Delay Doppler Spread Function

$R_{xx}(\tau) = \mathbf{E}[X^*(t)X(t + \tau)]$ — Autocorrelation function of $X(t)$

$R_{xx}(t, t\prime) = \mathbf{E}[X^*(t)X(t\prime)]$ — Autocorrelation of input signal of $X(t)$

$R_{yy}(t, t\prime) = \mathbf{E}[Y^*(t)Y(t\prime)]$ — Autocorrelation of output signal of $Y(t)$

$R_{xy}(\tau) = \mathbf{E}[X(t + \tau)Y^*(t)]$ — Cross-correlation function of $X(t)$ and $Y(t)$

ρ — Correlation coefficient

$\mathbf{r}(k)$ — MIMO receive signal vector

$\mathbf{r}_u$ — Receive signal vector of user u

$\mathbf{R}_{HH}$ — Covariance matrix of $\mathbf{H}$

$\mathbf{R}_{HH}^{\frac{1}{2}}$ — Square root matrix of $\mathbf{R}_{HH}$

$\mathbf{R}_{nn}$ — Covariance matrix of noise vector $\mathbf{n}$

$\mathbf{R}_{ss}$ — Covariance matrix of signal vector $\mathbf{s}$

$\mathbf{R}_{sn}$ — Cross-correlation matrix of $\mathbf{s}$ and $\mathbf{n}$

$\mathbf{R}_{tx}, \mathbf{R}_{rx}$ — Transmit, receive correlation matrix, resp.

s — Instant of Dirac impulse $\delta(t - s)$

$\mathrm{sinc}(x) = \frac{\sin(\pi x)}{\pi x}$ — sinc function

$\mathrm{sinc}(f_c t) \longmapsto \frac{1}{f_c} rect(\frac{f}{f_c})$ — Fourier transform of sinc $(f_c t)$

$s_j(k)$ — Transmit symbol of antenna j

$S_{gc}(f_{t\prime}; \tau, \tau\prime) \overset{\Delta t}{\longleftarrow} R_{gg}(\Delta t; \tau, \tau\prime)$ — Cross Power Spectral Density of $g(t, \tau)$

$S_{gD}(f_t, \tau) \overset{\Delta t}{\leftharpoondown} P_g(\Delta t; \tau)$ — Doppler Power Spectrum of $g(t, \tau)$

$S_{n_C n_C}(f) = \begin{cases} \frac{N_0}{2} & ; f_0 - f_I \le |f| \le f_0 + f_I \\ 0 & ; else \end{cases}$ — Power spectral density of receiver noise

$S_{xx}(f)$ — Power spectral density of $X(t)$

$S_{xx}(f_t)$ — Input Power Spectral Density

$S_{yy}(f_t)$ — Output Power Spectral Density

σ_x^2 — Variance of X

$\mathbf{s}(k)$ — MIMO transmit signal vector

$\hat{\mathbf{s}}$ — Estimate of $\mathbf{s}$

$\tilde{\mathbf{s}}, \tilde{\mathbf{r}}, \tilde{\mathbf{n}}$ — MIMO eigenmodes

$\mathbf{S}$ — Space-time coding matrix

t — Continuous time

T — Symbol interval

$\mathrm{tr}(\mathbf{W})$ — Trace of matrix $\mathbf{W}$

$\mathcal{T}[...]$ — Linear transmission operator

τ — Delay variable

$\tau_v(t)$ — Delay of channel path v

$\mathbf{t}_u$ — Beamforming vector of user u

$\mathbf{T}, \mathbf{T}_u$ — Multi-user downlink precoding matrix

$u(n)$ — Input of FIR channel

$u_j(t)$ — Output signal of transmit antenna j

u_{nk} — Sample n of channel input vector (block) k

$u_T(t)$ — Transmit signal

$U\left(e^{j2\pi f/f_s}\right)$ — Multicarrier transmit spectrum

$\mathbf{u} = \left(u_0 \; u_1 \; \cdots \; u_{N-1}\right)^T$ — Input signal vector of FIR channel

$\mathbf{u}(k) = \left(u_{0k} \; u_{1k} \; \cdots \; u_{N-1,k}\right)^T$ — Input signal vector (block) k of FIR channel

$\mathbf{u}_0(k) = \begin{pmatrix} \mathbf{u}(k) \\ \mathbf{0}_L \end{pmatrix}$ — Input vector k with zero padding

$\mathbf{u}_{pre}(k) = \left(u_{-G,k} \ \cdots \ u_{-1,k} \right)^{T}$

Vector k of prefix samples

$\mathbf{u}_{c}(k) = \begin{pmatrix} \mathbf{u}_{pre}(k) \\ \mathbf{u}_{0}(k) \end{pmatrix}$

Compound input vector k

$\underline{\mathbf{u}}(k) = \mathbf{F}\mathbf{u}(k)$

DFT of vector $\mathbf{u}(k)$

$\mathbf{U}$

Left hand singular value decomposition matrix

$\text{vec}(\mathbf{H})$

Column vector stack of $\mathbf{H}$

$v(n)$

Output of discrete-time equivalent baseband FIR channel

v_{B}

Bitrate (bit/s)

$v_{S} = \frac{v_{B}}{\kappa}$

Symbol rate (Baud rate, symbol/s)

v_{0}

speed of mobile terminal

$v_{i}(t)$

Receive signal of branch i

v_{nk}

Sample n of channel output vector (block) k

$\mathbf{v}(k) = \left(v_{0k} \ v_{1k} \ \cdots \ v_{N-1+L,k} \right)^{T}$

Output vector k of FIR channel

$\mathbf{v}_{pre}(k) = \left(v_{-G,k} \ \cdots \ v_{-1,k} \right)^{T}$

Prefix part of output vector k

$\mathbf{v}_{c}(k) = \begin{pmatrix} \mathbf{v}_{pre}(k) \\ \mathbf{v}(k) \end{pmatrix}$

Compound output vector k

$\mathbf{v}_{N}(k) = \left(v_{0k} \ v_{1k} \ \cdots \ v_{N-1,k} \right)^{T}$

Output vector k reduced to N samples

$\|\mathbf{v}\|^{2} = \mathbf{v}^{H}\mathbf{v}$

Squared norm of a vector $\mathbf{v}$

$\mathbf{V}^{H}$

Right-hand singular value decomposition matrix

$w(t, s)$

Impulse response of time-variant system

$w_{C,ij}(t, s)$

Time-Variant Channel Impulse Response between transmitter j and receiver i

$w_{e}(t, s)$

Overall impulse response of equivalent time-variant baseband system

$\mathbf{W}(k)$

MIMO receiver matrix

$\mathbf{W}^{(v)}$

Receiver matrix at iteration step v

$\mathbf{W}_u^{ZF}$	Multi-user zero-forcing uplink receive matrix of user u
$y(t)$	Output signal
$\mathbf{y}(k)$	Receiver output vector
z	Variable

Part I
Digital Communications over Single Input Single Output Channels

Chapter 1
Transmission System with Quadrature Amplitude Modulation

1.1 Introduction

This chapter presents an overview on the principles of digital communications. We focus on a system with one transmitter and one receiver, i.e. for a channel with a single input and a single output (SISO). This will also provide the necessary basics for multiple input multiple output (MIMO) systems investigated in Part III. Depending on the characteristics of the transmission medium we have to differentiate between a wire-line and a wireless connection. Both channel types exhibit different properties and therefore will be treated separately. We start with the wire-line transmission link and in Chap. 4 the wireless system will be discussed in detail. Also the transfer functions of the transmission media differ in general. They can have a lowpass or a bandpass characteristic. An electrical line, e.g.., a twisted pair or a coaxial cable exhibits a lowpass magnitude response, because the d.c. current can travel from the input to the output. In contrast, a wireless channel is characterized by a bandpass transfer function, because only high frequency spectral components can be emitted and received by the antennas. The optical transmission on a glass or a plastic fiber is similar but the transmission spectrum lies in the multi-THz frequency region. However, it should be noted that a connection over an electrical line may contain transformers at the transmitter or at the receiver side for galvanic isolation between the transmission line and the electronic equipment. Then the overall transfer function becomes a bandpass characteristic. The same is true, if decoupling capacitors are connected in series to the electrical transmission line. As the source signal at the transmitter normally has a lowpass spectrum, a frequency shift into the passband of the channel by dedicated pulse shaping or modulation is required. In the following we focus on a bandpass channel with a passband around the mid frequency f_0 and employ a modulator with a carrier signal

$$e^{j2\pi f_0 t} = \cos(2\pi f_0 t) + j\sin(2\pi f_0 t) \tag{1.1}$$

© Springer Nature Switzerland AG 2021
J. Speidel, *Introduction to Digital Communications*,
Signals and Communication Technology,
https://doi.org/10.1007/978-3-030-67357-4_1

at the transmitter, where the carrier frequency is f_0. This complex carrier contains an "in-phase" component with $\cos(2\pi f_0 t)$ and a "quadrature" component with $\sin(2\pi f_0 t)$, which are orthogonal. Therefore the scheme is called quadrature amplitude modulation (QAM) or I-Q modulation. We will introduce an equivalent baseband (or lowpass) system model providing the overall characteristic of a lowpass transfer function. With this approach lowpass and bandpass transmission schemes can be treated elegantly in a uniform way and the transmission scheme, which does not require modulation, is included for $f_0 = 0$. Then, only real-valued transmit symbols are applicable also called (multi-level) pulse amplitude modulated (PAM) symbols. We will also see that the principle methods for wire-line and wireless transmission have quite a lot in common and we focus in the next chapters on the basic principles of both.

1.2   The Transmitter

Figure 1.1 shows the principle block diagram of a transmission system. The output of the data source at the transmitter is a sequence of bits $b_S(l')$. The forward error correction (FEC) encoder allocates redundant bits to the input, which are thereafter temporally interleaved to prepare for burst errors. The resulting output bit sequence is denoted as $b(l)$. l and l' are discrete-time variables. The amount of redundancy can be defined by the temporal code rate $r_t \leq 1$, which is the ratio between the number of input and output bits in the same time frame. Then the output bit sequence contains $(1 - r_t) \cdot 100\%$ redundancy with respect to the input. The large number of methods and codes for forward error correction are not considered here. The reader is referred to the dedicated literature, e.g., [1–10]. The mapper periodically allocates tuples of κ successive bits to a symbol $a(k)$ and outputs the complex symbol sequence

$$a(k) = \mathrm{Re}[a(k)] + \mathrm{jIm}[a(k)] \tag{1.2}$$

at periodic intervals of duration T with the symbol rate $v_S = \frac{1}{T}$. The Fourier spectrum of the discrete-time signal $a(k)$ can be calculated with the help of the z-transform of $a(k)$, which is

$$A(z) = \sum_{k=-\infty}^{\infty} a(k)z^{-k} \tag{1.3}$$

and for $z = \mathrm{e}^{\mathrm{j}2\pi fT}$ the Fourier spectrum

$$A\left(\mathrm{e}^{\mathrm{j}2\pi fT}\right) = \sum_{k=-\infty}^{\infty} a(k)\mathrm{e}^{-\mathrm{j}2\pi fTk} \tag{1.4}$$

Transmitter

Fig. 1.1 Principle block diagram for digital transmission over a bandpass channel

is obtained. Obviously, $A\left(e^{j2\pi fT}\right)$ is a periodic function of the natural frequency f with period $\frac{1}{T}$, because the argument $e^{j2\pi fT}$ is periodic with $f = \frac{1}{T}$. Consequently, the spectrum of the symbol sequence $a(k)$ exhibits an infinite bandwidth, which has to be limited by a transmit lowpass filter with cut-off frequency f_I. This filter shall be linear and time-invariant with the impulse response $g_I(t)$ and can be used to shape the output impulses $u_I(t)$, for which reason the filter is called impulse shaper or pulse shaper. As is well known, the transfer function of a linear, time-invariant system is defined by the Fourier transform of its impulse response and in general the spectrum of a continuous-time signal can be obtained by the Fourier transform, [11]. In the following we denote the transform by the symbol $\rightarrowtail$ and assume that the reader is familiar with the Fourier transform calculus.

Thus, we obtain the spectrum of $g_I(t)$ as

$$g_I(t) \rightarrowtail G_I(f) = \int_{-\infty}^{\infty} g_I(t)e^{-j2\pi ft}dt \qquad (1.5)$$

and the inverse transform is given by

$$G_I(f) \leftarrowtail g_I(t) = \int_{-\infty}^{\infty} G_I(f)e^{j2\pi ft}df \qquad (1.6)$$

To check the existence of the integral (1.5) one of the sufficient Dirichlet conditions can be applied

$$\int_{-\infty}^{\infty} |g_I(t)|\, dt \le M_1 < \infty \quad \text{or} \quad \int_{-\infty}^{\infty} |g_I(t)|^2\, dt \le M_2 < \infty \qquad (1.7)$$

where M_1 and M_2 are finite real numbers. The second condition confirms that all signals with finite energy are equipped with a Fourier spectrum. $G_I(f)$ is also called the transfer function of the pulse shaper and shall have the lowpass property

$$G_I(f) \begin{cases} \ne 0 \,;\, |f| \le f_I \\ = 0 \,;\quad else \end{cases} \qquad (1.8)$$

where f_I is the cut-off frequency. The continuous-time signal at the input of the pulse shaper is described as

$$u_a(t) = \sum_{k=-\infty}^{\infty} a(k)\delta(t - kT) \qquad (1.9)$$

where $\delta(t)$ is the Dirac impulse. Then follows for the signal at the pulse shaper output

$$u_I(t) = u_a(t) * g_I(t) = \sum_{k=-\infty}^{\infty} a(k)g_I(t - kT) \qquad (1.10)$$

where $*$ denotes the convolution

$$u_a(t) * g_I(t) = \int_{-\infty}^{\infty} u_a(\tau)g_I(t - \tau)d\tau \qquad (1.11)$$

We obtain the output signal of the modulator with (1.10)

$$u_M(t) = u_I(t)e^{j2\pi f_0 t} = \left(\sum_{k=-\infty}^{\infty} a(k)g_I(t - kT) \right) e^{j2\pi f_0 t} \qquad (1.12)$$

where f_0 is the carrier frequency. Any physical channel exhibits a real-valued impulse response denoted as $g_C(t)$ in Fig. 1.1. If not otherwise stated, the channel shall be characterized as a linear, time-invariant system. In Chap. 4 we will focus on the wireless channel, which is characterized as linear and time-variant.[1]

If the complex signal $u_M(t)$ is directly input to the channel, a separation of the real and imaginary part at the receiver is not possible. Thus, only the real or the

---

[1] In case of a baseband transmission system with a lowpass channel, no modulation is required, thus we set $f_0 = 0$. Figure 1.1 still holds, wherein the modulator, the real-part operator Re[...], the demodulator, and the gain factors $\sqrt{2}$ are dropped. All signals including the symbols $a(k)$ take on real values only.

Fig. 1.2 Implementation of a QAM transmitter with exclusively real signals

imaginary part of $u_M(t)$ is feasible as the channel input signal. In Fig. 1.1 the real part is selected and $\sqrt{2}$ is just an amplification factor to achieve an overall amplification of one between the transmitter and the receiver. In the following we apply the general property for complex numbers

$$\text{Re}[u_M] = \frac{1}{2}\left(u_M + u_M^*\right) \tag{1.13}$$

where the superscript denotes the conjugate complex operation. With (1.12), (1.13), and assuming a real impulse response $g_I(t)$ a straightforward calculation yields the transmit signal

$$u_T(t) = \sqrt{2}\left(\sum_{k=-\infty}^{\infty} \text{Re}[a(k)] g_I(t - kT)\right) \cos(2\pi f_0 t) \\ -\sqrt{2}\left(\sum_{k=-\infty}^{\infty} \text{Im}[a(k)] g_I(t - kT)\right) \sin(2\pi f_0 t) \tag{1.14}$$

From (1.14) we recognize that the transmit signal can carry two independent symbol sequences $\text{Re}[a(k)]$ and $\text{Im}[a(k)]$. Furthermore, as depicted in Fig. 1.2, we can implement an alternative QAM transmitter, which contains only real-valued signals and provides the same output signal $u_T(t)$.

The reader can convince oneself easily that the spectrum $U_T(f)$ of $u_T(t)$ in (1.14) satisfies

$$U_T(f) \begin{cases} \neq 0 \; ; & f_0 - f_I \leq |f| \leq f_0 + f_I \\ = 0 \; ; & else \end{cases} \tag{1.15}$$

1.3 Signal Constellation Diagrams

The mapper in Figs. 1.1 and 1.2 assigns to each tuple of κ incoming bits
$b(l), b(l-1), ..., b(l-\kappa+1)$ a complex symbol $a(k)$. The allocation of these symbols in the complex plane constitutes the signal constellation diagram. The $a(k)$ is also denoted as signal point and encoded by a codeword with κ bits. Hence the number of different symbols is

$$L = 2^{\kappa} \tag{1.16}$$

and therefore this scheme is referred to as L-ary quadrature amplitude modulation or L-QAM. There is an infinite number of possible distributions of L signal points in the complex plane. The example in Fig. 1.3 shows the 4-level phase shift keying (4-PSK), where the signal points are distributed on a circle. In the following we present some more examples of important constellation diagrams used for digital transmission. The relation between the bitrate v_B and the symbol rate $v_S = \frac{1}{T}$ obviously is

$$v_S = \frac{v_B}{\kappa} = \frac{v_B}{\log_2 L} \tag{1.17}$$

and is measured in "symbols per second", which is also referred to as "Baud" according to the French engineer Baudot. Consequently, with L-QAM the clock rate is by factor $\frac{1}{\log_2 L}$ lower than the bitrate, which also holds for the required transmission bandwidth.

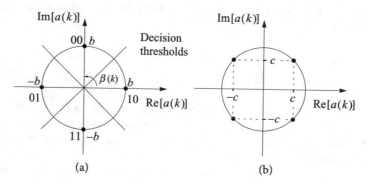

(a)   (b)

Fig. 1.3 **a** Constellation diagram of 4-PSK. Pairs of bits are allocated to four signal points **b** Constellation turned by $\frac{\pi}{4}$ and with $c = \frac{b}{\sqrt{2}}$

Fig. 1.4 Constellation
diagram of 16-QAM

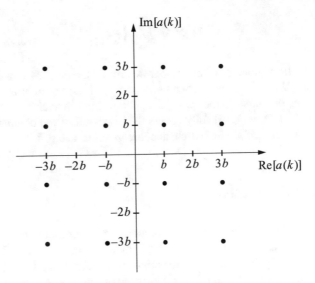

4-QAM and 4-PSK

Figure 1.3 shows the constellation diagram of 4-PSK as a special case of 4-QAM.
The four signal points are located on a circle and thus they differ only in their
angles $\beta(k)\epsilon\left\{0, \frac{\pi}{2}, \pi, \frac{3\pi}{2}\right\}$, which justifies the name phase shift keying. For illustration, if the binary sequence $b(l)$ shall be 0100110110..., the mapper output $a(k)$ is
$-b, jb, -jb, -b, b, \ldots$. Another special case is 2-PSK, where only two signal points
are used. If they are allocated on the real axis of the complex plane, the scheme is
very simple to implement. The symbol alphabet then is $\mathcal{B} = \{-b, b\}$ and the angles
are $\beta(k)\epsilon\{0, \pi\}$. Consequently, for the transmit signal in (1.14) follows

$$u_T(t) = \sqrt{2} \sum_{k=-\infty}^{\infty} \text{Re}[a(k)]\, g_I(t - kT)\, \cos(2\pi f_0 t) \; ; \; \text{with Re}\,[a(k)]\,\epsilon\,\{-b, b\}$$

$$(1.18)$$

16-QAM

Figure 1.4 shows another constellation diagram, the 16-QAM, where 4-tuples of bits
$b(l), b(l-1), b(l-2), b(l-3)$ are allocated to the sixteen different signal points
$a(k)$. The scheme is frequently used in many applications for wire-line and wireless
digital transmission.

1.4 Transmission Channel

The transmission channel shall be characterized by its time-invariant and real-valued
impulse response $g_C(t)$. The bandpass shaped transfer function $G_C(f)$ is obtained
by the Fourier transform of $g_C(t)$ and characterized as

$$g_C(t) \multimap G_C(f) \begin{cases} \neq 0\,; & f_0 - f_I \leq |f| \leq f_0 + f_I \\ = 0\,; & else \end{cases} \tag{1.19}$$

The channel passband is located around the center frequency $\pm f_0$ and the bandwidth is $2f_I$. Thus, the carrier frequency f_0 of the modulator and the cut-off frequency f_I of the pulse shaper have to be determined in such a way that the transmit spectrum $U_T(f)$ in (1.15) fully covers the passband of the channel transfer function $G_C(f)$. Models of a wireless channel are given in Chap. 4.

1.5   Receiver

In Fig. 1.1 the receiver is composed of a demodulator, which multiplies the receive signal $u_R(t)$ with the complex demodulation carrier $e^{-j(2\pi f_0 t + \varphi_0)}$. The frequency f_0 is exactly the same as for the modulating carrier and the phase φ_0 of the demodulation carrier is constant. Thus, the receiver operates synchronously with the transmitter. In the following we assume $\varphi_0 = 0$. The lowpass filter with real impulse response $\sqrt{2}g_R(t)$ selects the baseband out of the demodulated signal $u_D(t)$ resulting in the complex baseband signal $q_R(t)$. At the receiver there is no interest in the complete waveform of this analog signal. Only the samples taken with the symbol rate $\frac{1}{T}$ are required. Therefore a sampling device provides the sequence

$$q(k) = q_R(t_0 + kT) \tag{1.20}$$

where t_0 is the signal delay between the transmitter and receiver, which has to be estimated at the receiver. The sampling clock with frequency $\frac{1}{T}$ has to be synchronous with the symbol rate v_S at the transmitter. It is extracted by a special clock recovery circuit from the receive signal to guarantee synchronism and to control deviations. Synchronization methods [12] as well as sampling rates higher than $\frac{1}{T}$, which can meet the sampling theorem for advanced digital signal processing at the receiver, are not considered here.

The receiver input signal is corrupted by real-valued additive noise $n_C(t)$ coming from the channel and the electronic equipment of the receiver. We will just call it channel noise. Using the superposition principle we obtain the total receive signal

$$u_R(t) = u_T(t) * g_C(t) + n_C(t) \tag{1.21}$$

We will show that the output signal of the receive lowpass filter with the impulse response $\sqrt{2}g_R(t)$ and the transfer function

$$\sqrt{2}G_R(f) \begin{cases} \neq 0\,; & |f| \leq f_R \\ = 0\,; & else \end{cases} \tag{1.22}$$

is given by

$$q_R(t) = u_a(t) * h_e(t) + n_R(t) \tag{1.23}$$

where

$$h_e(t) = g_I(t) * \left(g_C(t)e^{-j2\pi f_0 t}\right) * g_R(t) \tag{1.24}$$

and

$$n_R(t) = \sqrt{2}\left(n_C(t)e^{-j2\pi f_0 t}\right) * g_R(t) \tag{1.25}$$

$n_R(t)$ represents the demodulated and lowpass filtered noise. (1.23) can be recognized as the input-output relation of the signals between the node (I) and the node (II) in Fig. 1.1. Consequently, $h_e(t)$ is the equivalent impulse response between (I) and (II) in case of no noise, $n_C(t) = 0$. Although this section is time-variant on the first glance due to the synchronous modulation and demodulation the convolution operation $u_a(t) * h_e(t)$ still holds, as will be proven in the following. This kind of convolution operation will change, if the phase of the demodulation carrier varies with time due to some phase noise or in case of a time-varying wireless channel, as outlined in Chap. 4. Please note that $u_I(t)$, $u_M(t)$, $h_e(t)$, and $n_R(t)$ are complex-valued signals and noise in general, whereas $g_I(t)$, $g_C(t)$, $g_R(t)$, and $n_C(t)$ are real-valued, respectively.

We also show that the spectrum $Q_R(f)$ of the signal $q_R(t)$ at the output of the receiver lowpass is given in case of no noise as

$$Q_R(f) = U_I(f)G_C(f + f_0)G_R(f) \tag{1.26}$$

where $u_I(t) \rightarrowtail U_I(f)$ is the lowpass spectrum of the output signal $u_I(t)$ of the pulse shaper

$$U_I(f) = G_I(f)U_a(f) \tag{1.27}$$

where $u_a(t) \rightarrowtail U_a(f)$ holds. Obviously, $Q_R(f)$ is a lowpass spectrum. From (1.26) follows with (1.27)

$$H_e(f) = \frac{Q_R(f)}{U_a(f)} = G_I(f)G_C(f + f_0)G_R(f) \tag{1.28}$$

which is the overall transfer function between the nodes (I) and (II). We recognize that $H_e(f)$ represents a lowpass and is called the transfer function of the *equivalent baseband (or lowpass) system*.[2]

---

[2]sometimes referred to as "equivalent baseband (or lowpass) channel".

Proof of (1.24) and (1.26)

We start with the output $u_I(t) \rightarrowtail U_I(f)$ of the pulse shaper, which has a lowpass spectrum with the cut-off frequency f_I given by (1.8). Then we obtain the modulator output signal

$$u_M(t) = u_I(t)e^{j2\pi f_0 t} \tag{1.29}$$

and with (1.13) the transmit signal

$$u_T(t) = \frac{\sqrt{2}}{2}\left[u_I(t)e^{j2\pi f_0 t} + u_I^*(t)e^{-j2\pi f_0 t}\right] \tag{1.30}$$

We proceed in the frequency domain and get with

$$u_I^*(t) \rightarrowtail U_I^*(-f) \tag{1.31}$$

and with the frequency shifting property of the Fourier transform the transmit signal

$$u_T(t) \rightarrowtail U_T(f) = \frac{\sqrt{2}}{2}\left[U_I(f - f_0) + U_I^*(-f - f_0)\right] \tag{1.32}$$

As expected, the spectrum $U_T(f)$ is located in the passband of the channel transfer function around the mid frequencies $-f_0$ and f_0.

The spectrum of the receive signal in case of no noise at the output of the channel can be found as

$$u_R(t) \rightarrowtail U_R(f) = \frac{\sqrt{2}}{2}\left[U_I(f - f_0) + U_I^*(-f - f_0)\right]G_C(f) \tag{1.33}$$

After demodulation the spectrum at the input of the receiver lowpass filter in case of no noise is $U_D(f) = U_R(f + f_0)$ and finally the spectrum of $q_R(t)$ follows

$$Q_R(f) = \sqrt{2}U_R(f + f_0)G_R(f) \tag{1.34}$$

Plugging in (1.33) yields

$$Q_R(f) = \left[U_I(f)G_C(f + f_0) + U_I^*(-f - 2f_0)G_C(f + f_0)\right]G_R(f) \tag{1.35}$$

The second term $U_I^*(-f - 2f_0)G_C(f + f_0)$ is a spectrum with passband around the center frequency $-2f_0$. Multiplied with the lowpass transfer function $G_R(f)$ the product is zero and we obtain from (1.35) $Q_R(f) = U_I(f)G_C(f + f_0)G_R(f)$, which finalizes the proof of (1.26).

We recognize that the overall amplification factor for $Q_R(f)$, excluding the noise, is $\sqrt{2}\sqrt{2}\frac{1}{2} = 1$, which justifies the introduction of the amplification factor $\sqrt{2}$ both at transmitter and receiver.

The inverse Fourier transform of $H_e(f)$ in (1.28) results in
$h_e(t) = g_I(t) * (g_C(t)e^{-j2\pi f_0 t}) * g_R(t)$ and the proof of (1.24) is finished.

Proof of (1.23) and (1.25)

We directly conclude from the left-hand side of (1.28) $Q_R(f) = H_e(f)U_a(f)$ and
the inverse Fourier transform yields $q_R(t) = u_a(t) * h_e(t)$. This finalizes the proof
of (1.23) in case of no noise.

The proof of (1.25) is straightforward. If the transmitter sends no signal, i.e.
$u_T(t) = 0$, then the output signal of the receiver lowpass is the filtered noise
$q_R(t) = n_R(t) = \sqrt{2}\,(n_C(t)e^{-j2\pi f_0 t}) * g_R(t)$. If $n_C(t) = 0$, then $q_R(t) = u_a(t) * h_e(t)$
holds. In the general case that the transmit signal and the noise are present the superposition principle can be applied to this linear system yielding $q_R(t) = u_a(t) * h_e(t) +$
$n_R(t)$, which completes the proof of (1.23).

1.6 Equivalent Baseband System Model

The block diagram in Fig. 1.1 is rather detailed. We take advantage of (1.23) and
find the much simpler structure in Fig. 1.5a. Details of modulation, demodulation
and filtering does not show up anymore. The input-output relation of the "black box"
between node (I) and (II) is given by (1.23) to (1.25). $h_e(t)$ is denoted as the impulse
response of the continuous-time equivalent baseband system model. Finally, we are
only interested in the discrete-time relation between the input sequence $a(k)$ and the
sampled output sequence $q(k)$ defined already in (1.20).

To this end, we insert (1.9) into (1.23) and obtain

Fig. 1.5 a Continuous-time
equivalent baseband system
model between nodes (I) and
(II) with reference to Fig. 1.1
b Discrete-time equivalent
baseband system model
between nodes (I) and
(III) with reference to
Fig. 1.1

$$q_R(t) = \sum_{m=-\infty}^{\infty} a(m)h_e(t - mT) + n_R(t) \tag{1.36}$$

Now we take samples at $t = t_0 + kT$ and get

$$q_R(t_0 + kT) = \sum_{m=-\infty}^{\infty} a(m)h_e\left(t_0 + (k - m)T\right) + n_R(t_0 + kT) \tag{1.37}$$

With

$$h(k) = h_e(t_0 + kT) = \left\{g_I(t) * \left(g_C(t)e^{-j2\pi f_0 t}\right) * g_R(t)\right\}_{t=t_0+kT} \tag{1.38}$$

and

$$n(k) = n_R(t_0 + kT) \tag{1.39}$$

follows from (1.37) with (1.20)

$$q(k) = \sum_{m=-\infty}^{\infty} a(m)h\left(k - m\right) + n(k) \tag{1.40}$$

which is the discrete-time input-output relation of the "black box" between nodes (I) and (III) in Fig. 1.1. This leads to the discrete-time equivalent baseband system (or channel) model depicted in Fig. 1.5b, which is very helpful, because it focuses our consideration on only a few characteristic parameters of the system. $h(k)$ is called the discrete-time impulse response of the equivalent baseband system model. The noises $n_R(t)$ and $n(k)$ are investigated in quite some detail in Sect. 2.6.

References

1. Shannon, C.: A mathematical theory of communication. Bell Syst. Tech. J. **27** (1948)
2. Gallager, R.: Low-density parity-check codes. IRE Trans. Inf., Theory (1962)
3. Hagenauer, J., Offer, E., Papke, L.: Iterative decoding of binary block and convolutional codes. IEEE Trans. Info. Theory **42** (1996)
4. Bossert, M.: Channel Coding for Telecommunications. Wiley (1999)
5. ten Brink, S.: Design of concatenated coding schemes based on iterative decoding convergence. Ph.D. thesis, University of Stuttgart, Institute of Telecommunications, Shaker Publ. (2001). ISBN 3-8322-0684-1
6. Costello, D., Forney, G.: Channel coding: The road to channel capacity. Proc, IEEE (2007)
7. Richardson, T., Urbanke, R.: Modern Coding Theory. Cambridge University Press (2008)
8. Ryan, W., Lin, S.: Channel Codes - Classical and Modern. Cambridge University Press (2009)
9. Declercq, D., Fossorier, M., Biglieri, E.: Channel Coding: Theory, Algorithms, and Applications. Academic (2014)
10. Tomlinson, M., Tjhai, C.J., Ambroze, M.A., Jibril, M.: Error-Correction Coding and Decoding. Springer (2017)

11. Papoulis, A.: The Fourier Integral and Its Applications. McGraw Hill (1976)
12. Simon, M., Hinedi, S., Lindsey, W.: Digital Communications: Synchronization. Prentice Hall (2000)

Chapter 2
Intersymbol Interference and Noise

2.1 Intersymbol Interference

With the help of the discrete-time equivalent baseband system model we can now get insight into the two major impairments a signal incurs from the transmitter to the receiver, namely intersymbol interference and noise. For that purpose we separate the term for $m = k$ from the sum in (1.40) and obtain

$$q(k) - a(k)h(0) + \sum_{\substack{m = -\infty \\ m \neq k}}^{\infty} a(m)h(k - m) + n(k) \qquad (2.1)$$

We see that the receive sample $q(k)$ is composed of the transmit symbol $a(k)$ multiplied by $h(0)$ of the discrete-time impulse response $h(k)$, the distortion term

$$I(k) = \sum_{\substack{m = -\infty \\ m \neq k}}^{\infty} a(m)h(k - m) \qquad (2.2)$$

and the noise $n(k)$. $I(k)$ is called intersymbol interference, because it consists of the previous and in case of a non-causality of $h(k)$ the future $a(m)$ weighted by the samples $h(k - m)$ of the equivalent baseband impulse response.

2.2 Nyquist's First Criterion in the Time Domain

To remove the intersymbol interference $I(k)$ defined in (2.2) we may not impose any constraint on the symbol sequence $a(k)$, because the system design should hold for any sequence given by the user at the transmitter. Therefore we can only touch

© Springer Nature Switzerland AG 2021
J. Speidel, *Introduction to Digital Communications*,
Signals and Communication Technology,
https://doi.org/10.1007/978-3-030-67357-4_2

Fig. 2.1 Example of a real-valued impulse $h_e(t)$ satisfying Nyquist's first criterion (2.3)

upon the impulse response $h(k)$. Looking at (1.38) the system is prepared already with two degrees of freedom, $g_I(t)$ and $g_R(t)$. Hence, for a given impulse response $g_C(t)$ of the physical channel we can design the overall impulse response in such a way that

$$h(k-m) = h_e(t_0 + (k-m)T) = \begin{cases} 0 & ; \ m \in \mathbb{Z} \ ; \ m \neq k \\ h(0) = h_e(t_0) \neq 0 \ ; & m = k \end{cases}$$

(2.3)

(2.3) is called Nyquist's first criterion in the time domain [1] and the corresponding impulse is referred to as Nyquist impulse. An example of a real-valued impulse response satisfying (2.3) is depicted in Fig. 2.1. Obviously, $h_e(t)$ owns equidistant zeros except at $t = t_0$.

Inserting the Nyquist condition (2.3) into (2.2) yields

$$I(k) = 0 \ ; \ \forall k \in \mathbb{Z}$$

(2.4)

and we obtain from (2.1)

$$q(k) = a(k)h(0) + n(k)$$

(2.5)

As expected, the signal $q(k)$ at the receiver output suffers not anymore from inter-symbol interference and the symbol sequence $a(k)h(0)$ is only corrupted by additive noise $n(k)$.

2.3   Nyquist's First Criterion in the Frequency Domain

An interesting question is: "How does the spectrum $H_e(f)$ of a Nyquist impulse look like?" We give the solution for the case of a real-valued spectrum $H_e(f)$ with the consequence that the corresponding impulse is real and an even function, $h_e(-t) = h_e(t)$.

Fig. 2.2 Real-valued transfer function $H_e(f)$ of a Nyquist lowpass. Roll-offs exhibit odd symmetry with respect to points P_1 $(f_N, 0.5)$ and P_2 $(-f_N, 0.5)$

Thus, in the following $t_0 = 0$ is assumed. Given these prerequisites the necessary and sufficient condition for the spectrum $H_e(f)$ is

$$\sum_{m=-\infty}^{\infty} H_e\left(f - m\frac{1}{T}\right) = h_e(0)T \; ; \; \forall f \; ; \; h_e(0)T = H_e(0) \qquad (2.6)$$

which is called Nyquist's first criterion in the frequency domain and will be proven in the following. (2.6) requires that the sum of all periodic repetitions of $H_e(f)$ is a constant, $H_e(0) = h_e(0)T$. A lowpass satisfying the condition (2.6) is also called a Nyquist lowpass. From (2.6) directly follows the solution for a lowpass spectrum with cut-off frequency

$$f_c = f_N + \Delta f \qquad (2.7)$$

$$H_e(f) \begin{cases} = A & ; \; |f| \le f_N - \Delta f \\ H_e\left(|f_N| - x\right) + H_e\left(|f_N| + x\right) = A ; & 0 < x \le \Delta f \\ = 0 & ; \; |f| > f_N + \Delta f \end{cases} \qquad (2.8)$$

where we have set $A = H_e(0)$. Obviously, $H_e(f)$ is an even function, $H_e(-f) = H_e(f)$. The principle spectrum of $H_e(f)$ is depicted in Fig. 2.2

The relation between the Nyquist frequency f_N and the symbol interval T is given by

$$f_N = \frac{1}{2T} \qquad (2.9)$$

The second condition in (2.8) requires a roll-off function from the passband to the stop-band, which exhibits odd symmetry with respect to the points P_1 and P_2. Δf defines half of the roll-off bandwidth

$$0 \le \Delta f \le f_N \qquad (2.10)$$

For $\Delta f = 0$ the spectrum $H_e(f)$ is an ideal lowpass filter and shows the minimal cut-off frequency $f_c = f_N$. On the contrary, the maximum cut-off frequency is $2f_N$ and is obtained for $\Delta f = f_N$. In that case the roll-off starts already at $f = 0$ and covers the whole passband. For a given symbol rate $v_S = \frac{1}{T}$ the Nyquist frequency (2.9) results in

$$f_N = \frac{1}{2} v_S \tag{2.11}$$

For example, if we strive for a symbol rate of 1 Gsymbol/s a minimal cut-off frequency of 500 MHz is required to satisfy Nyquist's criterion.

For the filter design the following definition of a roll-off factor is helpful

$$\alpha = \frac{\Delta f}{f_N} \tag{2.12}$$

From (2.10) we see that $0 \leq \alpha \leq 1$ holds and the cut-off frequency becomes $f_c = (1 + \alpha) f_N$.

If we drop the previous prerequisite and consider an impulse with $t_0 > 0$, as depicted in Fig. 2.1, then $H_e(f)$ turns into $H_e(f)e^{-j2\pi f t_0}$ according to the shifting property of the Fourier transform. Then the spectrum is equipped with a linear phase term $e^{-j2\pi f t_0}$.

As a conclusion, we can achieve the transmission of the symbol sequence $a(k)$ without intersymbol interference, if the overall transfer function $H_e(f)$ of the system is a Nyquist lowpass defined in (2.8). An adequate design is

$$f_I = f_R = (1 + \alpha) f_N \tag{2.13}$$

where f_R is the cut-off frequency of the receive lowpass filter $G_R(f)$ and $2f_I$ the bandwidth of the channel.

It is worth mentioning that also transfer functions with other than lowpass characteristics can be deduced from (2.6). For illustration the following example of a bandpass

$$B(f) = \frac{1}{2} H_e(f + 2f_N) + \frac{1}{2} H_e(f - 2f_N) \tag{2.14}$$

is given, which is composed of the Nyquist lowpass $H_e(f)$ according to (2.8). The reader can assure oneself easily that $\sum_{m=-\infty}^{\infty} B(f - m2f_N) = H_e(0) \; ; \; \forall f$ holds and thus $B(f)$ fulfills the first Nyquist criterion (2.6).

Proof of (2.6)

In the following we assume $t_0 = 0$ and first prove that (2.3) is a sufficient condition for (2.6), i.e. that (2.6) follows from (2.3). Ideal sampling of $h_e(t)$ yields $h_e(t) \sum_{k=-\infty}^{\infty} \delta(t - kT)$ and with (2.3) follows $h_e(t) \sum_{k=-\infty}^{\infty} \delta(t - kT) = h_e(0)\delta(t)$. Applying the Fourier transform on both sides results in $H_e(f) * \frac{1}{T} \sum_{m=-\infty}^{\infty} \delta(f - m\frac{1}{T}) = h_e(0)$, where we have used the transform pairs $\sum_{k=-\infty}^{\infty} \delta(t - kT) \longmapsto \frac{1}{T} \sum_{m=-\infty}^{\infty} \delta(f - m\frac{1}{T})$ and $\delta(t) \longmapsto 1$. With the convolution integral follows $H_e(f) * \frac{1}{T} \sum_{m=-\infty}^{\infty} \delta(f - m\frac{1}{T}) = \frac{1}{T} \sum_{m=-\infty}^{\infty} \int_{-\infty}^{\infty} H_e(\tau)\delta(f - m\frac{1}{T} - \tau)d\tau = \frac{1}{T} \sum_{m=-\infty}^{\infty} H_e(f - m\frac{1}{T})$. Hence, we end up with $\sum_{m=-\infty}^{\infty} H_e(f - m\frac{1}{T}) = h_e(0)T$, which validates (2.6).

Next we prove that (2.3) is a necessary condition for (2.6), i.e. we have to show that from (2.6) follows (2.3). This is easily done by starting from (2.6) and executing the steps done before in reverse direction using the inverse Fourier transform, which finally results in (2.3). This finalizes the proof that (2.3) and (2.6) are necessary and sufficient conditions.

2.4 Raised Cosine Nyquist Lowpass Filter

Now we consider a special Nyquist filter $H_e(f)$, which is frequently used as a design goal for digital communication systems

$$\frac{H_e(f)}{H_e(0)} = \begin{cases} 1 & ; & |f| \leq f_N(1 - \alpha) \\ \frac{1}{2}\left[1 + \cos\left(\frac{\pi}{2} \frac{|f| - f_N(1-\alpha)}{\alpha f_N}\right)\right] & ; & f_N(1 - \alpha) < |f| < f_N(1 + \alpha) \\ 0 & ; & |f| \geq f_N(1 + \alpha) \end{cases}$$

(2.15)

and portrait in Fig. 2.3. For an overall gain of one we have to determine $H_e(0) = 1$. The function $1 + \cos(\ldots)$ is called raised cosine. Significant are the roll-offs, which possess an odd symmetry with respect to the points $(\pm f_N, \frac{1}{2})$. The corresponding Nyquist impulses are obtained by inverse Fourier transform

$$\frac{h_e(t)}{h_e(0)} = \text{sinc}\left(\frac{t}{T}\right) \frac{\cos\left(\pi\alpha\frac{t}{T}\right)}{1 - \left(2\alpha\frac{t}{T}\right)^2}$$

(2.16)

where $h_e(0) = \frac{1}{T}H_e(0)$ and

$$\text{sinc}(x) = \frac{\sin(\pi x)}{\pi x}$$

(2.17)

Fig. 2.3 Raised cosine transfer function $H_e(f)$ of a Nyquist lowpass filter with various roll-off factors α

$H_e(f)/H_e(0)$

f/f_N

Fig. 2.4 Impulse response $h_e(t)$ of raised cosine Nyquist lowpass filter with various roll-off factors α

$h_e(t)/h_e(0)$

t/T

In Fig. 2.3 $H_e(f)$ is shown for various roll-off factors α and the corresponding impulse responses $h_e(t)$ are depicted in Fig. 2.4. As expected from the properties of the Fourier transform, the smoother the roll-off of $H_e(f)$ the smaller the magnitudes of the over- and under-shoots of the impulse responses are. The ideal lowpass filter with $\alpha = 0$ exhibits a step function at the transition from the passband to the stopband and therefore the over- and under-shoots are the largest. In this case (2.16) yields the well known sinc-function as the impulse response

$$\frac{h_e(t)}{h_e(0)} = \text{sinc}\left(\frac{t}{T}\right) \tag{2.18}$$

However, this response is not well suited for digital communications, because even a small deviation of the sampling phase at the receiver introduces strong intersymbol interference, which even approaches infinity theoretically. $\alpha = 1$ yields the maximal cut-off frequency $f_c = 2f_N$ and the roll-off is very smooth. Consequently, $h_e(t)$ is almost zero for $|t| > T$ and therefore is often approximated by a symmetrical triangular impulse with duration $2T$. For all α the periodic zero crossings of $h_e(t)$ for $|t| \geq T$, also called "Nyquist zeros" are clearly visible, which avoid the intersymbol interference completely.

Fig. 2.5 Receiver output signal $q_R(t)$ according to (1.36) with raised cosine Nyquist impulses $h_e(t)$ from (2.16), roll-off factor $\alpha = 0.3$, without noise, and binary transmit symbols $a(k) \in \{-1, +1\}$. No intersymbol interference is present at the sampling instants $t/T = 0, 1, 2, \ldots$. Source: Online platform "webdemo" [2]

Fig. 2.5 shows an example of the receiver output signal $q_R(t)$ in (1.36) without noise, simulated with the online platform "webdemo" [2], which also enables to adjust various system parameters online. The impulse response $h_e(t)$ of the equivalent baseband system is chosen as a raised cosine Nyquist impulse (2.16) with roll-off factor $\alpha = 0.3$. The transmit symbol sequence $a(k)$ is binary with the alphabet $\{-1, +1\}$ and indicated with red color. Each transmit symbol $a(m)$ arrives at the receiver as an impulse $a(m)h_e(t - mT)$ and $q_R(t)$ is the superposition of the received impulses. Apparently, the Nyquist impulse guarantees that no intersymbol interference is present in $q_R(t)$ at the sampling instants $t/T = 0, 1, 2, \ldots$. Hence, the transmit symbol sequence can be recovered without distortion in case of no noise.

2.5 Eye Diagram

Figure 2.6 portraits the *eye diagram*, which is an important measure for diagnosis of the receive signal quality for data transmission. Symbols $a(k) \in \{-1, +1\}$ are sent. At the receiver an oscilloscope with memory is triggered with the symbol rate $1/T$ and one receive impulse upon the other is recorded. On the display the shape of an eye can be seen in $|t| \leq T/2$. If the eye is vertically open, then intersymbol interference is low. The vertical eye opening of Nyquist impulses is even 100% at the sampling instant $t = 0$ and thus the margin of additive noise is maximal. If a jitter of the sampling clock phase moves the sampling instant aside from zero, the vertical eye opening reduces. Thus, the larger the vertical and horizontal eye openings are, the greater the tolerance for additive noise and phase jitter will be, respectively. Furthrtmore, the difference between the maximum and the minimum of $q_R(t)$ is a measure of the dynamic range of the receive signal.

Fig. 2.6 Eye diagram of raised cosine Nyquist impulse with roll-off factor $\alpha = 0.3$, without noise, and binary transmit symbols $a(k) \in \{-1, +1\}$. Source: Online platform "webdemo" [2]

2.6 Characterization of the Noise at the Receiver

Before we are going to discuss various detection algorithms for the receiver output signal $q(k)$ we have to characterize the noise $n_C(t)$ in Fig. 1.1. The main sources of noise are the resistors and the electronic components, such as transistors or the photo diode in an optical receiver. In a wireless system the receive antenna collects noise coming from the channel. In most cases the first stage of a receiver is composed of an amplifier associated with a bandpass filter to limit the noise spectrum to the passband of the transmit signal, which is given by the channel transfer function in (1.19). The resulting real-valued noise $n_C(t)$ is demodulated and lowpass filtered with $\sqrt{2}G_R(f)$ yielding the complex-valued noise $n_R(t)$ with a lowpass spectrum. The noise sequence $n(k)$ results after sampling and is depicted in the discrete-time equivalent baseband system in Fig. 1.5 (b). In the following we analyze the noise at the receiver step by step. For the basics on stochastic processes we refer to the Appendix A.

2.7 Channel Noise $n_C(t)$

A very likely model for the noise $n_C(t)$ at the receiver is the zero mean additive Gaussian noise with a bandpass shaped power spectral density. $n_C(t)$ is regarded as a sample function of a stationary stochastic bandpass process. In the Appendix A we show that this noise can be characterized by the following properties: [1]

In general the real-valued stationary bandpass noise is given in the time-domain by

$$n_C(t) = x_1(t) \cos(2\pi f_0 t) - x_2(t) \sin(2\pi f_0 t) \qquad (2.19)$$

where $x_1(t)$ and $x_2(t)$ are real-valued lowpass noises with the same Gaussian probability density function

---

[1] Different from Appendix A we denote the stochastic processes and their sample functions with the same lowercase letters to simplify the notation.

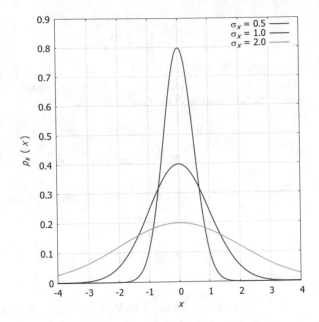

Fig. 2.7 Gaussian probability density function $p_x(x) = \frac{1}{\sqrt{2\pi}\sigma_x}e^{-\frac{x^2}{2\sigma_x^2}}$ with various standard deviations σ_x

$$p_{x_i}(x_i) = \frac{1}{\sqrt{2\pi}\sigma_x}e^{-\frac{x_i^2}{2\sigma_x^2}} \; ; \; i = 1, 2 \tag{2.20}$$

also called normal distribution. Fig. 2.7 shows the typical bell-shaped curve for three different standard deviations σ_x. The half-width (at 50% of the maximum) is known as $\Delta x = 2\sqrt{2\ln(2)}\sigma_x \approx 2.4\sigma_x$. Thus, the larger σ_x the wider the density function is. Please note, $\int_{-\infty}^{\infty} p_x(x)dx = 1$ holds in general.

It goes without saying that the probability density function holds for every fixed time instant t of the stationary stochastic process. $n_C(t)$, $x_1(t)$, and $x_2(t)$ possess zero mean

$$\mathbf{E}[n_C(t)] = \mathbf{E}[x_1(t)] = \mathbf{E}[x_2(t)] = 0 \tag{2.21}$$

and the same mean power

$$\sigma_x^2 = \mathbf{E}[n_C^2(t)] = \mathbf{E}[x_1^2(t)] = \mathbf{E}[x_2^2(t)] \tag{2.22}$$

Moreover we assume $x_1(t)$ and $x_2(t)$ as uncorrelated, i.e. the cross-correlation function is

$$R_{x_1x_2}(\tau) = \mathbf{E}[x_1(t)x_2(t+\tau)] = \mathbf{E}[x_1(t)]\,\mathbf{E}[x_2(t)] \tag{2.23}$$

and because of (2.21) $R_{x_1x_2}(\tau) = 0$ holds. As shown in the Appendix A, Gaussian processes with this property are even statistically independent.

The power spectral densities of $x_1(t)$ and $x_2(t)$ are identical. They are given by

$$S_{x_i x_i}(f) = \begin{cases} N_0 ; & |f| \leq f_I \\ 0 ; & else \end{cases} ; i = 1, 2 \tag{2.24}$$

where $2f_I$ is the bandwidth of the channel transfer function $G_C(f)$ in (1.19). The noise spectrum is step-wise constant. Although misleading, it is therefore sometimes called "band-limited white" Gaussian noise. The relation between $S_{x_i x_i}(f)$ and the mean noise power is

$$\sigma_x^2 = \int_{-\infty}^{\infty} S_{x_i x_i}(f) df = 2 f_I N_0 ; i = 1, 2 \tag{2.25}$$

As shown in the Appendix A, the power spectral density of $n_C(t)$ can be determined with $S_{x_i x_i}(f)$ as

$$S_{n_C n_C}(f) = \frac{1}{2} \left[S_{x_i x_i}(f - f_0) + S_{x_i x_i}(f + f_0) \right] = \begin{cases} \frac{N_0}{2} ; & f_0 - f_I \leq |f| \leq f_0 + f_I \\ 0 ; & else \end{cases} \tag{2.26}$$

Thus, we have a strictly band-limited bandpass shaped noise spectrum with the passband in the region of the transmission channel (1.19). This noise can be simulated by a frequency flat broadband noise at the input of an ideal bandpass filter with cut-off frequencies given in (1.19). Sometimes $n_c(t)$ is also called "band-limited white" noise, although misleading.

2.8 Noise After Demodulation and Lowpass Filtering

According to Fig. 1.1 after demodulation and lowpass filtering with $\sqrt{2} g_R(t)$ the real-valued channel noise $n_C(t)$ turns into a complex noise $n_R(t)$ given by (1.25). Rewriting (2.19) using $\cos(2\pi f_0 t) = \frac{1}{2} \left[e^{j2\pi f_0 t} + e^{-j2\pi f_0 t} \right]$ and $\sin(2\pi f_0 t) = \frac{1}{2j} \left[e^{j2\pi f_0 t} - e^{-j2\pi f_0 t} \right]$ we obtain from (1.25)

$$n_R(t) = \frac{\sqrt{2}}{2} \left[x_1(t) + j x_2(t) \right] * g_R(t) + \\ + \frac{\sqrt{2}}{2} \left\{ \left[x_1(t) - j x_2(t) \right] e^{-j4\pi f_0 t} \right\} * g_R(t) \tag{2.27}$$

The term in the second line of (2.27) represents the convolution of a bandpass noise in the frequency range around $f = 2f_0$ with a lowpass impulse response $g_R(t)$ with a cut-off frequency $f_R \ll 2f_0$. Thus, the result is zero. Consequently we obtain

$$n_R(t) = \frac{\sqrt{2}}{2} \left[x_1(t) + j x_2(t) \right] * g_R(t) \tag{2.28}$$

Now we assume that $g_R(t)$ is the impulse response of an ideal lowpass filter with cut-off frequency f_R. From the viewpoint of noise reduction we wish to make f_R small. However, please note that in this case we limit the bandwidth of the signal, which is f_I, at the same time. As this is not adequate, we have to accept

$$f_R = f_I \qquad (2.29)$$

according to (2.13), where $2f_I$ is the bandwidth of the bandpass channel transfer function. Let

$$G_R(f) = \begin{cases} 1 \; ; \; |f| \leq f_I \\ 0 \; ; \quad else \end{cases} \qquad (2.30)$$

be the ideal lowpass receive filter. Then its output is the complex-valued lowpass noise

$$n_R(t) = \frac{\sqrt{2}}{2} \left[x_1(t) + j x_2(t) \right] \qquad (2.31)$$

$n_R(t)$ and thus $x_1(t)$ as well as $x_2(t)$ remain Gaussian, because a complex Gaussian process passing through a linear system remains Gaussian, only the variance may change. Noting that the real and the imaginary part of $n_R(t)$ are statistically independent $n_R(t)$ has the mean power

$$\mathbf{E}\left[|n_R|^2\right] = \sigma_x^2 \qquad (2.32)$$

To simplify the notation we introduce new components of the complex noise incorporating the factor $\frac{\sqrt{2}}{2}$

$$\tilde{n}_i(t) = \frac{\sqrt{2}}{2} x_i(t) \; ; \; i = 1, 2 \qquad (2.33)$$

Then the output noise of the receive lowpass can be written as

$$n_R(t) = \tilde{n}_1(t) + j\tilde{n}_2(t) \qquad (2.34)$$

Obviously, $\tilde{n}_1(t)$ and $\tilde{n}_2(t)$ have identical Gaussian probability density functions

$$p_{\tilde{n}_i}(\tilde{n}_i) = \frac{1}{\sqrt{2\pi}\sigma_n} e^{-\frac{\tilde{n}_i^2}{2\sigma_n^2}} \; ; \; i = 1, 2 \qquad (2.35)$$

and the variance σ_n^2 of $\tilde{n}_i(t)$ is obtained from (2.33) with (2.22)

$$\sigma_n^2 = \mathbf{E}\left[\tilde{n}_i^2\right] = \frac{1}{2}\mathbf{E}\left[x_i^2\right] = \frac{1}{2}\sigma_x^2 \; ; \; i = 1, 2 \qquad (2.36)$$

The power spectral density of the noise at the output of the receive lowpass then follows from (2.24)

$$S_{\tilde{n}_i\tilde{n}_i}(f) = \begin{cases} \frac{N_0}{2} \; ; \; |f| \le f_I \\ 0 \; ; \quad else \end{cases} \; ; \; i = 1,2 \qquad (2.37)$$

As shown in the Appendix A, the autocorrelation function follows from the power spectral density by applying the inverse Fourier transform. Thus, we obtain the autocorrelation function $R_{\tilde{n}_i\tilde{n}_i}(\tau)$ of $\tilde{n}_i(t)$

$$S_{\tilde{n}_i\tilde{n}_i}(f) \multimap R_{\tilde{n}_i\tilde{n}_i}(\tau) = N_0 f_I \text{sinc}\,(2f_I\tau) \; ; \; i = 1,2 \qquad (2.38)$$

2.9 Noise After Sampling

For the detection of the QAM symbols the statistics of the real and imaginary part of the noise after sampling is important to know. At the output of the sampling device in Fig. 1.1 we obtain the noise sequence from (2.34)

$$n_R(t_0 + kT) = \tilde{n}_1(t_0 + kT) + j\tilde{n}_2(t_0 + kT) \qquad (2.39)$$

To simplify the notation we introduce

$$n(k) = n_R(t_0 + kT) \qquad (2.40)$$

and

$$n_i(k) = \tilde{n}_i(t_0 + kT) \; ; \; i = 1,2 \qquad (2.41)$$

Then (2.39) turns into

$$n(k) = n_1(k) + jn_2(k) \qquad (2.42)$$

As the Gaussian noise is stationary, the probability density function is independent of any time instant. Consequently, the samples $n_i(k)$ possess the same probability density functions as $\tilde{n}_i(t)$; $i = 1,2$, which are given in (2.35) and (2.36)

$$p_{n_i}(n_i) = \frac{1}{\sqrt{2\pi}\sigma_n} e^{-\frac{n_i^2}{2\sigma_n^2}} \; ; \; \mathbf{E}\left[n_i^2\right] = \sigma_n^2 = \frac{1}{2}\sigma_x^2 \; ; \; i = 1,2 \qquad (2.43)$$

Furthermore, both noise components have zero mean

$$\mathbf{E}\left[n_i\right] = 0 \; ; \; i = 1,2 \qquad (2.44)$$

The probability density function of $n(k)$ can be obtained with the following consideration. As $x_1(t)$ and $x_2(t)$ are statistically independent, this also holds for $n_1(k)$ and $n_2(k)$. Consequently, we obtain the density function of the complex noise $n(k)$ as

$$p_n(n) = p_{n_1}(n_1)p_{n_2}(n_2) = \frac{1}{2\pi\sigma_n^2}e^{-\frac{n_1^2+n_2^2}{2\sigma_n^2}} = \frac{1}{2\pi\sigma_n^2}e^{-\frac{|n|^2}{2\sigma_n^2}} \tag{2.45}$$

As shown in the Appendix A, sampling of a stochastic process with the sampling rate $\frac{1}{T}$ results in a discrete-time autocorrelation function with samples at $\tau = mT$; $m \in \mathbb{Z}$.[2] Thus, the autocorrelation function of $n_i(k)$ is $R_{\tilde{n}_i\tilde{n}_i}(mT) = R_{n_in_i}(m)$ and with (2.38) we obtain

$$R_{n_in_i}(m) = N_0 f_I \text{sinc}(2f_I T m) ; \quad i = 1, 2 ; \quad m \in \mathbb{Z} \tag{2.46}$$

If we choose the cut-off frequency $f_I = (1+\alpha)f_N = \frac{1+\alpha}{2T}$ as discussed in (2.13) we get

$$R_{n_in_i}(m) = N_0(1+\alpha)f_N \text{sinc}((1+\alpha)m) ; \quad i = 1, 2 ; \quad m \in \mathbb{Z} \tag{2.47}$$

The reader can convince oneself easily that for $\alpha = 0$ and $\alpha = 1$ the autocorrelation function is zero for $m = \pm 1, \pm 2, ...$, thus,

$$R_{n_in_i}(m) = \begin{cases} N_0(1+\alpha)f_N ; & m = 0 \\ 0 & ; m = \pm 1, \pm 2, ... \end{cases} \tag{2.48}$$

We conclude from (2.48) that two samples $n_i(k)$ and $n_i(k+m)$, which can be considered as random variables, are uncorrelated for $m \neq 0$. Hence, the sampling operation with the sampling frequency $\frac{1}{T}$ acts as a decorrelation. Because the uncorrelated random variables $n_i(k)$ and $n_i(k+m)$ are also Gaussian, they are even statistically independent. However, $\alpha = 0$ is not feasible for a practical implementation, as discussed earlier.

According to Appendix A the corresponding power spectral density of a discrete-time stochastic process is periodic and given by

$$S_{n_in_i}(f) = \sum_{\nu=-\infty}^{\infty} S_{\tilde{n}_i\tilde{n}_i}(f - \nu\frac{1}{T}) \tag{2.49}$$

Example 1

One expects that the power spectral density of an uncorrelated stochastic process is white. Consequently, $S_{n_in_i}(f) = const.$ must hold for $\alpha = 0$ and $\alpha = 1$.

---

[2] As mentioned in the footnote of Appendix A, the sampling function is defined as $T\sum_{k=-\infty}^{\infty}\delta(t-kT)$.

Solution

For $\alpha = 0$ the cut-off frequency of $S_{\tilde{n}_i \tilde{n}_i}(f)$ in (2.37) is $f_I = \frac{1+\alpha}{2T} = \frac{1}{2T}$. The periodic repetition of $S_{\tilde{n}_i \tilde{n}_i}(f)$ in (2.49) yields $S_{n_i n_i}(f) = \frac{N_0}{2} = const$. In case of $\alpha = 1$ follows $f_I = \frac{1}{T}$ and the spectral parts in (2.49) overlap resulting in $S_{n_i n_i}(f) = N_0 = const$.

2.10 Summary

We summarize the results to complete the noise properties for the equivalent baseband system models given in Sect. 1.6. The original block diagram and that of the equivalent system are shown in Figs. 1.1 and 1.5, respectively.

Continuous-Time Equivalent Baseband System Model
The bandpass noise $n_C(t)$ at the input of the receiver is characterized by (2.19). The lowpass in-phase $x_1(t)$ and quadrature component $x_2(t)$ are assumed to be statistically independent Gaussian noises with the same probability density function defined in (2.20), zero mean (2.21), and identical mean power (2.22). After demodulation and lowpass filtering $n_C(t)$ turns into a complex-valued Gaussian noise $n_R(t) = \tilde{n}_1(t) + j\tilde{n}_2(t)$ according to (2.34). Due to the amplification factor $\sqrt{2}$ of the receive lowpass the noise components possess the mean power $\mathbf{E}\left[\tilde{n}_i^2\right] = \frac{1}{2}\mathbf{E}\left[x_i^2\right]$ in (2.36) and the power spectral density given in (2.37) with lowpass characteristics. $\tilde{n}_1(t)$ and $\tilde{n}_2(t)$ are statistically independent with Gaussian probability density function (2.35).

Discrete-Time Equivalent Baseband System Model
The complex-valued, discrete-time noise $n(k) = n_1(k) + jn_2(k)$ is given by (2.42), in which $n_i(k) = \tilde{n}_i(t_0 + kT)$; $i = 1, 2$. The $n_1(k)$ and $n_2(k)$ exhibit the same zero mean Gaussian probability density function (2.43) and they are statistically independent. If the cut-off frequency of an ideal lowpass filter (2.30) at the receiver is chosen as $f_I = \frac{(1+\alpha)}{2T}$ with $\alpha = 0$ or $\alpha = 1$ and the noise input spectrum is flat, then the output samples $n_i(k)$ and $n_i(k+m)$ are statistically independent ($m \neq 0$) and the power spectral density $S_{n_i n_i}(f)$ in (2.49) is white ($i = 1, 2$).

References

1. H. Nyquist. Certain topics in telegraph transmission theory (reprint from Transactions of the A.I.E.E., febr. 1928). *Proceedings of the IEEE*, 2002
2. S. ten Brink. Pulse shaping, webdemo. Technical report, Online available: http://webdemo.inue. uni-stuttgart.de, Institute of Telecommunications, University of Stuttgart, Germany, 2018

Chapter 3
Detection Methods

3.1 Receive Signal Under Detection

In the following a survey on the most important detection methods is presented. We differentiate in principle between the symbol-by-symbol and the sequence or sequential detection. With the first method the receive signal $q(k)$ in Figs. 1.1 and 1.5 is decided at every time instant k. The sequential detection scheme takes decisions periodically after the observation of K past samples, e.g., after $q(0), q(1), \ldots, q(K-1)$. In this section we illustrate the key detection methods and consider 4-PSK depicted in Fig. 3.1 as an example. Assume that the intersymbol interference is completely removed and that the signal at the input of the detector is $q(k)$ given by (2.5). For simplification let $h(0) = 1$. Then we obtain from (2.5) the signal under decision as

$$q(k) = a(k) + n(k) \tag{3.1}$$

which is composed of the symbol sequence $a(k)$ sent by the transmitter and the additive Gaussian noise $n(k) = n_1(k) + jn_2(k)$ with zero mean and probability density function $p_{n_i}(n_i)$; $i = 1, 2$ in (2.43). Each $a(k)$ can represent one signal point out of the symbol alphabet $\mathcal{B} = \{a_1, a_2, a_3, a_4\}$. Figure 3.1 illustrates (3.1) in the complex plane for the case that the symbol $a(k) = a_1$ was sent.

3.2 Maximum Likelihood Symbol-by-Symbol Detection

3.2.1 Maximum Likelihood Detection

To show the principle let us assume that the symbol $a(k) = a_\nu$ was sent at time instant k. In the following we drop k to simplify notation. For maximum likelihood detection a likelihood function is defined, which is the conditional probability density function

© Springer Nature Switzerland AG 2021
J. Speidel, *Introduction to Digital Communications*,
Signals and Communication Technology,
https://doi.org/10.1007/978-3-030-67357-4_3

Fig. 3.1 QAM signal points $a_1, a_2, \ldots, a_4$, receive signal $q(k)$, and noise $n(k)$ for 4-PSK

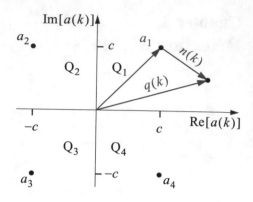

$p_L (q \mid a_\nu)$ or a monotonic function of it, e.g., the logarithmic function. $p_L (q \mid a_\nu)$ relates the density function of q under the condition that the symbol a_ν was sent. With $p_n(n)$ in (2.45) and with (3.1) follows

$$p_L (q \mid a_\nu) = p_n(q - a_\nu) = \frac{1}{2\pi\sigma_n^2} e^{-\frac{|q-a_\nu|^2}{2\sigma_n^2}} \; ; \; \nu = 1, 2, \ldots, 4 \qquad (3.2)$$

Please note that (3.2) describes for the signal constellation in Fig. 3.1 a set of four density functions, each one shifted by a_ν. The maximum likelihood detection (or decision) rule is:

Select that symbol a_ν, which is associated with the largest $p_L (q \mid a_\nu)$. Thus, the output of the detector is

$$\hat{a} = \arg \left\{ \max_{\nu=1,\ldots 4} [p_L (q \mid a_\nu)] \right\} \qquad (3.3)$$

Looking at (3.2) the maximal density function results for minimal $|q - a_\nu|^2$, as the exponent of the exponential function is negative. Consequently, (3.3) is equivalent to

$$\hat{a} = \arg \left\{ \min_{\nu=1,\ldots 4} \left[|q - a_\nu|^2 \right] \right\} \qquad (3.4)$$

and the probabilistic approach turns into the calculation of the squared Euclidean distance between two complex numbers, namely q and a_ν for $\nu = 1, 2, \ldots, 4$. It is also worth noting that the solution (3.4) does not depend on the mean power σ_n^2 of the noise n.

3.2.2 Threshold Detection

From the regular signal constellation given in Fig. 3.1 we see that the distance between the various a_ν and a given q is minimal in the first quadrant Q_1 of the complex plane. Consequently, we can formulate a detection criterion alternative to (3.4) by introducing decision regions in the complex plane. In our example these are the four quadrants $Q_1, Q_2, \ldots, Q_4$, if the symbols have equal a-priori probabilities. The decision regions are separated by decision thresholds, which are the real and the imaginary axis in the example of Fig. 3.1. The decision rule for a correct decision then is

$$\hat{a} = a_\nu \text{ if } q \, \epsilon \, Q_\nu \; ; \;\; \nu = 1, 2, \ldots, 4 \tag{3.5}$$

otherwise the decision is in error. If q is located on the decision threshold, no unique result is achieved. For low magnitudes of n we recognize a more reliable decision the closer q is located to a symbol. Therefore, advanced detection techniques take this fact into consideration and introduce "soft decision" as opposed to the described "hard decision" in particular together with forward error correction encoding [1–3].

3.2.3 Symbol Error Probability for Threshold Detection

The Gaussian Q-Function

The symbol error probability can be formulated with the Gaussian Q-function, which is introduced in the following. Let X be a random variable with Gaussian probability density function $p_x(x)$ given in (2.20), where we set $x_i = x$. As shown in the Appendix A, the probability that the random variable X is larger than a given real number α is

$$P[X > \alpha] = \int_\alpha^\infty p_x(x)dx = \frac{1}{\sqrt{2\pi}\sigma_x} \int_\alpha^\infty e^{-\frac{x^2}{2\sigma_x^2}} dx \tag{3.6}$$

where P$[\ldots]$ denotes the probability operator. With the substitution $u = \frac{x}{\sigma_x}$ follows

$$P[X > \alpha] = \frac{1}{\sqrt{2\pi}} \int_{\frac{\alpha}{\sigma_x}}^\infty e^{-\frac{u^2}{2}} du = Q\left(\frac{\alpha}{\sigma_x}\right) \tag{3.7}$$

with the Q-function

$$Q(\alpha) = \frac{1}{\sqrt{2\pi}} \int_\alpha^\infty e^{-\frac{u^2}{2}} du \tag{3.8}$$

It is straightforward to show the following properties of the Q-function

Fig. 3.2 Q-function $Q(\alpha) = \frac{1}{\sqrt{2\pi}} \int_\alpha^\infty e^{-\frac{x^2}{2}} dx$ together with approximations and upper bounds

$$Q(-\infty) = 1; \quad Q(0) = \frac{1}{2}; \quad Q(\infty) = 0; \quad Q(-\alpha) = 1 - Q(\alpha) \qquad (3.9)$$

and the relation with the error function $\mathrm{erf}(x) = \frac{2}{\sqrt{\pi}} \int_0^x e^{-u^2} du$ is

$$Q(\alpha) = \frac{1}{2}\left[1 - \mathrm{erf}(\frac{\alpha}{\sqrt{2}})\right] = \frac{1}{2}\mathrm{erfc}(\frac{\alpha}{\sqrt{2}}) \qquad (3.10)$$

where $\mathrm{erfc}(x) = 1 - \mathrm{erf}(x)$ is defined. There is no closed form solution for the integral $Q(\alpha)$ for arbitrary α. Hence, we have to rely on numerical calculations. The result is depicted in Fig. 3.2 together with some helpful approximations and upper bounds. Obviously, the Q-function is declining strongly with increasing α.

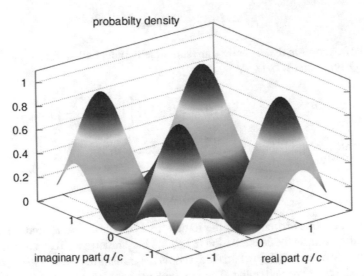

probabilty density

1

0.8

0.6

0.4

0.2

0

1

0

imaginary part q/c -1   -1   real part q/c

Fig. 3.3 Superposition of the Gaussian conditional probability density functions $p_L(q \mid a_\nu)$ in (3.2) of the receiver output signal q for the 4-PSK symbols a_ν with equal a-priori probabilities ($\nu = 1, 2, 3, 4$) and $10 \log\left(\frac{1}{2\sigma_n^2}\right) = 5$ dB. *Source* Online platform "webdemo" [4]

Example 2: Symbol Error Probability for 4-PSK

We consider Fig. 3.1, which depicts the signal points $a_1, \ldots, a_4$ for 4-PSK in the constellation diagram. Obviously, these points are located on a circle with radius $b = \sqrt{2}c$ around the origin. Figure 3.3 shows the surface diagram of the superposition of the conditional probability density functions $p_L(q \mid a_\nu)$ in (3.2) of the receiver output signal q for the 4-PSK symbols a_ν with equal a-priori probabilities ($\nu = 1, 2, 3, 4$). In this diagram $10 \log\left(\frac{1}{2\sigma_n^2}\right) = 5$ dB is assumed. It is generated with the online platform "webdemo" [4], which also enables to adjust different parameters. Apparently, the individual probability density functions centered around each PSK symbol overlap and hence we expect quite some decision errors.

Assume for the moment that only the symbol $a_1 = c + jc$ is transmitted, which shall be known at the receiver. Then the signal (3.1) under decision will be with (2.42)

$$q(k) = a_1 + n(k) = c + n_1(k) + j(c + n_2(k)) \tag{3.11}$$

at time instant k.

According to the decision rule (3.5) the decision for a_1 is correct, if $q(k)$ is located in the first quadrant Q_1 of the complex plane. The following equivalent relations are true (k is dropped to simplify notation)

$$q \in Q_1 \Leftrightarrow \text{Re}\,[q] > 0\ ;\ \text{Im}\,[q] > 0 \Leftrightarrow n_1 > -c\ ;\ n_2 > -c \tag{3.12}$$

Then follows for the probability P_{a_1} that a_1 is decided correctly

$$P_{a_1} = P[n_1 > -c, \, n_2 > -c] \tag{3.13}$$

Using the assumption that n_1 and n_2 are statistically independent follows with (2.43) and the Q-function (3.7), (3.9)

$$P_{a_1} = P[n_1 > -c] \, P[n_2 > -c] = \left[1 - Q\left(\frac{c}{\sigma_n}\right)\right]^2 \tag{3.14}$$

Consequently, the probability of a wrong decision of a_1 is

$$P_{S,1} = 1 - P_{a_1} = 2Q\left(\frac{c}{\sigma_n}\right) - Q^2\left(\frac{c}{\sigma_n}\right) \tag{3.15}$$

We can execute this procedure for the remaining symbols a_2, a_3, a_4 and achieve the same results, because the constellation in Fig. 3.1 is symmetrical. To get an idea about the symbol error probability when all symbols are transmitted over a long period of time we take the expected value. The symbols are sent by the transmitter with the a-priori probabilities $P[a_\nu]$, for which in general

$$\sum_{\nu=1}^{4} P[a_\nu] = 1 \tag{3.16}$$

holds. Then we obtain the mean symbol error probability

$$P_S = \sum_{\nu=1}^{4} P[a_\nu] \, P_{S,1} = P_{S,1} = 2Q\left(\frac{c}{\sigma_n}\right) - Q^2\left(\frac{c}{\sigma_n}\right) \tag{3.17}$$

Obviously, the minimal distance c from the decision threshold of each PSK symbol in Fig. 3.1 and σ_n determine the mean symbol error probability. It can be shown that this also holds in principle for higher order PSK or QAM constellations.

For high signal-to-noise ratios, $\frac{c}{\sigma_n} \gg 1$, we conclude $Q^2\left(\frac{c}{\sigma_n}\right) \ll 2Q\left(\frac{c}{\sigma_n}\right)$ and the mean symbol error probability approximately is

$$P_S = 2Q\left(\frac{c}{\sigma_n}\right) \tag{3.18}$$

P_S is depicted as a function of $\frac{c}{\sigma_n}$ in Fig. 3.4, which shows only minor differences between the exact solution given by (3.17) and the approximation (3.18).

For stochastic transmit data each symbol a_ν is a discrete random variable with probability of occurrence $P[a_\nu]$. The mean power of the symbol alphabet $\mathcal{B} = \{a_1, \ldots, a_4\}$ is $\sum_{\nu=1}^{4} |a_\nu|^2 P[a_\nu] = \sum_{\nu=1}^{4} \left(\sqrt{2}c\right)^2 P[a_\nu] = 2c^2$, in which we have used (3.16). Hence, the mean symbol error probability can also be expressed as $P_S = 2Q\left(\sqrt{\frac{\gamma}{2}}\right) - Q^2\left(\sqrt{\frac{\gamma}{2}}\right)$ where $\gamma = \frac{2c^2}{\sigma_n^2}$ is the signal-to-noise power ratio.

Fig. 3.4 Symbol error probability P_S for 4-PSK; "exact" according to (3.17), "approximation" (3.18)

3.3 Maximum A-Posterior Symbol-by-Symbol Detection

The signal under decision at time instant k is q given in (3.1) and depicted in Fig. 3.1 for the example of 4-PSK. We define the a-posterior probability

$$P_{APP}\,(a_\nu \mid q) = P\,[a_\nu \mid q] \; ; \; \nu = 1, 2, \ldots, 4 \tag{3.19}$$

as the conditional probability of the symbol a_ν to be detected under the condition that the signal q is observed. Equation (3.19) defines a set of probabilities. The maximum a-posterior probability (MAP) detection rule is as follows,

Select that symbol a_ν, which is associated with the largest $P_{APP}\,(a_\nu \mid q)$. Then the output of the MAP detector is

$$\hat{a} = \arg \left\{ \max_{\nu=1,\ldots 4} [P_{APP}\,(a_\nu \mid q)] \right\} \tag{3.20}$$

According to the decision theory this method provides the best symbol-by symbol detection of unknown events $a_1, a_2, \ldots$ in the stochastic signal q, because the detector deduces the cause a_ν from the effect q, which is expressed by the term a-posterior or "a posteriori". Obviously, the likelihood decision strategy in (3.3) argues with $p_L\,(q \mid a_\nu)$ the other way round.

Using a special form of the Bayes rule given in the Appendix A we can rewrite (3.19) as

$$P_{APP}\,(a_\nu \mid q) = \frac{p_L\,(q \mid a_\nu)\,P\,[a_\nu]}{p_q(q)} \; ; \; \nu = 1, 2, \ldots, 4 \tag{3.21}$$

where $p_q(q)$ is the probability density function of q. For the maximization of $P_{APP}(a_\nu \mid q)$ we have to find the maximal numerator in (3.21), because $p_q(q)$ does not depend on the decision of the detector. Consequently, we can rewrite the decision rule (3.20) as

$$\hat{a} = \arg\left\{ \max_{\nu=1,\dots4} [p_L(q \mid a_\nu) P[a_\nu]] \right\} \qquad (3.22)$$

Unlike the maximum likelihood detection rule in (3.3) we have to look for the maximal product composed of the likelihood probability density function $p_L(q \mid a_\nu)$ and the a-priori probabilities $P[a_\nu]$. Consequently, the detector at the receiver has to know all $P[a_\nu]$, which are normally only available at the transmitter, if at all. This is the main hurdle for the application of a maximum posterior probability detector. However, in many cases the transmit symbols have equal a-priori probabilities $P[a_\nu] = \frac{1}{L}$ for an L-ary QAM. Then the maximum posterior turns into the maximum likelihood criterion (3.3), because from (3.22) follows

$$\hat{a} = \arg\left\{ \frac{1}{L} \max_{\nu=1,\dots4} [p_L(q \mid a_\nu)] \right\} \qquad (3.23)$$

where the factor $\frac{1}{L}$ can be dropped, as it does not depend on ν. Another critical point is that the maximum posterior detection requires the knowledge of the mean power σ_n^2 of the noise n, which has to be estimated at the receiver. This will be evident with an example using the Gaussian noise. To this end we replace p_L in (3.22) by (3.2) yielding the rather complex maximum posterior decision rule

$$\hat{a} = \arg\left\{ \max_{\nu=1,\dots4} \left[\frac{1}{2\pi\sigma_n^2} e^{-\frac{|q-a_\nu|^2}{2\sigma_n^2}} P[a_\nu] \right] \right\} \qquad (3.24)$$

To execute (3.24) we can use the fact that the maximization is equivalent to the minimization of the reciprocal. In addition we take a monotonic function, such as the natural logarithm $\ln(\dots)$ and obtain

$$\hat{a} = \arg\left\{ \min_{\nu=1,\dots4} \left[\frac{|q-a_\nu|^2}{2\sigma_n^2} - \ln(P[a_\nu]) \right] \right\} \qquad (3.25)$$

where we have dropped the term $\ln(2\pi\sigma_n^2)$, because it does not affect the result. Equation (3.25) clearly reveals that the result $\hat{a}$ of the minimization procedure depends on the a-priori probabilities $P[a_i]$ and on the mean noise power σ_n^2. The MAP detector is optimal in the sense that it can minimize the symbol error probability [5].

3.4 Maximum Likelihood Sequence Detection

A sequence detector takes a decision after considering K receive samples $q(0), q(1), \ldots, q(K-1)$. The algorithm for maximum likelihood sequence detection will be best explained with the help of an example as follows.

3.4.1 System Model

Example 3
Given the simplified discrete-time transmission system depicted in Fig. 3.5 with the real-valued symbol sequence $a(k)$ and the symbol alphabet $\mathcal{B} = \{-1, 1\}$. Hence, the modulation scheme is 2-PSK, which allocates symbols only on the real axis of the complex plain. The equivalent baseband system is modeled as a finite impulse response (FIR) filter with real-valued impulse response

$$h(k) = \begin{cases} 1 & ; \quad k = 0 \\ -\frac{1}{2} & ; \quad k = 1 \\ 0 & ; \ k = -1, \pm 2, \pm 3, \ldots \end{cases} \quad (3.26)$$

The input signal to the sequential detector is

$$q(k) = w(k) + n(k) \quad (3.27)$$

where $n(k)$ is real-valued Gaussian noise with zero mean and probability density function

$$p_n(n) = \frac{1}{\sqrt{2\pi}\sigma_n} e^{-\frac{n^2}{2\sigma_n^2}} \quad (3.28)$$

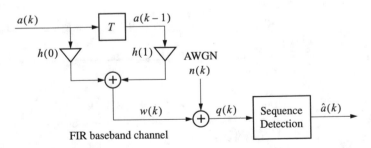

Fig. 3.5 Discrete-time equivalent baseband model of a digital transmission system with sequence detection (Viterbi equalizer)

The noise samples are statistically independent. The output of the decoder is referred to as $\hat{a}(k)$. The FIR system outputs

$$w(k) = a(k) * h(k) = a(k) - \frac{1}{2}a(k-1) \tag{3.29}$$

which contains strong intersymbol interference owing to the term $-\frac{1}{2}a(k-1)$.

3.4.2 State Space Trellis Diagram

Now we are going to describe the input-output relation in the state space. For that purpose a trellis diagram depicted in Fig. 3.6 is used, which is a state transition diagram annotated by the discrete time k. The key component is the state variable, which is always associated with the output of the system memories. If each state variable can take on L different values and if M is the number of independent memory elements of the system model, then the number of states is $N_S = L^M$. In our example the state variable is $a(k-1)$, as there is only a single memory. Consequently, $M = 1$ and with $L = 2$ we have $N_S = 2$ states, which we indicate freely as S_ν, $\nu = 1, 2$.

In general, the trellis diagram consists of nodes and transitions. The nodes represent the states and the transitions, which are labeled by $(a(k) \mid w(k))$, indicate the change of the output signal $w(k)$ depending on the input signal $a(k)$. As already defined, $w(k)$ is the output of the equivalent channel model without noise. The detector has to know the channel impulse response $h(k)$ in (3.26). To illustrate the trellis diagram in Fig. 3.6 assume that the system is in state S_1, thus $a(k-1) = 1$. If the input $a(k)$ keeps to be 1 for the next time instances $k = 1, 2, \ldots$, then the system remains in state S_1 indicated by the horizontal arrows and $w(k)$ remains $\frac{1}{2}$. In case the input $a(k)$ changes to -1 the system moves from state S_1 to state S_2 and provides the output $w(k) = -\frac{3}{2}$. The system remains in state S_2, when the input keeps going to be -1 until the input changes to 1 resulting in a state transition from S_2 to S_1

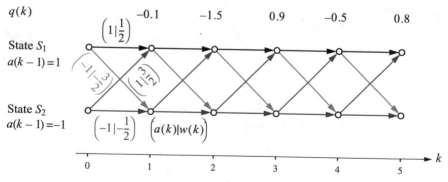

Fig. 3.6 Trellis diagram for the input-output relation of the equivalent baseband system in Fig. 3.5

and the output will be $w(k) = \frac{3}{2}$. We recognize that the trellis diagram in Fig. 3.6 is periodic with time period 1 and the same annotation repeats for $k > 1$, which is not shown.

3.4.3 Maximum Likelihood Sequence Detection

The detector estimates the transmit symbol sequence $a(0), a(1), \ldots, a(K-1)$ from the receive sequence $q(0), q(1), \ldots, q(K-1)$ in the maximum likelihood sense. Hence, the detection method is also called maximum likelihood sequence estimation (MLSE). As the receiver knows the channel coefficients $h(k)$, it can calculate the sequence $w(0), w(1), \ldots, w(K-1)$ for all possible symbols $a(k)$ of the symbol alphabet $\mathcal{B}$. This is indicated in the trellis diagram in Fig. 3.6, where $w(k)$ can take on four different values $\frac{1}{2}, -\frac{3}{2}, \frac{3}{2}, -\frac{1}{2}$. For a given symbol sequence $a(0), a(1), \ldots, a(K-1)$ the trellis illustrates the sequence $w(0), w(1), \ldots, w(K-1)$ as a path through the diagram from $k = 0$ to $K-1$ and the trellis shows the set of all such sequences. In the example we have chosen $K - 1 = 5$.

As for the symbol-by-symbol detection described previously, we also define a likelihood probability density function

$$p_K \left(q(0), q(1), \ldots, q(K-1) \mid w(0), w(1), \ldots, w(K-1) \right) \qquad (3.30)$$

which is the multivariate density function of the observation $q(0), q(1), \ldots, q(K-1)$ conditioned on $w(0), w(1), \ldots, w(K-1)$. There is a combinatorial multitude of sequences in the argument of (3.30). For the detection algorithm the set of density functions is relevant, which reflects all possible paths through the trellis diagram. Then the sequence detector is looking for that sequence $w(0), w(1), \ldots, w(K-1)$, which is maximal likely to the observation $q(0), q(1), \ldots, q(K-1)$ and from that dedicated path in the trellis diagram the sequence of optimal estimates $\hat{a}(0), \hat{a}(1), \ldots, \hat{a}(K-1)$ is derived. Assuming $q(0), q(1), \ldots$ as statistically independent and also independent of $w(0), w(1), \ldots$ yields the density function

$$p_K \left(q(0), q(1), \ldots, q(K-1) \mid w(0), w(1), \ldots, w(K-1) \right) = \prod_{k=0}^{K-1} p_n \left(q(k) - w(k) \right)$$
$$(3.31)$$

with p_n given in (3.28). Then follows

$$p_K \left(\ldots \mid \ldots \right) = \left(\frac{1}{\sqrt{2\pi}\sigma_n} \right)^K e^{-\frac{1}{2\sigma_n^2} \sum_{k=0}^{K-1} [q(k)-w(k)]^2} \qquad (3.32)$$

Now we introduce the branch metric

$$d_k = [q(k) - w(k)]^2 \qquad (3.33)$$

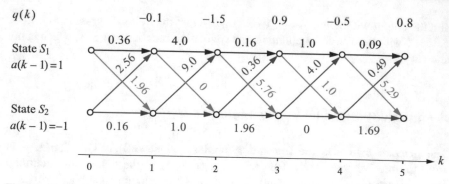

Fig. 3.7 Trellis diagram with branch metrics $d_k = [q(k) - w(k)]^2$; $k = 1, 2, \ldots, 5$

and the path metric

$$D_K = \sum_{k=0}^{K-1} d_k \tag{3.34}$$

and obtain

$$p_K (\ldots \mid \ldots) = \left(\frac{1}{\sqrt{2\pi}\sigma_n} \right)^K e^{-\frac{1}{2\sigma_n^2} D_K} \tag{3.35}$$

The calculated numbers for the branch metrics d_k are indicated in the trellis diagram in Fig. 3.7. Consequently, the path metrics D_K for all possible paths through the trellis can be found with (3.34) and also the set of conditional probability density functions in (3.35). Please note that D_K is positive.

The criterion for maximum likelihood detection is to select the maximal density function p_K out of the set. Equivalently, we can search for that path in the trellis, which exhibits minimal path metric D_K owing to the negative exponent in (3.35). Thus, the most likely symbol sequence $\hat{a}(0), \hat{a}(1), \ldots, \hat{a}(K-1)$ is deduced from the sequence of transitions in the trellis in Fig. 3.6

$$\{(a(0) \mid w(0)), (a(1) \mid w(1)), \ldots, (a(K-1) \mid w(K-1))\}_{\min[D_K]} \tag{3.36}$$

where the procedure is executed over all paths starting at $k = 0$ and ending at $k = K - 1$.

3.4.4 Solution Using the Viterbi Algorithm

A straightforward way to find the solution (3.36) is the calculation of all branch metrics d_k and all path metrics D_K of the trellis. Then we take the path with the smallest metric and deduce the estimated symbol sequence $\hat{a}(0), \hat{a}(1), \ldots, \hat{a}(K-1)$ using the trellis in Fig. 3.6. However, the computational amount with this "brute search" can be very large and this procedure is not effective, because we calculate all path metrics although we just require one, the minimal one. Consequently, more effective procedures have been the topic of research. A task similar to sequential detection is known from operations research as the "traveling salesman problem", where a salesman has to visit customers at various places spending minimal total traveling cost. The solution for the salesman problem was first given by Bellman in 1957. He formulated an optimality principle saying that "an optimal policy has the property that whatever the initial state and initial decision are, the remaining decisions must constitute an optimal policy with regard to the state resulting from the first decision" [6, 7]. The procedure to get the solution is called "dynamic programming". Similar tasks are known from computer science and solved with the algorithm proposed by Dijkstra [8].

For digital communications A. Viterbi has found a similar algorithm for the decoding of data, which are encoded with convolutional codes [9]. The Viterbi algorithm can be applied for many optimization problems based on a trellis diagram to find the most probable sequence of a hidden Markov model. For example it is applied for pattern recognition in bio-informatics and artificial intelligence. Here we focus on the maximum likelihood sequence detection. There are several formulations of the mathematical algorithm and as a program. Let us indicate the nodes in the trellis by its state S_ν and the time instant k, i.e. by (S_ν, k); $\nu = 0, 1, \ldots, N_S$, in which N_S is the number of states, in our example $N_S = 2$. All paths, which start at the beginning of the trellis and are entering that node are called partial paths. The principle strategy of the Viterbi algorithm is not to calculate all paths through the trellis up to the end $k = K - 1$, but to dynamically dismiss all those partial paths entering the node (S_ν, k), which exhibit larger partial path metrics compared to the partial path with minimal metric at that node. If there are more than one partial paths with the same minimum, the final decoding result will be ambiguous. The discarded "dead paths" are skipped from the trellis diagram. This procedure has to be executed for all nodes (S_ν, k); $\nu = 1, 2 \ldots, N_S$ and $k = 1, 2, \ldots, K - 1$ step by step. For the initialization at $k = 0$ the partial path metrics are set to zero at all nodes $(S_\nu, 0)$. Of course, the partial path metrics can be calculated recursively by adding the new branch metric in each step. At the end all branches are traced back from the end of the trellis and the "surviving" branches provide the optimal path (or paths), from which the sequence (3.36) is deduced.

Programs are adequately explained with an example. Therefore we go back to the trellis in Fig. 3.7. The various steps of the algorithm are explained in Fig. 3.8. In step 1 the branch metrics at nodes $(S_1, 1)$ and $(S_2, 1)$ are calculated. We notice that the partial path with metric 2.56 can be discarded at node $(S_1, 1)$, because $2.56 > 0.36$. With the similar argument we discard at node $(S_2, 1)$ the partial path with metric 1.96. In step 2 the partial path metrics are updated by adding the respective branch metrics of the second segment of the trellis. At node $(S_1, 2)$ the partial path with metric 9.16 can be dismissed as dead path. Similarly at node $(S_2, 2)$ we drop the partial path with metric 1.16. This procedure is continued until step 5, in which we cancel two partial paths with metrics 2.21 and 7.01, respectively. The algorithm finally reveals

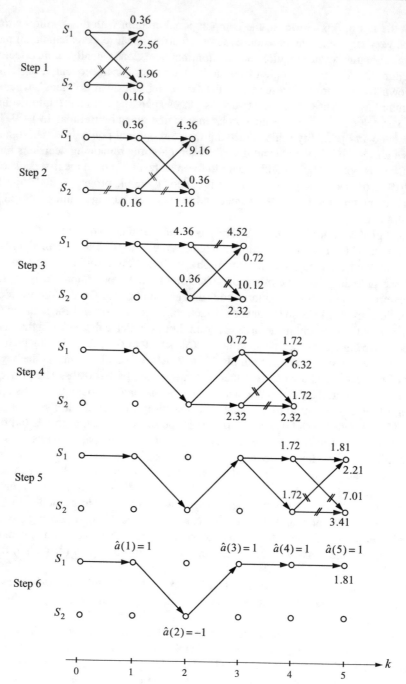

Fig. 3.8 Steps of the Viterbi algorithm with partial paths and associated metrics. Step 6 shows path with minimal metric and the detected sequence $\hat{a}(k)$

the result in step 6 from the path with minimal path metric 1.81 terminating in node $(S_1, 5)$. From the trellis diagram in Fig. 3.6 we allocate the input symbols $a(k)$ to the nodes crossed by the final path from $k = 1, \ldots, 5$ resulting in the estimated symbol sequence $\hat{a}(k)$

$$1, -1, 1, 1, 1 \tag{3.37}$$

For comparison a symbol-by-symbol threshold detector applied on $q(k)$ with threshold level $q = 0$ and decision rule

$$\hat{a}(k) = \begin{cases} 1 & ; q(k) \geq 0 \\ -1 & ; q(k) < 0 \end{cases} \tag{3.38}$$

shall be employed, which yields the sequence $\hat{a}(k)$ after the decision

$$-1, -1, 1, -1, 1 \tag{3.39}$$

Both results differ in the first and next to last symbol indicating that the symbol-by-symbol and the sequence detection can provide different results.

3.4.5 Viterbi Equalizer

In several practical applications the impulse response $g_R(t)$ of the receive lowpass in Fig. 1.1 is designed in such a way that the overall impulse response of the equivalent baseband system satisfies the first Nyquist criterion to eliminate the intersymbol interference. In this case the receive lowpass is called an *equalizer*. In the example of Fig. 3.5 the overall impulse response $h(k)$ shows a strong post cursor $h(1) = -\frac{1}{2}$, which causes severe intersymbol interference. No separate equalizer is present. However, the sequence detector takes this intersymbol interference together with the noise $n(k)$ into account and provides an optimal decision. Consequently, the tasks of equalization and detection under the impact of noise and intersymbol interference have been merged with this approach and the presented sequence detector is therefore often denoted as a *Viterbi equalizer*.

References

1. Hagenauer, J., Hoeher, P.: A Viterbi algorithm with soft-decision outputs and its applications. In: IEEE International Conference on Global Communications (GLOBECOM) (1989)
2. Hagenauer, J., Offer, E., Papke, L.: Iterative decoding of binary block and convolutional codes. IEEE Trans. Inf. Theory **42** (1996)
3. Hagenauer, J.: The turbo principle: tutorial introduction and state of the art. In: Proceedings of 1st International Symposium on Turbo Codes (1997)

4. Bernhard, M.: QAM, webdemo. Technical report. Institute of Telecommunications, University of Stuttgart, Germany (2018). http://webdemo.inue.uni-stuttgart.de
5. Barry, J.R., Lee, E.A., Messerschmitt, D.G.: Digital Communication. Springer, Berlin (2012)
6. Bellman, R.E.: Dynamic Programming. Princeton University Press, Princeton (1957)
7. Bellman, R.E.: Dynamic Programming. Princeton University Press, Princeton (2010)
8. Dijkstra, E.W.: A note on two problems in connexion with graphs. Numer. Math. **1** (1959)
9. Viterbi, A.: Error bounds for convolutional codes and an asymptotically optimum decoding algorithm. IEEE Trans. Inf. Theory (1967)

Chapter 4
Digital Transmission over Wireless, Time-Variant Channels

4.1 Transmission System with Time-Variant Channel

Digital signal transmission over a wireless channel has become an important field due to the high flexibility and comfort of wireless connections for many users with the motto "telecommunications anytime and anywhere". Therefore, in the following chapters we describe the principles of such systems and their design in some detail. A significant part is devoted to the wireless channel. The parameters of electrical cables or optical fibers are approximately constant over time. Consequently, they have been characterized as time-invariant and described by an impulse response $g_c(t)$ in Fig. 1.1 of Sect. 1.2. As we will see in detail, a wireless channel is significantly different and multifaceted. The transmit signal travels on a multitude of different paths from the transmitter to the receiver, undergoes reflections and scattering at objects, such as buildings. Moreover, if the transmitter or the receiver are moving, the signal suffers by the Doppler effect. There is quite a lot of knowledge on the basis of wave propagation and electromagnetic field theory. We will build upon these findings and emphasize a system-theoretical approach, which allows together with adequate channel models the effective design of algorithms for signal transmission and reception. We focus on single-input single output (SISO) channels in this Part I and prepare the ground for the wireless multiple input multiple output (MIMO) systems discussed in Part III well knowing that the MIMO channel is composed of many channels of the SISO type.

We start with the block diagram for the wireless transmission scheme and emphasize on the mathematical description of the input-output relation. The wireless channel is described as a time-variant system with the two-dimensional impulse response $w(t, s)$ or the delay spread function $g_C(t, \tau)$, also called modified impulse response. We will use and resume the principle results from Part II on the theory of time-variant systems. Part II is self-contained and readers not familiar with time-variant systems are encouraged to switch to that part beforehand or on demand during the study of the following chapters, because we shall merely quote those results here.

© Springer Nature Switzerland AG 2021
J. Speidel, *Introduction to Digital Communications*,
Signals and Communication Technology,
https://doi.org/10.1007/978-3-030-67357-4_4

Fig. 4.1 Block diagram of a transmission system with a time-variant channel with impulse response $w_C(t, s)$ and delay spread function $g_C(t, \tau)$

Figure 4.1 shows the block diagram of a system for wireless transmission. We recognize the strong similarity to Fig. 1.1 in Sect. 1.2. However, the differences are the time-variant channel and the different notations for the impulse responses. As outlined in Part II, linear and time-variant systems are completely characterized by their response $w(t, s)$ at observation time t to an input Dirac impulse $\delta(t - s)$, which is active at the time instant $s \leq t$. One denotes $w(t, s)$ as time-variant impulse response. s and t are independent variables. It can be observed that w as a function of t exhibits different shapes depending on the initial time instant s of the Dirac impulse. This is quite in contrast to a linear time-invariant system, where the shape does not change and the response is just shifted by s on the t-axis. Thus, a time-invariant system is characterized by the impulse response $w(t, s) = w(t - s)$.

As outlined in Part II, the Fourier transform of $w(t, s)$ does not provide meaningful system functions in the frequency domain, such as a transfer function. Therefore, the transformation of variables $s = t - \tau$ was proposed, which yields the delay spread function also called modified impulse response

$$w(t, t - \tau) = g(t, \tau) \tag{4.1}$$

This function can be regarded as the response of the time-variant system at observation time t to an input Dirac impulse effective at $t - \tau \leq t$.

For time-invariant systems $w(t - s) = w(\tau) = g(\tau)$ holds. $g(\tau)$ and not $g(t)$ is formally required, if a time-invariant system is described in the framework of time-variant systems, e.g., in case of a cascade of such systems. If the system is composed solely of time-invariant building blocks, τ can be replaced by t. Of course, $g(t)$ and $g(\tau)$ exhibit the same shape, because mathematically we can use any variable to describe a function. In Part II various input-output relations and Fourier spectra are derived in quite some detail, which we will use also in the following.

4.2   Overall Time-Variant Impulse Response

As in Sect. 1.6 for the time-invariant channel, we are now going to derive the impulse response between the nodes (I) and (II) in Fig. 4.1. For that purpose we excite the system at the input of the pulse shaper by $\delta(t - s)$ and allocate to all blocks impulse responses. The time-variant systems get impulse responses with the argument t, s and the time-invariant systems with $t - s$. Thus, the impulse responses of the time-invariant pulse shaper and receive lowpass filter are $w_I(t, s) = w_I(t - s)$ and $w_R(t, s) = w_R(t - s)$, respectively. The reader assures oneself easily of the time-variant impulse response of the modulator

$$w_M(t, s) = \delta(t - s)e^{j2\pi f_0 t} \qquad (4.2)$$

and the demodulator

$$w_D(t, s) = \delta(t - s)e^{-j2\pi f_0 t} \qquad (4.3)$$

We merely quote some more properties from Part II of time-variant systems before going into the details of the Fig. 4.1. Let $x(t)$ be the input signal of a time-variant system with impulse response $w(t, s)$. Then the output signal $y(t)$ is given by the "generalized convolution integral" or "time-variant convolution"

$$y(t) = x(t) \circledast w(t, s) = \int_{-\infty}^{\infty} x(s)w(t, s)ds \qquad (4.4)$$

Equation (4.4) actually incorporates the "conventional" or "time-invariant" convolution, if we set $w(t, s) = w(t - s)$ resulting in

$$y(t) = x(t) \circledast w(t - s) = \int_{-\infty}^{\infty} x(s)w(t - s)ds \qquad (4.5)$$

The overall time-variant impulse response of a cascade of systems with $w_1(t, s)$ and $w_2(t, s)$ is given by

$$w_1(t, s) \circledast w_2(t, s) = \int_{-\infty}^{\infty} w_1(\zeta, s) w_2(t, \zeta) d\zeta \tag{4.6}$$

If one system is time-invariant, e.g., $w_1(t, s)$, then we just write $w_1(t, s) = w_1(t - s)$ and replace the comma by the minus sign. In Part II we also show that the time-variant convolution is non-commutative in general. Hence, we have to respect the sequential arrangement of time-variant systems. This is in contrast to the time-invariant convolution.

Furnished with these basics we can focus on the system in Fig. 4.1. The cascade of the pulse shaper and the modulator with (4.2) owns the time-variant impulse response

$$w_I(t - s) \circledast w_M(t, s) = \int_{-\infty}^{\infty} w_I(\zeta - s) w_M(t, \zeta) d\zeta = w_I(t - s) e^{j2\pi f_0 t} \tag{4.7}$$

which we could have directly concluded from the Fig. 4.1. Then follows the time-variant impulse response at the output of the real part operator $\sqrt{2}\mathrm{Re}[...]$

$$w_1(t, s) = \frac{\sqrt{2}}{2} \left[w_I(t - s) e^{j2\pi f_0 t} + w_I^*(t - s) e^{-j2\pi f_0 t} \right] \tag{4.8}$$

Next we determine the impulse response of the cascade of the channel $w_C(t, s)$ and the demodulator $w_D(t, s)$ yielding with (4.3) and (4.6)

$$w_2(t, s) = w_C(t, s) \circledast w_D(t, s) = w_C(t, s) e^{-j2\pi f_0 t} \tag{4.9}$$

Finally, the overall time-variant impulse response between the nodes (I) and (II) in Fig. 4.1 is

$$w_e(t, s) = w_1(t, s) \circledast w_2(t, s) \circledast w_R(t - s) \sqrt{2} \tag{4.10}$$

where we have used the associative property of the time-variant convolution. Plugging in (4.8) we get

$$w_e(t, s) = w_{e1}(t, s) + w_{e2}(t, s) \tag{4.11}$$

with

$$\begin{aligned} w_{e1}(t, s) &= \left(w_I(t - s) e^{j2\pi f_0 t} \right) \circledast \left(w_C(t, s) e^{-j2\pi f_0 t} \right) \circledast w_R(t - s) \\ w_{e2}(t, s) &= \left(w_I^*(t - s) e^{-j2\pi f_0 t} \right) \circledast \left(w_C(t, s) e^{-j2\pi f_0 t} \right) \circledast w_R(t - s) \end{aligned} \tag{4.12}$$

In Sect. 4.4.2 we consider the spectrum $G_{e2}(f_t, f_\tau)$ of the delay spread function $g_{e2}(t, \tau)$ corresponding to $w_{e2}(t, s)$ and show that $G_{e2}(f_t, f_\tau)$ is zero for usual system parameters. As a consequence one can assume $g_{e2}(t, \tau) = 0$ and thus $w_{e2}(t, s) = 0$ in the following. Then we obtain the overall time-variant impulse response between the nodes (I) and (II) in Fig. 4.1 using (4.11) and (4.12)

$$w_e(t, s) = \left(w_I(t - s) e^{j2\pi f_0 t} \right) \circledast \left(w_C(t, s) e^{-j2\pi f_0 t} \right) \circledast w_R(t - s) \tag{4.13}$$

4.3 Overall Delay Spread Function

Now we are interested in the overall delay spread function between the nodes (I) and (II) in Fig. 4.1. As pointed out in Part II, the time-variant impulse response has no meaningful Fourier spectrum for signal processing. This can be overcome with the transformation of variables

$$s = t - \tau \qquad (4.14)$$

which turns the time-variant impulse responses into delay spread functions as follows

$$w_e(t, s) = g_e(t, \tau), \quad w_I(t - s) = g_I(\tau), \quad w_c(t, s) = g_C(t, \tau), \quad w_R(t - s) = g_R(\tau) \qquad (4.15)$$

We show at the end of this section that from (4.11) and (4.12) the overall delay spread function follows as

$$g_e(t, \tau) = g_{e1}(t, \tau) + g_{e2}(t, \tau) \qquad (4.16)$$

with

$$
\begin{aligned}
g_{e1}(t, \tau) &= g_I(\tau) \circledast \left(g_C(t, \tau) e^{-j2\pi f_0 \tau} \right) \circledast g_R(\tau) \\
g_{e2}(t, \tau) &= g_I^*(\tau) \circledast \left(g_C(t, \tau) e^{-j4\pi f_0 t} e^{j2\pi f_0 \tau} \right) \circledast g_R(\tau)
\end{aligned}
\qquad (4.17)
$$

In Sect. 4.4 the associated Fourier spectra are discussed. There we show that a usual choice of the cut-off frequencies results in

$$g_{e2}(t, \tau) = 0 \qquad (4.18)$$

yielding from (4.16)

$$g_e(t, \tau) = g_I(\tau) \circledast \left(g_C(t, \tau) e^{-j2\pi f_0 \tau} \right) \circledast g_R(\tau) \qquad (4.19)$$

Please note, we use the same symbol $\circledast$ for the time-variant convolution of delay spread functions and for the time-variant impulse responses. However, their integral representations are different.

As shown in Part II, the time-variant convolution of two delay spread functions $g_1(\tau)$ and $g_2(t, \tau)$ is given by

$$g_{12}(t, \tau) = g_1(\tau) \circledast g_2(t, \tau) = \int_{-\infty}^{\infty} g_1(\tau - \zeta) g_2(t, \zeta) d\zeta = \int_{-\infty}^{\infty} g_1(\eta) g_2(t, \tau - \eta) d\eta \qquad (4.20)$$

and

$$g_{21}(t, \tau) = g_2(t, \tau) \circledast g_1(\tau) = \int_{-\infty}^{\infty} g_2(t - \zeta, \tau - \zeta) g_1(\zeta) d\zeta \qquad (4.21)$$

Both relations apply for (4.17) and apparently are non-commutative. Thus, the sequential arrangement may not be interchanged. For the input-output relation we obtain from (4.4) with $s = t - \tau$ and $w(t, t - \tau) = g(t, \tau)$

$$y(t) = x(t) \circledast g(t, \tau) = \int_{-\infty}^{\infty} x(t - \tau)g(t, \tau)d\tau \qquad (4.22)$$

Proof of (4.16) and (4.17)

We use the definition of the time-variant convolution (4.6) and obtain for the upper line in (4.12)

$w_{e1}(t, s) = \left(\int_{-\infty}^{\infty} w_I(\zeta - s)w_C(t, \zeta)e^{-j2\pi f_0(t-\zeta)}d\zeta \right) \circledast w_R(t - s)$. Introducing the substitution $\zeta = t - u$; $d\zeta = -du$ yields

$w_{e1}(t, s) = \left(\int_{-\infty}^{\infty} w_I(t - u - s)w_C(t, t - u)e^{-j2\pi f_0 u}du \right) \circledast w_R(t - s)$.

With $s = t - \tau$ and (4.15) follows

$w_{e1}(t, t - \tau) = g_{e1}(t, \tau) = \left(\int_{-\infty}^{\infty} g_I(\tau - u)g_C(t, u)e^{-j2\pi f_0 u}du \right) \circledast g_R(\tau)$. Using the definition (4.20) we obtain $g_{e1}(t, \tau) = g_I(\tau) \circledast \left(g_C(t, \tau)e^{-j2\pi f_0 \tau} \right) \circledast g_R(\tau)$ and the first equation in (4.17) is proven.

In a similar way we prove $g_{e2}(t, \tau)$. We start with the lower line in (4.12)

$w_{e2}(t, s) = \left(\int_{-\infty}^{\infty} w_I^*(\zeta - s)e^{-j2\pi f_0\zeta}w_C(t, \zeta)e^{-j2\pi f_0 t}d\zeta \right) \circledast w_R(t - s)$ and the substitution $\zeta = t - u$; $d\zeta = -du$ yields

$w_{e2}(t, s) = \left(\int_{-\infty}^{\infty} w_I^*(t - u - s)w_C(t, t - u) \ e^{-j4\pi f_0 t}e^{j2\pi f_0 u}du \right) \circledast w_R(t - s)$.

With $s = t - \tau$, $w_{e2}(t, t - \tau) = g_{e2}(t, \tau)$, $w_C(t, t - u) = g_C(t, u)$,

$g_C'(t, u) = g_C(t, u)e^{-j4\pi f_0 t}e^{j2\pi f_0 u}$, and $w_R(t - s) = g_R(\tau)$ follows

$g_{e2}(t, \tau) = \left(\int_{-\infty}^{\infty} g_I^*(\tau - u)g_C'(t, u)du \right) \circledast g_R(\tau)$ from which we obtain with (4.20) the symbolic notation

$g_{e2}(t, \tau) = g_I^*(\tau) \circledast g_C'(t, \tau) \circledast g_R(\tau) = g_I^*(\tau) \circledast \left(g_C(t, \tau) \ e^{-j4\pi f_0 t}e^{j2\pi f_0 \tau} \right) \circledast g_R(\tau)$
and the proof is finished.

4.4   Overall Doppler Spread Function

4.4.1   Fourier Transform of the Overall Delay Spread Function

To get inside into (4.17) and to define alternative system functions we consider the frequency domain. To this end we will apply the Fourier transform with respect to the variables t and τ. Therefore we define the corresponding variables as

$$t \leftrightarrow f_t \ ; \ \tau \leftrightarrow f_\tau$$

and use the symbol $\longmapsto$ for the transform. In wireless communications f_t is called Doppler frequency, as it illustrates the time-variance of the channel. f_τ is the "natural" frequency, also used for the ordinary frequency response of a time-invariant system or the spectrum of a signal. In Part II the delay spread functions and their Fourier transforms are discussed in detail, where the table in Sect. 11.6 presents a summary. We shall merely quote the results here without proof. We define with capital letters the transfer function of the pulse shaper

$$g_I(\tau) \overset{\tau}{\rightarrowtail} G_I(f_\tau) \tag{4.23}$$

the transfer function of the receive filter

$$g_R(\tau) \overset{\tau}{\rightarrowtail} G_R(f_\tau) \tag{4.24}$$

the Doppler spread function of the channel

$$g_C(t, \tau) \overset{t,\tau}{\rightarrowtail} G_C(f_t, f_\tau) \tag{4.25}$$

and the overall Doppler spread function

$$g_e(t, \tau) \overset{t,\tau}{\rightarrowtail} G_e(f_t, f_\tau) \tag{4.26}$$

On top of the arrows we indicate the direction of the Fourier transform with respect to the t- and the τ-coordinate.

We quote from Sect. 11.5 of Part II the following transform pairs:
Let $g_1(\tau) \overset{\tau}{\rightarrowtail} G_1(f_\tau)$ and $g_2(t, \tau) \overset{t,\tau}{\rightarrowtail} G_2(f_t, f_\tau)$, then

$$g_1(\tau) \circledast g_2(t, \tau) \overset{t,\tau}{\rightarrowtail} G_1(f_\tau)G_2(f_t, f_\tau) \tag{4.27}$$

$$g_2(t, \tau) \circledast g_1(\tau) \overset{t,\tau}{\rightarrowtail} G_2(f_t, f_\tau)G_1(f_t + f_\tau) \tag{4.28}$$

With these findings the following Fourier transforms of (4.16) and (4.17) are proven at the end of this section,

$$g_e(t, \tau) \overset{t,\tau}{\rightarrowtail} G_e(f_t, f_\tau) = G_{e1}(f_t, f_\tau) + G_{e2}(f_t, f_\tau) \tag{4.29}$$

with

$$g_{e1}(t, \tau) \overset{t,\tau}{\rightarrowtail} G_{e1}(f_t, f_\tau) = \quad G_I(f_\tau)G_C(f_t, f_\tau + f_0)G_R(f_t + f_\tau)$$
$$g_{e2}(t, \tau) \overset{t,\tau}{\rightarrowtail} G_{e2}(f_t, f_\tau) = G_I^*(-f_\tau)G_C(f_t + 2f_0, f_\tau - f_0)G_R(f_t + f_\tau) \tag{4.30}$$

In the following it will be shown that the cut-off frequencies of the pulse shaper and the receive filter can be determined in such a way that $G_{e2}(f_t, f_\tau) \approx 0$, which results in the overall Doppler spread function

$$G_e(f_t, f_\tau) = G_I(f_\tau)G_C(f_t, f_\tau + f_0)G_R(f_t + f_\tau) \tag{4.31}$$

Apparently, $G_e(f_t, f_\tau)$ is composed of the frequency shifted Doppler spread function of the time-variant channel, $G_C(f_t, f_\tau + f_0)$, filtered by $G_I(f_\tau)$ and $G_R(f_t + f_\tau)$.

4.4.2 Principal System Model Parameters

Cut-Off Frequencies

To bring the relevance of (4.29) and (4.30) to light, we impose the following model parameters. The impulse shaper and the receive lowpass shall possess transfer functions with ideal stop-bands

$$G_I(f_\tau) \begin{cases} \neq 0 \; ; \; |f_\tau| \leq f_I \\ = 0 \; ; \quad else \end{cases} \; ; \quad G_R(f_\tau) \begin{cases} \neq 0 \; ; \; |f_\tau| \leq f_R \\ = 0 \; ; \quad else \end{cases} \tag{4.32}$$

f_I and f_R are the cut-off frequencies of the pulse shaper and the receive filter, respectively. The Doppler spread function of the channel shall have a bandpass shape with a passband around the carrier frequency $f_\tau = \pm f_0$ and shall be strictly band-limited as follows

$$G_C(f_t, f_\tau) \begin{cases} \neq 0 \; ; \; f_0 - f_{\tau,c} \leq |f_\tau| \leq f_0 + f_{\tau,c} \\ \neq 0 \; ; \quad -f_{t,c} \leq |f_t| \leq f_{t,c} \\ 0 \; ; \qquad\qquad else \end{cases} \tag{4.33}$$

$2f_{\tau,c}$ defines the transmission bandwidth and $f_{t,c}$ the maximal Doppler frequency of the time-variant channel.

Overall Doppler Spread Function Without Receive Filter

For a better understanding we first define an intermediate delay spread function and a Doppler spread function from the input of the pulse shaper (node I) to the receive filter input in Fig. 4.1 as $g_{e,D}(t, \tau)$ and $G_{e,D}(f_t, f_\tau)$. With (4.16) and (4.17) we obtain

$$g_{e,D}(t, \tau) = g_I(\tau) \circledast \left(g_C(t, \tau) e^{-j2\pi f_0 \tau} \right) + g_I^*(\tau) \circledast \left(g_C(t, \tau) e^{-j4\pi f_0 t} e^{j2\pi f_0 \tau} \right) \tag{4.34}$$

With (4.29) and (4.30) follows

$$G_{e,D}(f_t, f_\tau) = G_I(f_\tau) G_C(f_t, f_\tau + f_0) + G_I^*(-f_\tau) G_C(f_t + 2f_0, f_\tau - f_0) \tag{4.35}$$

where we do not take account of the gain factor $\sqrt{2}$ of the receive filter at the moment. In Fig. 4.2 the top view of the spectral components of $G_{e,D}(f_t, f_\tau)$ in (4.35) is illustrated. This busy figure needs some comments. To show the principle we assume real-valued spectra. As defined in (4.33), the passbands of the channel Doppler spread function $G_C(f_t, f_\tau)$ are rectangles located around $f_\tau = \pm f_0$ and exhibit a bandwidth of $2f_{\tau,c}$ and $2f_{t,c}$. The passband of $G_I(f_\tau)$ is shown as a horizontal stripe. Please note, if a one-dimensional function $G_I(f_\tau)$ is plotted in a two-dimensional diagram we have to keep $G_I(f_\tau)$ constant along the f_t-axis. The inside of the stripe represents the passband and the outside the stop-band of this filter. $G_I^*(-f_\tau)$ in principle has the same shape.

The frequency shifted Doppler spread function $G_C(f_t + 2f_0, f_\tau - f_0)$ is illustrated in the upper left part of Fig. 4.2. For the defined cut-off frequencies we can substantiate that all spectral parts for $|f_\tau| > f_I$ are canceled by $G_I(f_\tau)$ and $G_I^*(-f_\tau)$. The important baseband $G_I(f_\tau)G_C(f_t, f_\tau + f_0)$ lies as a rectangle around the origin in the frequency range $|f_\tau| \le f_I$ and $|f_t| \le f_{t,C}$. Apparently, also a main part of the annoying two-dimensional spectrum $G_I^*(-f_\tau)G_C(f_t + 2f_0, f_\tau - f_0)$ around $(-2f_0, 0)$ contributes to $G_{e,D}(f_t, f_\tau)$. It does not overlap with the baseband, because normally the carrier frequency is much larger than the maximal Doppler shift of the channel, $2f_0 \gg f_{t,C}$.

Impact of the Receive Filter

Now we introduce the receive filter $g_R(\tau)$ and consider the overall delay spread function, which is

$$g_e(t, \tau) = g_{e,D}(t, \tau) \circledast g_R(\tau) \tag{4.36}$$

as well as the overall Doppler spread function

$$G_e(f_t, f_\tau) = G_{e,D}(f_t, f_\tau)G_R(f_t + f_\tau) \tag{4.37}$$

Figure 4.3 illustrates the remaining spectral components from Fig. 4.2. The transfer function of the receive lowpass is defined in (4.32) with the cut-off frequency f_R. In (4.37) it is represented as $G_R(f_t + f_\tau)$ and thus as a diagonal stripe in the top view of Fig. 4.3. The inside of the stripe indicates the passband and the outside is the stopband with $G_R(f_t + f_\tau) = 0$. Apparently, the receive filter cancels the remaining part of $G_I^*(-f_\tau)G_C(f_t + 2f_0, f_\tau - f_0)$ located around $(-2f_0, 0)$ and cuts even some corners of the channel baseband. As a result

$$G_I^*(-f_\tau)G_C(f_t + 2f_0, f_\tau - f_0)G_R(f_t + f_\tau) = G_{e2}(f_t, f_\tau) = 0 \tag{4.38}$$

and $G_e(f_t, f_\tau)$ in (4.31) is verified. From $G_{e2}(f_t, f_\tau) = 0$ follows $g_{e2}(t, \tau) = 0$ and (4.18) is proven as well.

The top view of $G_e(f_t, f_\tau)$ in Fig. 4.3 results as a baseband spectrum with a rectangle or a hexagon shape in our example depending on the cut-off frequencies and indicated by dashed lines. The interesting question is, whether or not the filter $G_R(f_t + f_\tau)$ can limit the maximal Doppler shift $f_{t,C}$ of the fading channel. We see from Fig. 4.3 that the border lines of $G_R(f_t + f_\tau)$ cut the f_t-axis at $\pm f_R$ as well as the f_τ-axis. A reduction of f_R is able to limit the Doppler frequency of the channel on the f_t-axis, however simultaneously also the transmission bandwidth along the f_τ-axis. Therefore a time-invariant receive lowpass is not effective in reducing the impact of the fading other than the spectral parts of $G_I^*(-f_\tau)G_C(f_t + 2f_0, f_\tau - f_0)$.

Finally, please note that f_I in Fig. 4.3 can not be considered as the final bandwidth of the receiver output signal $q_R(t)$. The reason is the Doppler shift $f_{t,C}$ of the fading channel, which increases the bandwidth of $q_R(t)$ beyond f_I, as will be discussed in the next section.

Fig. 4.2 Top view of spectral parts of the Doppler spread function $G_{e,D}(f_t, f_\tau)$ in (4.35) between node I and the input of the receive lowpass in Fig. 4.1. All spectra shall be real-valued. $G_C(f_t, f_\tau)$, Doppler spread function of time-variant channel; $G_C(f_t, f_\tau + f_0)$, version of $G_C(f_t, f_\tau)$ shifted on f_τ-axis; $G_C(f_t + 2f_0, f_\tau - f_0)$, version of $G_C(f_t, f_\tau)$ shifted on f_t- and f_τ-axis; $G_I(f_\tau)$, transfer function of the pulse shaper

Proof of (4.29) and (4.30)

We start with the first line in (4.17) and obtain with the frequency shifting property of the Fourier transform $g_C(t, \tau)\mathrm{e}^{-\mathrm{j}2\pi f_0\tau} \overset{t,\tau}{\longmapsto} G_C(f_t, f_\tau + f_0)$. Using (4.27) yields $g_I(\tau) \circledast \left(g_C(t, \tau)\mathrm{e}^{-\mathrm{j}2\pi f_0\tau}\right) \overset{t,\tau}{\longmapsto} G_I(f_\tau)G_C(f_t, f_\tau + f_0)$ and with the help of (4.28) we obtain $\left[g_I(\tau) \circledast \left(g_C(t, \tau)\mathrm{e}^{-\mathrm{j}2\pi f_0\tau}\right)\right] \circledast g_R(\tau) \overset{t,\tau}{\longmapsto}$ $[G_I(f_\tau)G_C(f_t, f_\tau + f_0)]\,G_R(f_t + f_\tau)$ which proves the first line of (4.30).

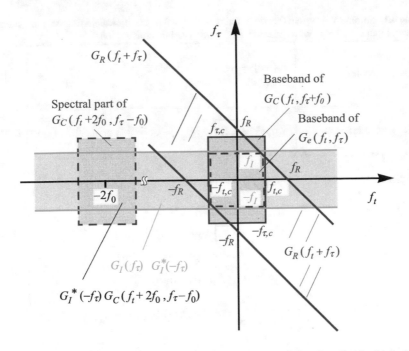

Fig. 4.3 Top view of spectral parts of the overall Doppler spread function $G_e(f_t, f_\tau)$ in (4.29) with (4.30). All spectra shall be real-valued. $G_R(f_t + f_\tau)$, Doppler spread function of receive filter; $G_I(f_\tau)$, transfer function of the pulse shaper

For the second line we go back to (4.17) and start with $g_C(t, \tau)e^{-j4\pi f_0 t}e^{j2\pi f_0 \tau} \overset{t,\tau}{\longmapsto}$ $G_C(f_t + 2f_0, f_\tau - f_0)$. With $g_I^*(\tau) \overset{\tau}{\longmapsto} G_I^*(-f_\tau)$ and using (4.27) results in $g_I^*(\tau) \circledast g_C(t, \tau)e^{-j4\pi f_0 t}e^{j2\pi f_0 \tau} \overset{t,\tau}{\longmapsto} G_I^*(-f_\tau)G_C(f_t + 2f_0, f_\tau - f_0)$ and with the help of (4.28) we obtain $\left[g_I^*(\tau) \circledast \left(g_C(t, \tau)e^{-j4\pi f_0 t}e^{j2\pi f_0 \tau}\right)\right] \circledast g_R(\tau) \overset{t,\tau}{\longmapsto}$ $\left[G_I^*(-f_\tau)G_C(f_t + 2f_0, f_\tau - f_0)\right]G_R(f_t + f_\tau)$, which proves the second line of (4.30).

The proof of (4.29) is straightforward.

4.5 Equivalent Time-Variant Baseband System and Receiver Output Signal

4.5.1 Equivalent Time-Variant Baseband System

Equipped with the overall impulse response $w_e(t, s)$, the delay spread function $g_e(t, \tau)$, and the overall Doppler spread function $G_e(f_t, f_\tau)$ summarized in Table 4.1 we can illustrate the previous results with the block diagram in Fig. 4.4. It shows the

Table 4.1 Summary—system functions and output signal of the equivalent time-variant baseband system, input signal $x(t)$ with spectrum $X(f_t)$, time-variant convolution ⊛

Impulse response	$w_e(t,s) = \left(w_I(t-s)e^{j2\pi f_0 t}\right) \circledast \left(w_C(t,s)e^{-j2\pi f_0 t}\right) \circledast w_R(t-s)$	(4.13)
Delay spread function	$g_e(t,\tau) = g_I(\tau) \circledast \left(g_C(t,\tau)e^{-j2\pi f_0\tau}\right) \circledast g_R(\tau)$	(4.19)
Doppler spread function	$G_e(f_t,f_\tau) = G_I(f_\tau)G_C(f_t, f_\tau + f_0)G_R(f_t + f_\tau)$	(4.31)
Output signal	$q_R(t) = x(t) \circledast g_e(t,\tau) + n_R(t)$	(4.40)
Sampled output signal	$q(k) = a(k)g(k,0) + \sum_{\substack{m=-\infty \\ m \neq k}}^{\infty} a(m)g(k, k-m) + n(k)$	(4.44)
Output spectrum	$Q_R(f_t) = \int_{-\infty}^{\infty} X(u)G_e(f_t - u, u)du$	(4.46)
Noise	$n_R(t) = \sqrt{2}\left(n_C(t)e^{-j2\pi f_0 t}\right) * g_R(t)$	(4.41)

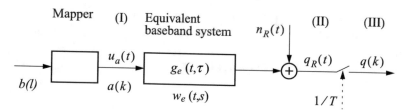

Fig. 4.4 Equivalent time-variant baseband system for wireless transmission with impulse response $w_e(t,s)$ and delay spread function $g_e(t,\tau)$

equivalent time-variant baseband system between nodes (I) and (II) of Fig. 4.1. In the last column of Table 4.1 the equation numbers are given as a reference. The noise $n_R(t)$ is investigated in quite some detail in Sect. 2.6. This equivalent model provides a condensed system description between the input and the output as a "black box" with just one system function, without details of filtering, modulation, and demodulation. We will use this scheme in the next section to determine the output signal $q_R(t)$ and its Fourier spectrum $Q_R(f_t)$.

4.5.2 Receiver Output Signal

Receiver Output Signal in the Time Domain

With the time-variant convolution (4.4) quoted from Part II and using the overall impulse response $w_e(t, s)$ of the equivalent time-variant baseband system model in Fig. 4.4 we obtain

$$q_R(t) = u_a(t) \circledast w_e(t, s) + n_R(t) = \int_{-\infty}^{\infty} u_a(s)w_e(t, s)ds + n_R(t) \qquad (4.39)$$

With the introduction of $s = t - \tau$ and the delay spread function $g_e(t, \tau) = w_e(t, t - \tau)$ follows

$$q_R(t) = u_a(t) \circledast g_e(t, \tau) + n_R(t) = \int_{-\infty}^{\infty} u_a(t - \tau)g_e(t, \tau)d\tau + n_R(t) \qquad (4.40)$$

where $n_R(t)$ is the lowpass noise given earlier in (1.25) as

$$n_R(t) = \sqrt{2}\left(n_C(t)e^{-j2\pi f_0 t}\right) * g_R(t) \qquad (4.41)$$

With the input signal $u_a(t)$ of the pulse shaper (1.9) follows from (4.40) $q_R(t) = \left(\sum_{m=-\infty}^{\infty} a(m)\delta(t - mT)\right) \circledast g_e(t, \tau) + n_R(t)$ and the time-variant convolution yields $q_R(t) = \sum_{m=-\infty}^{\infty} a(m) \int_{-\infty}^{\infty} \delta(t - \tau - mT)g_e(t, \tau)d\tau + n_R(t)$ resulting in

$$q_R(t) = \sum_{m=-\infty}^{\infty} a(m)g_e(t, t - mT) + n_R(t) \qquad (4.42)$$

As depicted in Fig. 4.4, $q_R(t)$ is sampled at $t = t_0 + kT$. With

$$g_e(t_0 + kT, t_0 + kT - mT) = g(k, k - m) \qquad (4.43)$$

$q_R(t_0 + kT) = q(k)$, and $n_R(t_0 + kT) = n(k)$ the receive signal at observation instant k is obtained as

$$q(k) = a(k)g(k,0) + \sum_{\substack{m=-\infty \\ m \neq k}}^{\infty} a(m)g(k,k-m) + n(k) \qquad (4.44)$$

where $g(k,m)$ is denoted as discrete-time delay spread function of the equivalent time-variant baseband system between node (I) and (III) in Fig. 4.4. Equation (4.44) is an interesting result and is quite similar to (2.1) of a transmission system with time-invariant channel. We recognize that $q(k)$ is composed of the transmit symbol $a(k)$, the intersymbol interference

$$I(k,k) = \sum_{\substack{m=-\infty \\ m \neq k}}^{\infty} a(m)g(k,k-m) \qquad (4.45)$$

and the additive noise $n(k)$. The receive sample $a(k)$ is weighted by $g(k,0)$. In contrast to $h(0)$ of the time-invariant system in (2.1), $g(k,0)$ changes with time due to the fading of the channel. Also the intersymbol interference $I(k,k,)$ is characterized by fading compared to $I(k)$ in (2.2).

In summary, the receive signal $q(k)$ of a wireless time-variant connection suffers from three impairments, namely the fading of the gain coefficient $g(k,0)$, the time-variant intersymbol interference $I(k,k)$, and the additive noise $n(k)$.

Spectrum of the Receiver Output Signal

Now we are interested to see the change of the transmit spectrum when passing through the time-variant channel. To this end we input the signal $u_a(t) = x(t)$ with spectrum $X(f_t)$ to the pulse shaper and determine the spectrum $Q_R(f_t)$ of the receive signal $q_R(t)$. Of course, we take advantage of the equivalent time-variant system model in Fig. 4.4 with the Doppler spread function $G_e(f_t, f_\tau)$.

Please note, in contrast to time-invariant systems $Q_R(f_t)$ is not the product of a "transfer function" $G_e(f_t, f_\tau)$ and the input spectrum $X(f_t)$. In Part II we show in quite some detail that the correct input-output relation is given by an integral as follows

$$Q_R(f_t) = \int_{-\infty}^{\infty} X(u)G_e(f_t - u, u)du \qquad (4.46)$$

This equation has some similarities with a convolution integral (except for the second argument u in G_e) and is the reason why the output spectrum $Q_R(f_t)$ can have a wider bandwidth than the input spectrum $X(f_t)$, as will be demonstrated in the Example 4. We insert (4.31) into (4.46)

$$Q_R(f_t) = \int_{-\infty}^{\infty} X(u)G_I(u)G_C(f_t - u, u + f_0)G_R(f_t - u + u)du \qquad (4.47)$$

and obtain

$$Q_R(f_t) = G_R(f_t) \int_{-\infty}^{\infty} X(u)G_I(u)G_C(f_t - u, u + f_0)du = G_R(f_t)U_D(f_t)$$

(4.48)

where

$$U_D(f_t) = \int_{-\infty}^{\infty} X(u)G_I(u)G_C(f_t - u, u + f_0)du$$ (4.49)

is the Fourier spectrum of the output signal $u_D(t)$ of the demodulator in Fig. 4.1. Thus, (4.46) boils down to (4.48), which clearly reveals that the receive lowpass $G_R(f_t)$ filters its input signal $U_D(f_t)$, as expected. Moreover, in contrast to time-invariant systems we cannot determine a quotient $\frac{Q_R(f_t)}{X(f_t)}$ from (4.48), such as the transfer function of a time-invariant system in (1.28). All findings are briefly summarized in Table 4.1.

Example 4

The top view of the real-valued Doppler spread function $G_e(f_t, f_\tau)$ of an *equivalent time-variant baseband system* is given in Fig. 4.5a. The time-variance is revealed by the maximum Doppler frequency (Doppler shift) $f_{t,G}$, which is the cut-off frequency on the f_t-axis. The higher the speed of the transmitter or receiver the larger $f_{t,G}$ will be. If the physical channel would be time-invariant and thus refrains from temporal fading, then $f_{t,G} = 0$. The cut-off frequency of the Doppler spread function in the f_τ-direction shall be $f_{\tau,G}$ and indicates the transmission bandwidth. We have assumed an input signal $x(t)$ of the pulse shaper with real-valued spectrum $X(f_\tau)$ and the cut-off frequency $f_{\tau,X} < f_{\tau,G}$. In principle, for any one-dimensional spectrum X we can allocate either frequency variable, f_t or f_τ, because the functions are mathematically the same. However, in a two-dimensional diagram the argument indicates in what frequency direction the filtering is effective. When we plot $X(f_\tau)$ as a two-dimensional function, we have to consider $X(f_\tau)$ as constant with respect to f_t. Consequently, $X(f_\tau)$ is a horizontal stripe in Fig. 4.5a. The magnitudes of the two spectra are unequal to zero in the shaded areas and outside they are zero with sharp transitions. For simplicity we assume real-valued spectra. Figure 4.5b shows $G_e(0, f_\tau)$ and $X(f_\tau)$ as a function of f_τ and Fig. 4.5c illustrates $G_e(f_t - u, u)$ in the integrand of (4.46) for various f_t to calculate $Q_R(f_t)$. Figure 4.4d depicts the resulting output spectrum $Q_R(f_t)$. Apparently, its cut-off frequency

$$f_{t,Q} = f_{\tau,X} + f_{t,G}$$ (4.50)

is by the quantity $f_{t,G}$ larger than $f_{\tau,X}$ of the input $X(f_\tau)$. The reason clearly is the time-variance of the channel, because for the case of time-invariance, $f_{t,G} = 0$ and no excess bandwidth can be observed. As a conclusion, the stronger the time variance of the channel is the larger $f_{t,G}$ and $f_{t,Q}$ will be.

Fig. 4.5 a Top view of the Doppler spread function $G_e(f_t, f_\tau)$ of the equivalent time-variant baseband system together with the spectrum $X(f_\tau)$ of its input signal **b** $G_e(0, f_\tau)$ and $X(f_\tau)$ as a function of f_τ **c** Steps to determine the integrand in (4.46) **d** Spectrum $Q_R(f_t)$ of the output signal of the equivalent time-variant baseband system

According to (4.48), the bandwidth f_R of the receive filter must cover $f_{\tau,X} + f_{t,G}$ and not just the bandwidth $f_{\tau,X}$ of the transmit signal, as in the case of a time-invariant system. Hence, the receive lowpass must be designed with the cut-off frequency $f_R \geq f_{\tau,X} + f_{t,G}$ otherwise the filter limits its input spectrum.

Chapter 5
Basic Parameters of Wireless Transmission and Multipath Propagation

In this chapter we summarize the main facts, which characterize a single input single output wireless link. Such channels can be partitioned into different segments. The inner part is the wave propagation channel, which is characterized by the free space between the output of the transmit antenna and the input to the receive antenna. The next level includes the characteristics of the transmit and the receive antenna, such as radiation pattern and antenna gains. Finally, the equivalent baseband system incorporates modulation, demodulation, and filtering, as described in Sect. 4.1. In the following we first characterize the main transmission effects of wireless and mobile channels with adequate models. There are several physical details for refinements, which are beyond our scope here, such as specific indoor and outdoor scenarios as well as details of electromagnetic field theory. The interested reader is referred to dedicated material, such as [1–4]. Finally, the input-output relation for a wireless system with multipath channel is outlined.

5.1 Path Loss

An electromagnetic wave traveling from the transmit to the receive antenna undergoes free space path loss, scattering, refraction and defraction from surfaces of buildings, hills, vegetation, rain, and various objects. These effects are well understood and investigated theoretically and by measurements using two- and tree-dimensional modeling of the landscape, ray tracing and Snell's law [2, 5]. Friis' law [6] for free space transmission relates the receive mean power P_r to the mean transmit power P_t as

$$\frac{P_r}{P_t} = G_t G_r \frac{1}{P_L} \tag{5.1}$$

© Springer Nature Switzerland AG 2021
J. Speidel, *Introduction to Digital Communications*,
Signals and Communication Technology,
https://doi.org/10.1007/978-3-030-67357-4_5

where G_t and G_r are the effective antenna gains for the transmit and receive antenna, respectively. $P_L = \left(\frac{4\pi d}{\lambda}\right)^2$ denotes the path loss. $\lambda = \frac{c}{f_0}$ is the wavelength of the electromagnetic waves, which are assumed to be planar waves in the far field. c denotes the speed of light in the air and f_0 is the carrier frequency of the modulator. The path loss P_L increases proportional to the square of the distance d between the transmitter and the receiver. However, for more complex propagation environments highly sophisticated and empirical models are required, such as the COST 231 Hata model or the Walfisch–Ikegami model, which are frequently used for the planning of cellular networks. Often, simple and experimental path loss models will do, in which the proportionality

$$P_L \sim \left(\frac{d}{d_0}\right)^n \tag{5.2}$$

is used approximately, where n is the path loss exponent and d_0 a reference. The range of n is reported to be about $2 \ldots 6$ and it depends on the environment for indoor and outdoor communications. As can be seen from (5.1) and (5.2), on a logarithmic scale the receive power P_r declines linearly with the distance d. On the basis of this path loss the necessary power budget between the transmit and the receive antenna can be calculated including antenna gains and also some margin for additional effects, such as small scale fading. This approach provides the bottom line of the system design and mainly reflects an area denoted as path loss region, which is beyond the multitude of local scattering and reflections close to the antennas. On top of that statistical fluctuations of the receive power is observed, which is called fading. The main source of fading are scattering and reflections of the propagating waves at objects on the transmitter and the receiver side as well as due to statistical movement of the transmitter and/or the receiver. This is briefly outlined in the next section.

5.2 Shadowing

An effect, which causes a fluctuation of the receive power is shadowing also called shadow fading. It can be characterized by a long-term power loss ψ and the main reason is the movement of the transmitters and/or receiver behind some shielding objects in the propagation path. Also changing surfaces of objects, scatterers, and reflectors contribute to this effect. Shadow fading normally is a stochastic effect and the resulting path loss $\psi > 0$ is often modeled satisfactorily by a log-normally distributed random variable with probability density function

$$p_0(\psi) = \frac{10}{\ln(10)\sqrt{2\pi}\sigma_0\psi} e^{-\frac{(10\log_{10}(\psi)-\mu_0)^2}{2\sigma_0^2}} \tag{5.3}$$

in which μ_0 and σ_0^2 are the mean and the variance, respectively [5]. μ_0 and σ_0 in dB and are the result of measurements or analytical models. Typical values are $\mu_0 = 0\,\mathrm{dB}$ and $\sigma_0 = 8\,\mathrm{dB}$ [3]. Shadow fading loss ψ and path loss P_L can be multiplied to get approximately the overall path loss.

In particular for broadband applications, where the signal bandwidth is a noticeable fraction of the carrier frequency and where sophisticated modulation and detection methods have to be used, the receiver does not only require a sufficient mean power but also a reasonable approximation of the impulse response $w_e(t, s)$ or delay spread function $g_e(t, \tau)$ of the equivalent time-variant baseband system. In the following we consider the link budget design as given and focus on additional impairments, such as the frequency and the time dependent fading. Practical examples are modern cellular networks, where the carrier frequency and the signal bandwidth are around 2 and 0.1 GHz, respectively. Given an omni-directional antenna pattern the electromagnetic waves are continuously distributed over the whole space. For a typical wave propagation area in a city with many buildings there is no line of sight between the base stations and the mobile stations. In this scenario reflections and scattering of the waves are favorable for sufficient signal reception. Other propagation scenarios, e.g., according to the COST 207 channel model already used for GSM and the COST 259 model differentiate between "rural area", "typical urban" with approximately no significant reflections and scattering, as well as "bad urban". The model "hilly terrain" accounts for significant reflections [1, 3, 7]. Difficult propagation conditions can occur, if there is just a narrow "key hole" between the transmitter and the receiver [8].

5.3 Multipath Model of Time-Invariant Channel

Let us now consider the propagation scenario depicted in Fig. 5.1, which is a model for finding the propagation channel parameters approximately by measurement with ray tracing. Assume that the transmitter emits an electromagnetic wave as a narrow beam received by the mobile station. Consequently, discrete paths ν from the transmitter to the receiver result, where the waves undergo path losses and all kinds of scattering and reflections, which are investigated using the wave propagation theory. Often there is also a direct path between transmitter and receiver, called line of sight path. Each ray ν can be characterized approximately by its path loss coefficient α_ν and its delay time τ_ν. The delay $\tau_\nu = \frac{l_\nu}{c}$ is given by the path length l_ν and the speed of light c in the air. The path loss coefficient $0 < \alpha_\nu \leq 1$ depends on l_ν as given by the path loss model in (5.2), thus $\alpha_\nu \sim \left(\frac{1}{\tau_\nu}\right)^n$ holds approximately with n as the path loss exponent. Consequently, a long path exhibits a high loss and thus can often be dropped for approximation. Now assume that the transmit antenna emits the signal $u_T(t)$, which is traveling on ray ν and arrives at the receive antenna as $\alpha_\nu u_T(t - \tau_\nu)$. At the receiver the signals of all paths superimpose yielding the receive signal

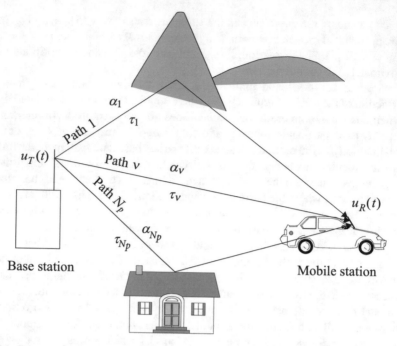

Fig. 5.1 Wireless downlink multipath propagation

$$u_R(t) = \sum_{\nu=1}^{N_P} \alpha_\nu u_T(t - \tau_\nu) \tag{5.4}$$

where in (5.4) only the N_P strongest paths are kept.

This model characterizes a linear and time-invariant system. Consequently it can be fully described by its impulse response. Let $u_T(t) = \delta(t)$ then the impulse response is

$$h(t) = \sum_{\nu=1}^{N_P} \alpha_\nu \delta(t - \tau_\nu) \tag{5.5}$$

We substantiate that the single Dirac impulse at the input results in a chain of impulses, which superimpose and the output can be considered as a broad impulse with a duration given approximately by the difference between the largest and the smallest delay, also called delay spread. Hence, the multipath channel is broadening the input signal in the time domain.

It should be mentioned that the presented model characterizes a channel with an infinite transmission bandwidth, which is not realistic. However, the pulse shaper at the transmitter and the lowpass filter at the receiver finally limit the bandwidth of the overall system.

5.4 Multipath Model of Time-Variant Channel

5.4.1 Delay Spread Function of the Time-Variant Multipath Channel

If the receiver moves and/or scattering, reflections, and shadowing effects are time variant, the path loss coefficients and the delays depend on time, $\alpha_\nu = \alpha_\nu(t)$ and $\tau_\nu = \tau_\nu(t)$, respectively. Hence, the change of the receive signal $u_R(t)$ is not only caused by the variations of the transmit signal $u_T(t)$ but also by the temporal variations of the channel parameters. Consequently, the channel model has to be refined and described as a time-variant system. As outlined in Part II, time-variant systems are characterized by the delay spread function $g_C(t, \tau)$. By definition $g_C(t, \tau)$ is the response at observation time t to a Dirac impulse at the time instant $t - \tau \leq t$. We first consider the input Dirac impulse $u_T(t) = \delta(t - (t_0 - \tau))$ active at the instant $t_0 - \tau$. Then we proceed with $t_0 \to t$. Hence, on the basis of the model in Fig. 5.1 we obtain the delay spread function

$$g_C(t, \tau) = \sum_{\nu=1}^{N_P} \alpha_\nu(t)\delta(t - (t_0 - \tau) - \tau_\nu(t))\,|_{t_0 \to t} = \sum_{\nu=1}^{N_P} \alpha_\nu(t)\delta(\tau - \tau_\nu(t)) \quad (5.6)$$

which shows some differences compared to (5.5). The multipath model in Fig. 5.1 with (5.6) is widely accepted for wireless communications. As all path loss coefficients $\alpha_\nu(t)$ are real-valued, also the delay spread function $g_C(t, \tau)$ shows this property and is in line with the fact that any physical channel exhibits a real-valued impulse response or delay spread function. It is readily appreciated that the delay spread function in (5.6) of the multipath channel is much simpler than the general form $g_C(t, \tau)$ treated in Sect. 4.1, because (5.6) is almost separated into the product of a function of t and a function of τ. The separation is perfect, if we can assume $\tau_\nu(t) = \tau_\nu = const.$ that is often fulfilled approximately. We follow this special case later and will see that the results are becoming rather simple.

The input-output relation is given by the time-variant convolution defined in (4.22). Consequently, we obtain the channel output signal $u_R(t)$ as a function of the input signal $u_T(t)$ with the help of (4.22)

$$u_R(t) = u_T(t) \circledast g_C(t, \tau) = \int_{-\infty}^{\infty} u_T(t - \tau)\, g_C(t, \tau)d\tau \quad (5.7)$$

Plugging in (5.6) yields $u_R(t) = \sum_{\nu=1}^{N_P} \int_{-\infty}^{\infty} u_T(t - \tau)\alpha_\nu(t)\delta(\tau - \tau_\nu(t))\, d\tau$ and the result is

$$u_R(t) = \sum_{\nu=1}^{N_P} \alpha_\nu(t)u_T(t - \tau_\nu(t)) . \quad (5.8)$$

Hence, the channel output signal for path ν is $\alpha_\nu(t)u_T\,(t - \tau_\nu(t))$ determined by the input signal, which is delayed by the time-variant path delay and attenuated by the time-variant path loss coefficient.

As a special case, if all path delays are constant, $\tau_\nu(t) = \tau_\nu$, the delay spread function of the path ν in (5.6) is

$$g_{C,\nu}(t, \tau) = \alpha_\nu(t)\delta(\tau - \tau_\nu) \; ; \quad \nu = 1, 2, \ldots, N_P \tag{5.9}$$

and is separated into the product of solely time and delay depending functions $\alpha_\nu(t)$ and $\delta(\tau - \tau_\nu)$, respectively.

5.4.2 Delay Spread Function of the Equivalent Time-Variant Multipath Baseband System

Now we are going to determine the delay spread function $g_e(t, \tau)$ of the equivalent baseband system between the nodes (I) and (II) in Figs. 4.1 and 4.4 for a wireless channel with time-variant multipath propagation. The channel delay spread function is given by (5.6). We prove at the end of this section that

$$g_e(t, \tau) = \sum_{\nu=1}^{N_P} g_{e,\nu}(t, \tau) \tag{5.10}$$

holds, in which

$$g_{e,\nu}(t, \tau) = \left[\tilde{\alpha}_\nu(t)g_I\,(\tau - \tau_\nu(t))\right] \circledast g_R(\tau) \; ; \quad \nu = 1, 2, \ldots, N_P \tag{5.11}$$

is the equivalent baseband delay spread function of path ν and

$$\tilde{\alpha}_\nu(t) = \alpha_\nu(t)e^{-j2\pi f_0\tau_\nu(t)} \; ; \quad \nu = 1, 2, \ldots, N_P \tag{5.12}$$

defines the complex path loss coefficient. As can be seen, $g_e(t, \tau)$ in (5.10) is the superposition of N_P delay spread functions $g_{e,\nu}(t, \tau)$ of the various paths, which are composed of the delay spread function $g_I\,(\tau - \tau_\nu(t))$ of the pulse shaper weighted by $\tilde{\alpha}_\nu(t)$ and filtered by the receive lowpass $g_R(\tau)$.

Proof of (5.10) and (5.11)

We plug (5.6) into the general form of the equivalent delay spread function (4.19) and obtain $g_e(t, \tau) = g_I(\tau) \circledast \left(\sum_{\nu=1}^{N_P} \alpha_\nu(t)\delta\,(\tau - \tau_\nu(t))\,e^{-j2\pi f_0\tau}\right) \circledast g_R(\tau)$. As the time-variant convolution is distributive, we obtain $g_e(t, \tau) = \sum_{\nu=1}^{N_P} g_{e,\nu}(t, \tau)$ with $g_{e,\nu}(t, \tau) = g_I(\tau) \circledast \left[\alpha_\nu(t)\delta\,(\tau - \tau_\nu(t))\,e^{-j2\pi f_0\tau}\right] \circledast g_R(\tau)$. To execute the first time-variant convolution we use (4.20) and obtain $g_I(\tau) \circledast \left[\alpha_\nu(t)\delta\,(\tau - \tau_\nu(t))\,e^{-j2\pi f_0\tau}\right]$

$= \int_{-\infty}^{\infty} g_I(\eta)\alpha_\nu(t)\delta\left(\tau - \eta - \tau_\nu(t)\right)e^{-j2\pi f_0(\tau-\eta)}d\eta = g_I(\tau - \tau_\nu(t))\tilde{\alpha}_\nu(t), \text{ with } \tilde{\alpha}_\nu(t)$
$= \alpha_\nu(t)e^{-j2\pi f_0\tau_\nu(t)}.$ Finally, $g_{e,\nu}(t,\tau) = \left[\tilde{\alpha}_\nu(t)g_I\left(\tau - \tau_\nu(t)\right)\right] \circledast g_R(\tau)$ follows and the proof is finished.

5.4.3 Doppler Spread Function of the Equivalent Time-Variant Multipath Baseband System

The Doppler spread function $G_e(f_t, f_\tau)$ of the equivalent time-variant multipath baseband system is obtained by the Fourier transform of $g_e(t, \tau)$ in (5.10) with respect to t and τ. To this end we make the following presuppositions. Firstly, all complex path loss coefficients $\tilde{\alpha}_\nu(t)$ exhibit the Fourier spectrum $\tilde{A}_\nu(f_t)$ with lowpass characteristic and secondly, all path delays are constant, $\tau_\nu(t) = \tau_\nu = const.$ ($\nu = 1, 2, \ldots, N_P$). We proof at the end of this section that $G_e(f_t, f_\tau)$ is determined by

$$g_e(t, \tau) \overset{t,\tau}{\longmapsto} G_e(f_t, f_\tau) = \sum_{\nu=1}^{N_P} G_{e,\nu}(f_t, f_\tau) \tag{5.13}$$

with

$$g_{e,\nu}(t, \tau) \overset{t,\tau}{\longmapsto} G_{e,\nu}(f_t, f_\tau) = G_I(f_t)\tilde{A}_\nu(f_t)e^{-j2\pi\tau_\nu f_\tau}G_R(f_t + f_\tau); \quad \nu = 1, 2, \ldots, N_P \tag{5.14}$$

where the $G_{e,\nu}(f_t, f_\tau)$ are the Doppler spread functions of the individual paths. The term $e^{-j2\pi\tau_\nu f_\tau}$ reflects the signal delay on each path.

Proof of (5.13) and (5.14)
We consider (5.11). Under the given prerequisite $\tau_\nu(t) = \tau_\nu = const.$ the Fourier transform is applicable and we get $\tilde{\alpha}_\nu(t)g_I(\tau - \tau_\nu) \overset{t,\tau}{\longmapsto} \tilde{A}_\nu(f_t)G_I(f_\tau)e^{-j2\pi\tau_\nu f_\tau}$. Next, we apply the transform pair (4.28) on (5.11) and obtain
$g_{e,\nu}(t, \tau) \overset{t,\tau}{\longmapsto} \tilde{A}_\nu(f_t)G_I(f_\tau)e^{-j2\pi\tau_\nu f_\tau}G_R(f_t + f_\tau) = G_{e,\nu}(f_t, f_\tau),$ which proves (5.14). The summation over all N_P paths, $\sum_{\nu=1}^{N_P} G_{e,\nu}(f_t, f_\tau)$, finalizes the proof.

5.4.4 Receiver Output Signal $q_R(t)$

With the help of $g_e(t, \tau)$ in (5.10), (5.11), and the general input-output relation (4.22) for a time-variant system we obtain the signal at the receiver output

$$q_R(t) = \sum_{\nu=1}^{N_P} q_{R,\nu}(t) + n_R(t) \tag{5.15}$$

with the signal component of path ν

$$q_{R,\nu}(t) = u_a(t) \circledast \left[\tilde{\alpha}_\nu(t) g_I \left(\tau - \tau_\nu(t) \right) \right] \circledast g_R(\tau) \; ; \; \nu = 1, 2, \ldots, N_P \qquad (5.16)$$

and the transmit signal $u_a(t)$.

5.4.5   Fourier Spectrum $Q_R(f_t)$ of the Receiver Output Signal

The Fourier spectrum $Q_R(f_t)$ of $q_R(t)$ is obtained with the general input-output relation (4.46) and the Doppler spread function (5.13) under the prerequisite $\tau_\nu(t) = \tau_\nu = const.$ With the spectrum $X(f_t)$ of the signal $u_a(t) = x(t)$ at the input of the pulse shaper we get without considering the receiver noise $Q_R(f_t) = \int_{-\infty}^{\infty} X(u) \sum_{\nu=1}^{N_P} G_{e,\nu}(f_t - u, u) du$. Interchanging integration and summation yields

$$Q_R(f_t) = \sum_{\nu=1}^{N_P} Q_{R,\nu}(f_t) \qquad (5.17)$$

where

$$Q_{R,\nu}(f_t) = \int_{-\infty}^{\infty} X(u) G_{e,\nu}(f_t - u, u) du \qquad (5.18)$$

is the receiver output spectrum allocated to path ν. With (5.14) follows

$$
\begin{aligned}
Q_{R,\nu}(f_t) &= \int_{-\infty}^{\infty} X(u) G_I(u) G_R(f_t) e^{-j2\pi\tau_\nu u} \tilde{A}_\nu(f_t - u) du \\
&= \left[\left(X(f_t) G_I(f_t) e^{-j2\pi\tau_\nu f_t} \right) * \tilde{A}_\nu(f_t) \right] G_R(f_t)
\end{aligned}
\qquad (5.19)
$$

We recognize that the spectrum $Q_{R,\nu}(f_t)$ of the νth receive signal is given by the convolution of $X(f_t) G_I(f_t) e^{-j2\pi\tau_\nu f_t}$ and the path loss spectrum $\tilde{A}_\nu(f_t)$ with respect to f_t, subsequently filtered by $G_R(f_t)$. As expected from the convolution, the temporal fading of the multipath channel results in an excess bandwidth of the input signal of the receive filter compared to the transmit signal. The difference is given by the bandwidth of the path loss $\tilde{A}_\nu(f_t)$. For more details please see Example 5. The reason why the time-variant convolution boils down to the time-invariant convolution in (5.19) is the fact that the Doppler spread function $G_{e,\nu}(f_t, f_\tau)$ in (5.14) excluding G_R is a product of isolated functions of f_t and f_τ, respectively, which holds for the given prerequisite of constant delays τ_ν. Again, we recognize that a quotient $\frac{Q_{R,\nu}(f_t)}{X(f_t)}$ like the transfer function of a time-invariant system cannot be defined owing to the integral in (5.18).

Example 5

(a) Let the cut-off frequencies of $X(f_t)$, $G_I(f_t)$, $\tilde{A}_\nu(f_t)$, and $G_R(f_t)$ be f_X, $f_I = f_X$, $f_{\tilde{A}_\nu}$, and $f_R > f_I + f_{\tilde{A}_\nu}$, respectively. Find the maximal cut-off frequency of $Q_{R,\nu}(f_t)$.

(b) Determine the output spectrum $Q_{R,\nu}(f_t)$ of path ν, if the channel is approximately static, i.e. its parameters do not change with time.

Solution:

(a) In general the convolution operation in (5.19) yields a maximal cut-off frequency, which is the sum of the cut-off frequencies of the spectra under convolution. Hence, $f_I + f_{\tilde{A}_\nu}$ is the cut-off frequency of $Q_{R,\nu}(f_t)$ as a maximum. If we determine $f_R > f_I + f_{\tilde{A}_\nu}$, the receive lowpass $G_R(f_t)$ is not cutting its input spectrum.

(b) If the channel is showing long periods, in which there is no fading the path loss is almost constant, say $\tilde{\alpha}_\nu(t) = 1$ and consequently $\tilde{A}_\nu(f_t) = \delta(f_t)$. Then follows from (5.19) $Q_{R,\nu}(f_t) = X(f_t)G_I(f_t)e^{-j2\pi\tau_\nu f_t}G_R(f_t)$, which is the output spectrum of a time-invariant system with channel transfer function $G_C(f_t) = e^{-j2\pi\tau_\nu f_t}$, similar to (1.26) of Sect. 1.5. Consequently, $f_R = f_I$ suffices in this case.

5.5 Multipath Channel and Mobile Receiver

5.5.1 System Model

An important practical case is considered in Fig. 5.2 for a wireless time-variant multipath channel. The fading of the parameters $\alpha_\nu(t)$ and $\tau_\nu(t)$ is caused by the movement of the receiver with the velocity $\mathbf{v}_0$. Only the path ν of the multipath propagation is depicted. The receiver starts at time instant $t_0 = 0$ at the location P_0 and moves with the velocity $\mathbf{v}_0$ to arrive at the location P at the time instant $t > 0$. We assume the speed $|\mathbf{v}_0| = v_0 = constant$. Within a short interval $t - t_0$ the receiver moves approximately on a straight line in the direction indicated by φ_ν.

Fig. 5.2 Downlink transmission over a multipath wireless channel to a receiver, which moves with velocity $\mathbf{v}_0$

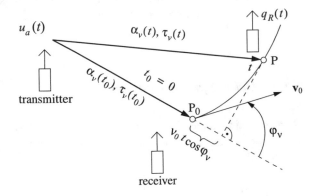

The delay of the receive signal at time instant t in location P is determined approximately by

$$\tau_\nu(t) = \tau_\nu(t_0) + \frac{v_0 \cos(\varphi_\nu)}{c} t \qquad (5.20)$$

where c is the speed of light in the air.

5.5.2   Doppler Shift

Plugging (5.20) into (5.11) and (5.12) results in the delay spread function of the equivalent baseband system for the path ν

$$g_{e,\nu}(t, \tau) = \left[\alpha_\nu(t) e^{-j2\pi f_0 \tau_\nu(t_0)} e^{-j2\pi f_{D,\nu} t} g_I \left(\tau - \tau_\nu(t_0) - \frac{v_0 \cos(\varphi_\nu)}{c} t \right) \right] \circledast g_R(\tau)$$

$$(5.21)$$

where

$$f_{D,\nu} = f_0 \frac{v_0}{c} \cos(\varphi_\nu) \qquad (5.22)$$

is denoted as the Doppler frequency or the Doppler shift for the path ν. From (5.22) we conclude the following,

- The Doppler shift $f_{D,\nu}$ is proportional to the speed v_0 of the receiver and to the carrier frequency f_0 of the modulator. Consequently, if the receiver stands still, then $f_{D,\nu} = 0$. The same holds, if no modulation is present, $f_0 = 0$.
- For $-\frac{\pi}{2} < \varphi_\nu < \frac{\pi}{2}$ the receiver moves away from the transmitter and $f_{D,\nu} > 0$. For $\varphi_\nu = 0$ the Doppler shift is maximal,

$$f_{D,max} = f_0 \frac{v_0}{c} \qquad (5.23)$$

- For $\frac{\pi}{2} < \varphi_\nu < \frac{3\pi}{2}$ the receiver moves towards the transmitter and $f_{D,\nu} < 0$.
- For $\varphi_\nu = \pm \frac{\pi}{2}$ the receiver does not change the distance to the transmitter and $f_{D,\nu} = 0$.

5.5.3   Delay Spread Function

For a small time interval $t - t_0$ and $\frac{v_0}{c} \ll 1$ we can neglect $\frac{v_0 \cos(\varphi_\nu)}{c} t$ and obtain from (5.21) approximately the delay spread function of the equivalent baseband system for the path ν

$$g_{e,\nu}(t, \tau) = \left[\alpha_\nu(t) e^{-j2\pi f_0 \tau_\nu(t_0)} e^{-j2\pi f_{D,\nu} t} g_I \left(\tau - \tau_\nu(t_0) \right) \right] \circledast g_R(\tau) \qquad (5.24)$$

Obviously, the delays $\tau_\nu(t_0) = \tau_\nu$ in (5.24) are constant. Plugging in (5.12) yields

$$g_{e,\nu}(t, \tau) = \left[\tilde{\alpha}_\nu(t) e^{-j2\pi f_{D,\nu} t} g_I(\tau - \tau_\nu) \right] \circledast g_R(\tau) \tag{5.25}$$

with $\tilde{\alpha}_\nu(t) = \alpha_\nu(t) e^{-j2\pi f_0 \tau_\nu}$.

Regarding (5.25) with (5.22) the movement of the receiver causes a modulation of the path loss coefficient $\tilde{\alpha}_\nu(t)$ with the "carrier" $e^{-j2\pi f_{D,\nu} t}$ and consequently the spectrum $\tilde{A}_\nu(f_t)$ of $\tilde{\alpha}_\nu(t)$ will experience a frequency shift, also called Doppler shift

$$\tilde{\alpha}_\nu(t) e^{-j2\pi f_{D,\nu} t} \overset{t}{\longmapsto} \tilde{A}_\nu(f_t + f_{D,\nu}) \tag{5.26}$$

As can be seen the main difference between (5.25) and (5.11) is the Doppler shift $e^{-j2\pi f_{D,\nu} t}$ and the constant delay $\tau_\nu(t) = \tau_\nu(t_0) = \tau_\nu$. Hence, we can use the terms for $g_{e,\nu}(t, \tau)$ and $q_{R,\nu}(t)$ derived earlier, if we just replace $\tilde{\alpha}_\nu(t)$ by $\tilde{\alpha}_\nu(t) e^{-j2\pi f_{D,\nu} t}$ and take $\tau_\nu(t_0) = \tau_\nu$. This will be done in the next section.

5.5.4 Receiver Output Signal $q_R(t)$ with Doppler Shift

For the receiver output signal $q_R(t)$ (5.15) is valid

$$q_R(t) = \sum_{\nu=1}^{N_P} q_{R,\nu}(t) + n_R(t) \tag{5.27}$$

and with (5.25) follows

$$q_{R,\nu}(t) = x(t) \circledast \left[\tilde{\alpha}_\nu(t) e^{-j2\pi f_{D,\nu} t} g_I(\tau - \tau_\nu) \right] \circledast g_R(\tau) \; ; \; \nu = 1, 2, \ldots, N_P \tag{5.28}$$

where $u_a(t) = x(t)$ is the transmit signal at the pulse shaper input. Compared to (5.16) the Doppler shift $e^{-j2\pi f_{D,\nu} t}$ is effective and $\tau_\nu(t) = \tau_\nu(t_0) = \tau_\nu$ holds.

From (5.24) and (5.28) we derive the equivalent time-variant baseband system model depicted in Fig. 5.3. Each path ν is composed of a delay τ_ν, a complex path loss coefficient $\alpha_\nu(t) e^{-j2\pi f_0 \tau_\nu}$ and a modulator with carrier $e^{-j2\pi f_{D,\nu} t}$, which represents the Doppler shift of the spectrum. In this tapped delay line model the signals at the input and at the output are filtered by the pulse shaper $g_I(\tau)$ and the receive lowpass filter $g_R(\tau)$, respectively. The resulting output signal $q_R(t)$ is given by (5.27) and (5.28) without considering the receiver noise.

Example 6
Find the Doppler spread function $G_{e,\nu}(f_t, f_\tau)$ of the equivalent baseband system of path ν for a receiver with velocity $\mathbf{v}_0$ and constant path delay in Fig. 5.2. How large is the maximal Doppler shift, if a person is moving with $v_0 = 1$ m/s (=3.6 km/h) and a fast train with 360 km/h?

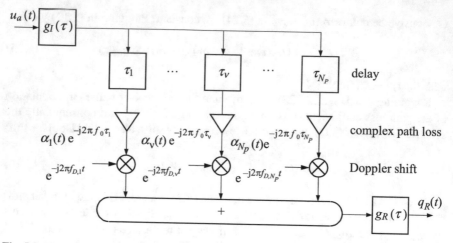

Fig. 5.3 Tapped delay line model of equivalent time-variant baseband system with multipath channel and moving receiver

Determine the receiver output spectrum $Q_R(f_t)$ associated to path ν.

Solution:
We can determine $G_{e,\nu}(f_t, f_\tau)$ from (5.25) by the Fourier transform with respect to t and τ or alternatively from (5.14), if we replace $\tilde{A}_\nu(f_t)$ by $\tilde{A}_\nu(f_t + f_{D,\nu})$. The result is

$$g_{e,\nu}(t,\tau) \overset{t,\tau}{\longmapsto} G_{e,\nu}(f_t, f_\tau) = G_I(f_\tau)e^{-j2\pi\tau_\nu f_\tau}\tilde{A}_\nu(f_t + f_{D,\nu})G_R(f_t + f_\tau) \quad (5.29)$$

Compared to the baseband of $G_e(f_t, f_\tau)$ in Fig. 4.3 the the spectrum $\tilde{A}_\nu(f_t + f_{D,\nu})$ is shifted to the left by $f_{D,\nu}$ on the f_t-axis and subsequently filtered by $G_R(f_t + f_\tau)$.
With (5.23) the maximal Doppler shifts are $f_{D,max} \approx 7$ Hz and 700 Hz.
Output spectrum: We apply (5.19) and replace $\tilde{A}_\nu(f_t)$ by $\tilde{A}_\nu(f_t + f_{D,\nu})$ yielding

$$Q_{R,\nu}(f_t) = \left[\left(X(f_t)G_I(f_t)e^{-j2\pi\tau_\nu f_t} \right) * \tilde{A}_\nu(f_t + f_{D,\nu}) \right] G_R(f_t) \quad (5.30)$$

Example 7
Consider a harmonic transmit signal $u_a(t) = x(t) = \hat{x}e^{j2\pi f_X t}$ with the frequency $f_X < f_I < f_R$, where f_I and f_R are the cut-off frequencies of the pulse shaper $g_I(t)$ and the receive lowpass $g_R(t)$, respectively. Find the spectrum $Q_{R,\nu}(f_t)$ at the receiver output for path ν, if the receiver moves with the velocity $\mathbf{v}_0$ given in Fig. 5.2.

Solution:
We take advantage of the frequency domain. Knowing that
$x(t) \overset{t}{\longmapsto} X(f_t) = \hat{x}\delta(f_t - f_X)$ we find with (4.46)
$Q_{R,\nu}(f_t) = \int_{-\infty}^{\infty} X(u)G_{e,\nu}(f_t - u, u)du = \hat{x}\int_{-\infty}^{\infty}\delta(u - f_X)G_{e,\nu}(f_t - u, u)du.$

With the sifting property of the Dirac impulse we obtain

$$Q_{R,\nu}(f_t) = \hat{x} G_{e,\nu}(f_t - f_X, f_X) \tag{5.31}$$

Plugging in (5.29) yields

$$Q_{R,\nu}(f_t) = \left[\hat{x} G_I(f_X) e^{-j2\pi\tau_\nu f_X} \right] \tilde{A}_\nu(f_t - f_X + f_{D,\nu}) G_R(f_t) \tag{5.32}$$

The term in brackets is a constant factor. The remaining part of $Q_{R,\nu}(f_t)$ represents a continuous spectrum rather than just a Dirac impulse as $X(f_t)$ owing to the temporal fading of $\tilde{\alpha}_\nu(t)$.

5.6 Frequency Selective Fading of Multipath Channel

In this section we consider the time-variant transfer function of the multipath wireless channel and investigate its frequency response for a constant time instant t. We will see that it can change significantly as a function of f_τ. This effect is called frequency-selectivity or frequency selective fading. With the transform pair $\delta(\tau - \tau_\nu(t)) \overset{\tau}{\longmapsto} e^{-j2\pi\tau_\nu(t)f_\tau}$ the Fourier transform yields from the delay spread function in (5.6) the time variant transfer function

$$g_C(\tau, t) \overset{\tau}{\longmapsto} G_{C,t}(t, f_\tau) = \sum_{\nu=1}^{N_P} \alpha_\nu(t) e^{-j2\pi\tau_\nu(t)f_\tau} \tag{5.33}$$

With the following simple example we show that this channel transfer function can be strongly frequency selective.

Example 8
Assume a two-path channel model with $N_P = 2$, $\alpha_1(t) = \alpha_2(t) = 1$, and $\tau_1(t) = 0$ for a fixed time instant t. We rename f_τ as f and obtain from (5.33)

$$G_{C,t}(t, f) = 2 \cos\left(\pi f \tau_2(t)\right) e^{-j\pi\tau_2(t)f} \tag{5.34}$$

We recognize deep notches $\left| G_{C,t}(t, f_m) \right| = 0$ for $f = f_m = \frac{1+2m}{2\tau_2(t)}$; $m = 0, \pm 1, \ldots$.
This is attributed as a fading effect and denoted as frequency selective fading or multipath fading. Hence, the term fading is not exclusively used for temporal fading.

Example 9: Time and Frequency Selective Fading
We consider a channel incorporating time and frequency selective fading and characterize this channel by its time-variant transfer function $G_{C,t}(t, f_\tau)$. Figure 5.4 shows the squared magnitude $\left| G_{C,t}(t, f_\tau) \right|^2$ of the time-variant transfer function of a typical channel for the wireless local area network standard WLAN IEEE 802.11 in the indoor office environment. The frequency range is at around 2400 MHz. The graph is generated by the open online simulation platform "webdemo" [9]. Apparently, for a fixed time instant the magnitude response is changing with respect to the

squared magnitude of time-variant transfer function

Fig. 5.4 Squared magnitude $\left|G_{C,t}(t, f_\tau)\right|^2$ of the time-variant transfer function of a typical indoor channel for the WLAN IEEE 802.11 standard in the office environment. *Source* Online platform "webdemo" [9]

frequency and thus is frequency selective. On the other hand, for a given frequency $\left|G_{C,t}(t, f_\tau)\right|^2$ varies with time. Hence, the channel is time and frequency selective often denoted as double selective.

5.7   Statistical Description of Wireless Multipath Channel

In the following we are going to describe the time-variance of a multipath channel by statistical parameters. We assume a multipath channel with delay spread function $g_c(t, \tau)$ given in (5.6) and neglect the receiver noise in the following. For the receive signal (5.15) then holds with (5.16) and $u_a(t) = x(t)$

$$q_R(t) = \sum_{\nu=1}^{N_P} x(t) \circledast \left[\tilde{\alpha}_\nu(t)g_I\left(\tau - \tau_\nu(t)\right)\right] \circledast g_R(\tau) \tag{5.35}$$

As can be seen, even for a constant transmit signal $x(t) = const.$ the receiver output signal changes with time, due to the temporal fading of the path loss coefficients $\tilde{\alpha}_\nu(t)$ and the path delays $\tau_\nu(t)$. Because there are many impacts, which are causing this fading, such as the polymorphic environment in the first place, a statistical description is adequate. Nevertheless, the deterministic approach outlined in the previous chapters is helpful to understand the interaction of the various building blocks of the wireless transmission system. For an introduction on stochastic processes the reader is referred to the Appendix A.

5.7.1 Complex Gaussian Multipath Model

To derive the complex Gaussian multipath model the following prerequisites shall apply,

- $\alpha_\nu(t)$ and $\tau_\nu(t)$ change independently at random,
- The N_P paths are independent,
- The number of independent paths grows to infinity, $N_P \rightarrow \infty$.

Then the conditions of the central limit theorem [10] are fulfilled resulting in the following properties:

- $q_R(t)$ in (5.35) becomes a complex-valued Gaussian process,
- $\text{Re}\,[q_R(t)]$ and $\text{Im}\,[q_R(t)]$ are statistically independent, real-valued Gaussian processes each with variance σ^2.

5.7.2 Channel Model with Rayleigh Fading

The multipath channel model with Rayleigh fading is determined under the prerequisites in Sect. 5.7.1.

If there is no line of sight or if there are no fixed scatterers or reflectors between the transmit and the receive antenna, then the Gaussian process $q_R(t)$ has zero mean, $\mathbf{E}\,[q_R] = 0$.

Let $z = |q_R(t)|$ and $\phi = \arg\,[q_R(t)]$, then the probability density function $p_z(z)$ is a Rayleigh density given by

$$p_z(z) = \begin{cases} \frac{z}{\sigma^2}e^{-\frac{z^2}{2\sigma^2}} & ; z \geq 0 \\ 0 & ; z < 0 \end{cases} \tag{5.36}$$

with

$p_z(z) = 0 \,\forall\, z < 0$, as z is the absolute value of $q_R(t)$.

$\mathbf{E}\,[z] = \sigma\sqrt{\frac{\pi}{2}}$ the expected value

$\mathbf{E}\,[z^2] = 2\sigma^2$ the mean power

$var\,[z] = \mathbf{E}\left[(z - \mathbf{E}\,[z])^2\right] = \sigma^2\left(2 - \frac{\pi}{2}\right)$ the variance of z.

The probability density function $p_\phi(\phi)$ of ϕ is uniform with

$$p_\phi(\phi) = \begin{cases} \frac{1}{2\pi} & ; -\pi \leq \phi \leq \pi \\ 0 & ; \quad else \end{cases} \tag{5.37}$$

It should be pointed out that the Rayleigh fading model is applicable to each branch of the tapped delay line model in Fig. 5.3, because in reality each branch can be regarded as a composition of an infinite number of independent sub-paths, for which the central limit theorem also applies. One of the first implementations of Rayleigh fading has been given by Clarke and Jakes, as described later in Sect. 5.7.5.

$p_z(z)$ is depicted in Fig. 5.5a. Obviously, the density function is zero for $z = 0$ and approaches zero for $z \to \infty$. For increasing values of σ the magnitudes decrease and we see a wider distribution.

5.7.3   Channel Model with Rice Fading

If there are line of sight or dominant fixed scatterers or reflectors present between the transmit and the receive antenna, the receive signal $q_R(t)$ exhibits a mean value unequal to zero. The probability density function of $z = |q_R(t)|$ is modeled by

$$p_z(z) = \begin{cases} \frac{z}{\sigma^2} e^{-\frac{z^2+s^2}{2\sigma^2}} I_0 \left(\frac{zs}{\sigma^2}\right) ; \ z \geq 0 \\ \qquad\qquad 0 \qquad\qquad ; \ z < 0 \end{cases} \tag{5.38}$$

where $s \in \mathbb{R}$ is called the non-centrality parameter. The term $I_0(x)$ defines the modified Bessel function of the first kind and zero order. $I_0(x)$ is monotonically increasing with x and not oscillating as the ordinary Bessel functions $J_m(x)$. Please note, $I_0(0) = 1$ and the approximation $I_0(x) \approx \frac{1}{\sqrt{2\pi x}} e^x$ holds for $x \gg 1$ [11] . It is straightforward to show that

$\mathbf{E}[z] = \sigma \sqrt{\frac{\pi}{2}} + s$ expected value

$\mathbf{E}[z^2] = 2\sigma^2 + s^2$ mean power

$p_z(z)$ is depicted in Fig. 5.5b. For $s = 0$ the Riceian density boils down to the Rayleigh density in Fig. 5.5a. Roughly speaking, the parameter $s > 0$ initiates a shift to the right combined with a change of the shape. This will be outlined in Example 10. Please note that s also increases the mean power of z.

Example 10
Show that the parameter $s > 0$ approximately shifts the Riceian probability density function in (5.38) to the right associated with a change of the shape.

Solution:
We decompose the dominant exponential function $e^{-\frac{z^2+s^2}{2\sigma^2}} = e^{-\frac{(z-s)^2}{2\sigma^2}} e^{-\frac{zs}{\sigma^2}}$ and see that the first term represents a shift and the second term changes the shape.

5.7.4   Channel Model with Nakagami-m Fading

An interesting and versatile statistical model for $z = |q_R(t)|$ is given by the Nakagami-m probability density function [12]

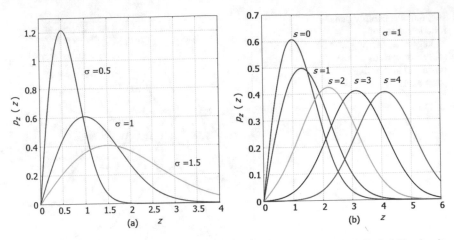

Fig. 5.5 **a** Probability density function of Rayleigh fading given by (5.36). **b** Probability density function of Rice fading given by (5.38) with $\sigma = 1$

$$
p_z(z) = \begin{cases} \frac{2}{\Gamma(m)} \left(\frac{m}{\Omega}\right)^m z^{2m-1} e^{-\frac{m}{\Omega}z^2} & ; z \geq 0 \\ 0 & ; z < 0 \end{cases} \quad ; \quad m \geq \frac{1}{2} \qquad (5.39)
$$

$\Gamma(m)$ is the gamma function defined by Euler's integral $\Gamma(m) = \int_0^\infty t^{m-1} e^{-t} dt$, where in general $m \in \mathbb{C}$, $\mathrm{Re}\,[m] > 0$. For real $m > 0$ this function can be calculated recursively by $\Gamma(m+1) = m\Gamma(m)$ with $\Gamma(1) = 1$. Then, $\Gamma(m)$ is positive and exhibits a bathtub shape with a pole at $m = 0$ while strongly approaching infinity for $m \to \infty$. In case of integer $m \geq 1$ the gamma function turns into $\Gamma(m) = (m-1)!$ and is monotonically increasing [11]. In (5.39) $\Omega = \mathbf{E}\left[z^2\right]$ defines the mean power of z and m is called the Nakagami fading parameter. $p_z(z)$ is depicted in Fig. 5.6 for various Ω and m. With the parameter m the impact of the fading is adjusted.

E.g., for $m = 1$ and $\Omega = 2\sigma^2$ the probability density function (5.39) turns into the density of the Rayleigh fading given by (5.36). With increasing m the fading strength is lowered and for $m \to \infty$ it can be shown that the Nakagami-m model resembles no fading, i.e. a time-invariant channel [13]. We observe from Fig. 5.6 that for $m > 0.5$ an increasing mean power Ω shifts the main lobe of the density function to the right and the Nakagami-m density can provide a reasonable approximation of the Riceian probability density function for many applications [14]. For $m = 0.5$, $\Gamma(\frac{1}{2}) = \sqrt{\pi}$, and $\Omega = \sigma^2$ the Nakagami density represents the one-sided Gaussian density $p_G(z) = 2\frac{1}{\sqrt{2\pi}\sigma} e^{-\frac{z^2}{2\sigma^2}}$ for $z \geq 0$ and $p_G(z) = 0$ else. In summary, the Nakagami-m fading model can be adapted rather flexible to a variety of realistic fading channels.

Fig. 5.6 Probability density function $p_z(z)$ of Nakagami-m fading given by (5.39) for various mean powers Ω and fading parameters m

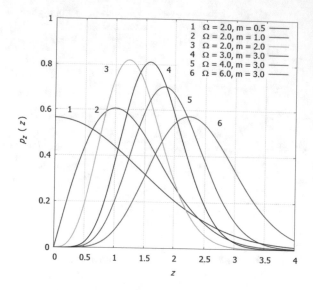

5.7.5 Channel Model of Clarke and Jakes

Clarke considers in his approach [15] Gaussian distributed I- and Q-signals at the output of the multipath channel. The result is a Rayleigh distributed magnitude of the envelope. It is one of the first implementations of Rayleigh fading channel models.

With the model of Jakes [16] the receiver is located in the middle of a circle and receives N_P signals from reflectors, which are uniformly distributed on the circle with an angular spacing of $\frac{2\pi}{N_P}$. Hence, the angles of arrival, equivalent to $\pi - \varphi_\nu$ in Fig. 5.2, are equally distributed. Each receive signal consists of the I- and Q-component of the complex envelope and all signals superimpose at the receiver. The I- and Q-component signals are equipped with constant amplitudes and their phases are assumed to be equally distributed in the interval from 0 to 2π. There is a variety of enhancements of this channel model. Some are using additional filters for a higher decorrelation of the signal components [17].

In the following we focus again on the scenario in Fig. 5.2 with a receiver moving at the velocity $\mathbf{v}_0$. For small noise the receiver output signal $q_R(t)$ in (5.27) is

$$q_R(t) = \sum_{\nu=1}^{N_P} q_{R,\nu}(t) \tag{5.40}$$

All delays τ_ν are assumed to be constant. From (5.28) follows with $x(t) \circledast g_I(\tau - \tau_\nu)$ $= x(t) * g_I(t - \tau_\nu) = u_I(t - \tau_\nu)$ the receiver output signal for path ν

$$q_{R,\nu}(t) = \left[\tilde{\alpha}_\nu(t)e^{-j2\pi f_{D,\nu}t}u_I(t - \tau_\nu)\right] \circledast g_R(\tau) \; ; \; \nu = 1, 2, \ldots, N_P \tag{5.41}$$

In the following $g_R(\tau)$ shall represent the delay spread function of an ideal lowpass receive filter with a cut-off frequency large enough to pass its input signal without any distortions to the output. Therefore we will drop $g_R(\tau)$ and get from (5.41) the receive signal for path ν

$$q_{R,\nu}(t) = \alpha_\nu(t)e^{j\Theta_\nu}e^{-j2\pi f_{D,\nu}t}u_I(t - \tau_\nu) \tag{5.42}$$

where we have substituted $\tilde{\alpha}_\nu(t) = \alpha_\nu(t)e^{j\Theta_\nu}$ with $\Theta_\nu = -2\pi f_0\tau_\nu$. The Doppler shift $f_{D,\nu}$ is defined by (5.22). The following prerequisites shall apply:

1. All signals are described as wide sense stationary (WSS) stochastic processes. As outlined in Appendix A, a wide sense stationary process owns a (joint) probability density functions and a mean value, which are independent of the time t. Furthermore, the auto- and cross-correlation functions R depend on a time difference $\zeta = t_2 - t_1$ rather than on the distinct time instants t_1 and t_2, thus $R(t_1, t_2) = R(\zeta)$ is true. The statistical description of the signals holds for any time instant t, which is skipped in some cases to simplify notation.
2. As with the Jakes model, a multitude of signals shall be present at the receiver in Fig. 5.2 with angles of arrival $\pi - \varphi_\nu$. All φ_ν, $\nu = 1, 2, \ldots, N_P$, are uniformly distributed with the probability density function

$$p_\varphi(\varphi) = \begin{cases} \frac{1}{2\pi} \; ; \; |\varphi| \leq \pi \\ 0 \; ; \quad else \end{cases} \tag{5.43}$$

   where φ stands for φ_ν.
3. Each Θ_ν shall be uniformly distributed with the density function (5.43), where φ stands for Θ_ν.
4. α_ν, $e^{j\Theta_\nu}$, $e^{-j2\pi f_{D,\nu}t}$, and $u_I(t - \tau_\nu)$ are uncorrelated $\forall\, \nu$.
5. $e^{j\Theta_\nu}$ and $e^{j\Theta_\mu}$ are uncorrelated $\forall\, \nu \neq \mu$.
6. The transmit signal at the output of the pulse shaper shall be $u_I(t)$ with autocorrelation function

$$R_{u_I u_I}(\zeta) = \mathbf{E}\left[u_I^*(t)u_I(t + \zeta)\right] \tag{5.44}$$

With these prerequisites we can determine the autocorrelation function of $q_R(t)$ and obtain

$$R_{q_R q_R}(\zeta) = \mathbf{E}\left[q_R^*(t)q_R(t + \zeta)\right] = R_{u_I u_I}(\zeta)J_0\left(2\pi f_{D,max}\zeta\right)\sum_{\nu=1}^{N_P} R_{\alpha_\nu\alpha_\nu}(\zeta) \tag{5.45}$$

where

$$R_{\alpha_\nu\alpha_\nu}(\zeta) = \mathbf{E}\left[\alpha_\nu(t)\alpha_\nu(t + \zeta)\right] \tag{5.46}$$

represents the autocorrelation function of $\alpha_\nu(t)$. If we can assume that the changes of the $\alpha_\nu(t)$ are small, $\alpha_\nu(t) \approx \alpha_\nu(t + \zeta)$, we obtain approximately

$$R_{q_R q_R}(\zeta) = P_\alpha R_{u_I u_I}(\zeta) J_0 \left(2\pi f_{D,max}\zeta\right) \tag{5.47}$$

where

$$P_\alpha = \sum_{\nu=1}^{N_P} \mathbf{E}\left[\alpha_\nu^2\right] \tag{5.48}$$

is the total mean power of the path loss coefficients. $J_0(x)$ denotes the Bessel function of the first kind and zero order defined as

$$J_0(x) = \frac{1}{2\pi} \int_{-\pi}^{\pi} e^{jx \cos(\beta)} d\beta \tag{5.49}$$

with $J_0(0) = 1$ and $J_0(x) \approx \sqrt{\frac{2}{\pi x}} \cos\left(x - \frac{\pi}{4}\right)$; $x \gg \frac{1}{4}$ [11]. The proof of (5.45) and (5.47) is given at the end of this section. $R_{q_R q_R}(\zeta)$ in (5.47) is determined approximately by the total mean power P_α of the path losses, the autocorrelation function $R_{u_I u_I}(\zeta)$ of the transmit signal at the output of the pulse shaper, and the Bessel function $J_0 \left(2\pi f_{D,max}\zeta\right)$. As outlined in Appendix A, the Fourier transform of the autocorrelation function provides the power spectral density. Thus, for the power spectral density $S_{u_I u_I}(f_t)$ of $u_I(t)$ holds

$$R_{u_I u_I}(\zeta) \overset{\zeta}{\longmapsto} S_{u_I u_I}(f_t) \tag{5.50}$$

and with the convolution operation in the frequency domain follows from (5.47) the power spectral density $S_{q_R q_R}(f_t)$ of $q_R(t)$

$$R_{q_R q_R}(\zeta) \overset{\zeta}{\longmapsto} S_{q_R q_R}(f_t) = \frac{P_\alpha}{\pi f_{D,max}} S_{u_I u_I}(f_t) * \frac{\text{rect}\left(\frac{f_t}{2 f_{D,max}}\right)}{\sqrt{1 - \left(\frac{f_t}{f_{D,max}}\right)^2}} \tag{5.51}$$

where we have used the Fourier transform pair

$$\pi a J_0 \left(2\pi a t\right) \overset{t}{\longmapsto} \frac{\text{rect}\left(\frac{f_t}{2a}\right)}{\sqrt{1 - \left(\frac{f_t}{a}\right)^2}} \quad ; \quad a \neq 0 \tag{5.52}$$

The rectangular function is defined as $\text{rect}\left(\frac{f_t}{2a}\right) = 1$ for $|f_t| \le a$, and zero else.

The second term of the convolution operation in (5.51) exhibits poles at $f_t = \pm f_{D,max}$. However, the power spectral density $S_{u_I u_I}(f_t)$ of realistic transmit signals $u_I(t)$ owns a lowpass characteristic and thus the result $S_{q_R q_R}(f_t)$ of the convolution shows a smooth transition at the band edges and no poles.

To get a rough idea about the correlation property of the receiver output signal $q_R(t)$ we discuss $R_{q_R q_R}(\zeta)$ in Example 11 for a simple input signal, which is constant,

$u_I(t) = 1$. Its autocorrelation function is $R_{u_I u_I}(\zeta) = 1$ and from (5.47) follows

$$R_{q_R q_R}(\zeta) = P_\alpha J_0\left(2\pi f_{D,max}\zeta\right) \tag{5.53}$$

Figure 5.7a portraits $R_{q_R q_R}(\zeta)$. The envelope of $J_0(x)$ declines approximately with $\sqrt{\frac{2}{\pi x}}$ for increasing x and we conclude that the sample $q_R(t)$ at time instant t and the sample $q_R(t + \zeta)$ at $t + \zeta$ are increasingly less correlated the larger ζ is. Hence, uncorrelated scattering of the samples of $q_R(t)$ is valid approximately. Computer simulations with the online platform "webdemo" [18], verify approximately the Rayleigh probability density function and the Bessel shaped autocorrelation function of the magnitude of a fading receive signal.

Example 11
We consider the special transmit signal $u_I(t) = 1$ with constant amplitude and look for the power spectral densities $S_{u_I u_I}(f_t)$ and $S_{q_R q_R}(f_t)$.

We obtain the autocorrelation function of $u_I(t)$ from (5.44) as $R_{u_I u_I}(\zeta) = 1$, from which the power spectral density follows $R_{u_I u_I}(\zeta) \overset{\zeta}{\mapsto} S_{u_I u_I}(f_t) = \delta(f_t)$. Then we get from (5.51)

$$S_{q_R q_R}(f_t) = \frac{P_\alpha}{\pi f_{D,max}} \frac{\mathrm{rect}\left(\frac{f_t}{2 f_{D,max}}\right)}{\sqrt{1 - \left(\frac{f_t}{f_{D,max}}\right)^2}} \tag{5.54}$$

which is shown in Fig. 5.7b. Apparently, $S_{q_R q_R}(f_t) \approx const.$ for $|f_t| \ll f_{D,max}$ and $S_{q_R q_R}(f_t) \to \infty$ for $f_t \to \pm f_{D,max}$. Measured power spectral densities verify the shape of $S_{q_R q_R}(f_t)$ approximately, however, with maxima rather than poles at $\pm f_{D,max}$ [2]. Anyhow, in reality, transmit signals $u_I(t)$ are not constant and their power spectral density $S_{u_I u_I}(f_t)$ in (5.51) shows a lowpass characteristic which makes $S_{q_R q_R}(f)$ more smooth at the edges of the frequency band.

Proof of (5.45)
For $R_{q_R q_R}(\zeta)$ follows with (5.40)

$$R_{q_R q_R}(\zeta) = \mathbf{E}\left[\sum_{\nu=1}^{N_P} \sum_{\mu=1}^{N_P} q_{R,\nu}^*(t) q_{R,\mu}(t + \zeta) \right] \tag{5.55}$$

As outlined in Appendix A, the cross-correlation function of two wide sense stationary and uncorrelated stochastic processes $X(t)$ and $Y(t)$ is $R_{XY}(\zeta) = \mathbf{E}[X^*(t)Y(t + \zeta)] = \mathbf{E}[X^*(t)]\mathbf{E}[Y(t)]$. As the expectation operator is linear we obtain with (5.42)

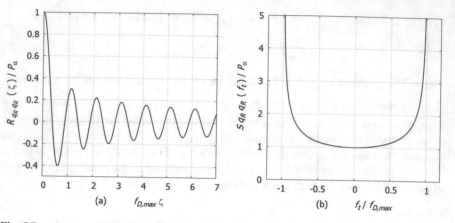

Fig. 5.7 a Autocorrelation function $R_{q_R q_R}(\zeta)$ in (5.53) of receive signal $q_R(t)$. **b** Power spectral density $S_{q_R q_R}(f_t)$ in (5.54) of receive signal $q_R(t)$ (normalized to $1/\pi f_{D,max}$)

$$R_{q_R q_R}(\zeta) = \sum_{\nu=1}^{N_P} \sum_{\mu=1}^{N_P} \left\{ \mathbf{E}\left[\alpha_\nu(t)\alpha_\mu(t+\zeta)\right] \mathbf{E}\left[e^{-j\Theta_\nu}e^{j\Theta_\mu}\right] \right\} \cdot$$
$$\left\{ \mathbf{E}\left[e^{j2\pi f_{D,\nu}t}e^{-j2\pi f_{D,\mu}(t+\zeta)}\right] \mathbf{E}\left[u_I^*(t-\tau_\nu)u_I\left(t-\tau_\mu+\zeta\right)\right] \right\} \tag{5.56}$$

in which we have accounted for the precondition 4 with the consequence that the expected value of the product is equal to the product of the expected values. Next, we make use of the prerequisite 5 and obtain

$$\mathbf{E}\left[e^{-j\Theta_\nu}e^{j\Theta_\mu}\right] = \begin{cases} \mathbf{E}\left[e^{-j\Theta_\nu}\right]\mathbf{E}\left[e^{j\Theta_\mu}\right] ; \nu \neq \mu \\ 1 \qquad\qquad ; \nu = \mu \end{cases} \tag{5.57}$$

Before we are going to determine $\mathbf{E}\left[e^{-j\Theta_\nu}\right]$ let us review some theorems from Appendix A.

Given a random variable X with probability density function $p_x(x)$. Then $\mathbf{E}[g(X)] = \int_{-\infty}^{\infty} g(x)p_x(x)dx$ holds, in which $g(\ldots)$ is a given function.

Using this property and prerequisite 3 for Θ_ν one can find

$$\mathbf{E}\left[e^{-j\Theta_\nu}\right] = \int_{-\infty}^{\infty} e^{-j\Theta_\nu} p_\varphi(\Theta_\nu)d\Theta_\nu = \frac{1}{2\pi}\int_{-\pi}^{\pi} e^{-j\Theta_\nu}d\Theta_\nu = 0 \tag{5.58}$$

yielding from (5.57)

$$\mathbf{E}\left[e^{-j\Theta_\nu}e^{j\Theta_\mu}\right] = \begin{cases} 0 ; \nu \neq \mu \\ 1 ; \nu = \mu \end{cases} \tag{5.59}$$

Then follows from (5.56) with (5.44) and (5.46)

$$R_{q_R q_R}(\zeta) = \sum_{\nu=1}^{N_P} R_{\alpha_\nu \alpha_\nu}(\zeta) \mathbf{E}\left[e^{-j2\pi f_{D,\nu}\zeta}\right] R_{u_I u_I}(\zeta) \tag{5.60}$$

With the Doppler frequency in (5.22) and the density function (5.43) for φ_ν we obtain $\mathbf{E}\left[e^{-j2\pi f_{D,\nu}\zeta}\right] = \mathbf{E}\left[e^{-j2\pi \zeta f_{D,max}\cos(\varphi_\nu)}\right] = \frac{1}{2\pi}\int_{-\pi}^{\pi} e^{-j2\pi \zeta f_{D,max}\cos(\varphi_\nu)} d\varphi_\nu$. With the Bessel function $J_0(x) = \frac{1}{2\pi}\int_{-\pi}^{\pi} e^{jx\cos(\beta)} d\beta = \frac{1}{2\pi}\int_{-\pi}^{\pi} \cos(x\cos(\beta)) d\beta = J_0(-x)$ follows $\mathbf{E}\left[e^{-j2\pi f_{D,\nu}\zeta}\right] = J_0(2\pi f_{D,max}\zeta)$, which is true for all ν. Then we get from (5.60) $R_{q_R q_R}(\zeta) = R_{u_I u_I}(\zeta) J_0(2\pi f_{D,max}\zeta) \sum_{\nu=1}^{N_P} R_{\alpha_\nu \alpha_\nu}(\zeta)$ and the proof of (5.45) ends. For the case $\alpha_\nu(t) \approx \alpha_\nu(t+\zeta)$ we conclude from (5.46) $R_{\alpha_\nu \alpha_\nu}(\zeta) = \mathbf{E}\left[\alpha_\nu^2\right]$. From (5.45) follows with (5.48) the result (5.47) and the proof is finished.

References

1. Molisch, A.F., Asplund, H., Heddergott, R., Steinbauer, M., Zwick, T.: The COST259 directional channel model - part i: Overview and methodology. IEEE Trans. Wireless Commun. **5** (2006)
2. Molisch, A.F.: Wireless Communications. Wiley and IEEE Press, New York (2009)
3. Physical layer aspects for evolved universal terrestrial radio access (UTRA). Technical report, 3GPP TR 25.814 v7.1.0 (2006)
4. Haykin, S., Moher, M.: Modern Wireless Communications. Pearson Prentice Hall, Upper Saddle River (2005)
5. Goldsmith, A.: Wireless Communications. Cambridge University Press, New York (2005)
6. Friis, H.T.: A note on a simple transmission formula. Proceed. IRE **34** (1946)
7. Physical channels and modulation, Technical Specifications. TS 36.211, V11.5.0. Technical report, 3GPP (2012)
8. Mueller, A., Speidel, J.: Performance limits of multiple-input multiple-output keyhole channels with antenna selection. Eur. Trans. Telecommun. **19** (2008)
9. Ziegler, M.: 802.11 WLAN spectrum analysis, webdemo. Technical report, Institute of Telecommunications, University of Stuttgart, Germany. http://webdemo.inue.uni-stuttgart.de (2018)
10. Papoulis, A., Pillai, S.U.: Probability, Random Variables, and Stochastic Processes, 4th edn. McGraw-Hill, Boston (2002)
11. Abramowitz, A., Stegun, I.A.: Handbook of Mathematical Functions. Dover Publications Inc, New York (1974)
12. Nakagami, M.: The m-distribution - a general formula for intensity distribution of rapid fading. In: Statistical Methods in Radio Wave Propagation, Pergamon (1960)
13. Alouini, M., Goldsmith, J.: A unified approach for calculating the error rates of linearly modulated signals over generalized fading channels. IEEE Trans. Commun. (1999)
14. Simon, K.M., Alouini, M.-S.: Digital Communication Over Fading Channels: A Unified Approach to Performance. Wiley, New York (2000)
15. Clarke, R.H.: A statistical theory of mobile radio reception. Bell Syst. Techn. J. (1968)
16. Jakes, W.: Microwave Mobile Communications. Wiley, New York (1975)
17. Dent, P., Bottomley, G.E., Croft, T.: Jakes fading model revisited. Electron. Lett. (1993)
18. ten Brink, S.: Rayleigh fading, webdemo. Technical report, Institute of Telecommunications, University of Stuttgart, Germany. http://webdemo.inue.uni-stuttgart.de (2018)

Chapter 6
Block-Wise Signals with/without Prefix over FIR Channels

6.1 Finite Impulse Response Channel

In the following we consider channels, which can be modeled by a causal impulse response with finite duration, what we call finite impulse response (FIR)

$$h(n) = \begin{cases} h_n \; ; \; n = 0, 1, \ldots, L \\ 0 \; ; \qquad else \end{cases} \; ; \quad h_L \neq 0 \qquad (6.1)$$

We say, $h(n)$ owns the *length* $L + 1$, which is also the dimension of a corresponding signal vector. The channel parameters h_n shall be time-invariant. n represents discrete time. All properties of FIR filters known from signal processing apply. Furthermore, we study discrete-time input signals with finite duration, which can be described by a signal vector also called block of samples. Long signals will be structured block-wise and the principle of block-wise transmission is discussed in quite detail. Finally, we introduce a guard interval between the signal blocks by means of a prefix or a cyclic prefix and study important properties.

6.2 Channel Input-Output Relation with Convolution

For the moment we neglect the noise $n_R(n)$ at the receiver. According to Fig. 6.1a we obtain with discrete-time convolution the output signal

$$v(n) = u(n) * h(n) = \sum_{m=-\infty}^{\infty} u(m)h(n-m) = \sum_{m=-\infty}^{\infty} u(n-m)h(m) \qquad (6.2)$$

Now, assume an input signal with length N

© Springer Nature Switzerland AG 2021
J. Speidel, *Introduction to Digital Communications*,
Signals and Communication Technology,
https://doi.org/10.1007/978-3-030-67357-4_6

(a)   (b)

Fig. 6.1 a Transmission link with FIR channel impulse response $h(n)$, **b** Block-wise transmission with channel matrix $\mathbf{H}_0$ and zero-forcing equalizer $\mathbf{E} = \mathbf{H}_0^{-1}$

$$u(n) = \begin{cases} u_n \; ; \; n = 0, 1, \ldots, N-1 \\ 0 \; ; \qquad\qquad else \end{cases} \; ; \quad u_{N-1} \neq 0 \qquad (6.3)$$

Then follows from (6.2)

$$v(n) = \sum_{m=0}^{N-1} u(m)h(n-m) = \begin{cases} v_n \; ; \; n = 0, 1, \ldots, N-1+L \\ 0 \; ; \qquad\qquad else \end{cases} \qquad (6.4)$$

Apparently, the length of $v(n)$ is $N + L$ and by L larger than the length N of the input $u(n)$.

6.3 Channel Input-Output Relation with Matrix Notation

To this end we introduce signal vectors $\mathbf{u}$, $\mathbf{v}$, and $\mathbf{h}$ for the finite length signals $u(n)$, $v(n)$, and $h(n)$, respectively.

$$\mathbf{u} = \left(u_0 \; u_1 \cdots u_{N-1}\right)^T \; ; \; \mathbf{v} = \left(v_0 \; v_1 \cdots v_{N-1+L}\right)^T \; ; \; \mathbf{h} = \left(h_0 \; h_1 \cdots h_L\right)^T \quad (6.5)$$

For block-wise transmission it is beneficial to have vectors for input $\mathbf{u}$ and output $\mathbf{v}$ with the same dimension (length of signal). Just to delete the last L samples from $v(n)$ would leave the output signal premature and different from the convolution result, which makes no sense. The general solution is to allocate L new elements $u_N = u_{N+1} \ldots = u_{N-1+L} = 0$ to $u(n)$, what we call *zero padding*. The result $v(n)$ is not affected, because the last L samples $v_{N+L}, v_{N+L+1}, \ldots, v_{N+2L-1}$ turn out to be always zero. The channel vector $\mathbf{h}$ is padded with $N-1$ zeros increasing the dimension to $L + N$ as well without any change of the output $v(n)$. In summary, after zero padding the vectors are

$$\mathbf{u}_0 = \begin{pmatrix} \mathbf{u} \\ \mathbf{0}_L \end{pmatrix} = \left(u_0 \; u_1 \cdots u_{N-1} \mid \mathbf{0}_L^T\right)^T \; ; \; \mathbf{h}_0 = \begin{pmatrix} \mathbf{h} \\ \mathbf{0}_{N-1} \end{pmatrix} = \left(h_0 \; h_1 \cdots h_L \mid \mathbf{0}_{N-1}^T\right)^T$$

$$(6.6)$$

where $\mathbf{0}_m$ is a column vector with m zeros. Please note, $u(n)$ and $h(n)$ still keep their original lengths, N and $L + 1$, respectively, because tail end zeros do not count. The following example illustrates the presented method.

Example 12

A channel impulse response $h(n)$ with $L = 2$ and an input signal $u(n)$ with length $N = 4$ shall be given. Calculate the output signal $v(n)$, (a) using convolution, (b) using matrix notation. (c) How can the receiver reconstruct the transmit signal $u_0, \ldots, u_3$? The noise at the receiver shall be small and be neglected.

Solution:

(a) From (6.4) follows $v(n) = u_0 h_n + u_1 h_{n-1} + u_2 h_{n-2} + u_3 h_{n-3}$ and $n = 0, 1,$ $\ldots, 5$ yields the result $v_0, \ldots, v_5$ in (6.7 left), which can also be expressed as $\mathbf{v} = \mathbf{Hu}$.

$$
\underbrace{\begin{pmatrix} h_0 & 0 & 0 & 0 \\ h_1 & h_0 & 0 & 0 \\ h_2 & h_1 & h_0 & 0 \\ 0 & h_2 & h_1 & h_0 \\ 0 & 0 & h_2 & h_1 \\ 0 & 0 & 0 & h_2 \end{pmatrix}}_{\mathbf{H}} \underbrace{\begin{pmatrix} u_0 \\ u_1 \\ u_2 \\ u_3 \end{pmatrix}}_{\mathbf{u}} = \underbrace{\begin{pmatrix} v_0 \\ v_1 \\ v_2 \\ v_3 \\ v_4 \\ v_5 \end{pmatrix}}_{\mathbf{v}} \Longleftrightarrow \underbrace{\begin{pmatrix} h_0 & 0 & 0 & 0 & 0 & 0 \\ h_1 & h_0 & 0 & 0 & 0 & 0 \\ h_2 & h_1 & h_0 & 0 & 0 & 0 \\ 0 & h_2 & h_1 & h_0 & 0 & 0 \\ 0 & 0 & h_2 & h_1 & h_0 & 0 \\ 0 & 0 & 0 & h_2 & h_1 & h_0 \end{pmatrix}}_{\mathbf{H}_0} \underbrace{\begin{pmatrix} u_0 \\ u_1 \\ u_2 \\ u_3 \\ 0 \\ 0 \end{pmatrix}}_{\mathbf{u}_0} = \underbrace{\begin{pmatrix} v_0 \\ v_1 \\ v_2 \\ v_3 \\ v_4 \\ v_5 \end{pmatrix}}_{\mathbf{v}}
$$

$$(6.7)$$

Apparently, $v(n)$ has length $N + L = 6$, as predicted by (6.4).
(b) We obtain from (6.2)
$v(n) = u_0 h_n + u_1 h_{n-1} + u_2 h_{n-2} + u_3 h_{n-3} + u_4 h_{n-4} + u_5 h_{n-5}$, where the zero padding elements are $u_4 = u_5 = 0$ and $h_3 = h_4 = h_5 = 0$. The matrix notation is derived with $n = 0, 1, \ldots, 5$ and shown in (6.7 right). The reader convinces oneself easily that both output vectors $\mathbf{v}$ in (6.7) provide the same result. Please note, $\mathbf{H}$ has dimension 6×4 whereas $\mathbf{H}_0$ is a 6×6 square matrix.
(c) Only the first four equations in (6.7) are sufficient to determine $u_0, \ldots, u_3$, resulting in

$$
\begin{pmatrix} h_0 & 0 & 0 & 0 \\ h_1 & h_0 & 0 & 0 \\ h_2 & h_1 & h_0 & 0 \\ 0 & h_2 & h_1 & h_0 \end{pmatrix} \begin{pmatrix} u_0 \\ u_1 \\ u_2 \\ u_3 \end{pmatrix} = \begin{pmatrix} v_0 \\ v_1 \\ v_2 \\ v_3 \end{pmatrix}
\qquad (6.8)
$$

The solution is straightforward, owing to the lower triangular matrix with determinant h_0^4. Thus, an inverse matrix exists for $h_0 \neq 0$.

Channel Matrix

From (6.7 right) we can generalize the input-output relation of an FIR channel as

$$\mathbf{H}_0 \mathbf{u}_0 = \mathbf{v}$$

$$
\begin{array}{cccccccccccc|c}
u_0 & u_1 & u_2 & \cdots & u_L & u_{L+1} & \cdots & u_N & u_{N-1} & 0 & \cdots & 0 \; 0 & = \\
\hline
h_0 & 0 & 0 & \cdots & 0 & 0 & \cdots & 0 & 0 & 0 & \cdots & 0 \; 0 & v_0 \\
h_1 & h_0 & 0 & \cdots & & & & & & & \cdots & 0 & v_1 \\
\cdots & \cdots & \cdots & \cdots & & \cdots & & & \cdots & & & & \vdots \\
 & & & & & \mathbf{0} & & & & & & & \\
h_L & h_{L-1} & & \cdots & h_0 & 0 & \cdots & & & & \cdots & 0 & v_L \\
0 & h_L & h_{L-1} & \cdots & h_1 & h_0 & \cdots & & & & \cdots & 0 & v_{L+1} \\
\vdots & 0 & h_L & & & \ddots & \ddots & & & & \cdots & & \vdots \\
 & & & \ddots & & & & & & & & & \\
\mathbf{0} & & & & & & \cdots & h_1 & h_0 & 0 & \cdots & 0 & v_{N-1} \\
 & \cdots & & & & & & \ddots & \ddots & \ddots & \cdots & & \vdots \\
 & & & & & & & & \ddots & \ddots & & & \\
0 & 0 & 0 & \cdots & 0 & 0 & & 0 & h_L & h_{L-1} & \cdots & h_1 \; h_0 & v_{N-1+L} \\
\end{array}
$$

$$\leftarrow \qquad\qquad\qquad \mathbf{H}_0 \qquad\qquad\qquad \rightarrow$$

(6.9)

where $\mathbf{u}_0$, $\mathbf{h}_0$, and $\mathbf{v}$ own the same dimensions $N + L$ according to (6.6). We use an alternative matrix notation in tabular form, in which the vector $\mathbf{u}_0$ is written on the top with entries straight above the corresponding matrix entries to be multiplied with. This allows an easy identification of the structure.

The $(N + L) \times (N + L)$ channel matrix $\mathbf{H}_0$ shows a lower triangular form. The entries along the main diagonal and along each off-diagonal are identical, which characterizes a Toeplitz matrix. We can construct $\mathbf{H}_0$ by filling the main diagonal with h_0, the lower next diagonal with h_1 etc. up to a last diagonal filled with h_L. The remaining entries are zero. Please note that $v_{N+L}, v_{N+L+1}, \ldots$ are always zero due to zero padding of $\mathbf{u}_0$ and $\mathbf{h}_0$, hence not shown. The determinant of a lower triangular matrix is the product of the main diagonal entries

$$\det(\mathbf{H}_0) = h_0^{N+L} \tag{6.10}$$

Thus, $\mathbf{H}_0^{-1}$ exists for $h_0 \neq 0$ and a solution $\mathbf{u}_0$ of (6.9) is guaranteed. Finally, we mention that in case of an infinite impulse response (IIR) channel, $L \to \infty$, the dimension of $\mathbf{H}_0$ tends to infinity, but the lower triangular Toeplitz structure is maintained.

6.4 Block-Wise Transmission

In reality, the channel input signal can be very long. Therefore, we consider a method to divide the message into blocks. "Block" is just the technical term for a vector. To this end we allocate to the blocks $\mathbf{u}$, $\mathbf{v}$, and $\mathbf{u}_0$ in (6.5) and (6.6) a block number k as

$$\mathbf{u}_0(k) = \begin{pmatrix} \mathbf{u}(k) \\ \mathbf{0}_L \end{pmatrix} ; \; \mathbf{u}(k) = \begin{pmatrix} u_{0k} \; u_{1k} \; \cdots \; u_{N-1,k} \end{pmatrix}^T$$

$$\mathbf{v}(k) = \begin{pmatrix} v_{0k} \; v_{1k} \; \cdots \; v_{N-1,k} \; v_{Nk} \; \cdots \; v_{N-1+L,k} \end{pmatrix}^T , \quad k = 0, 1, 2, \ldots \tag{6.11}$$

The entries are

$$u_{nk} \text{ and } v_{nk} \tag{6.12}$$

where the timing of the samples inside a block is denoted by n with time interval T_S. As illustrated in Fig. 6.2a of Example 13, consecutive blocks do not overlap, i.e. no inter-block interference is present at the channel output, if the input block length is made identical to the output block length by zero padding. The input-output relation for block-wise transmission then follows from (6.9)

$$\mathbf{v}(k) = \mathbf{H}_0 \mathbf{u}_0(k) \tag{6.13}$$

The receive signal with additive noise $\mathbf{n}(k) \in \mathbb{C}^{(N-1+L)\times 1}$ is $\mathbf{v}(k) + \mathbf{n}(k)$ and illustrated in Fig. 6.1b.

In case that no zeros $\mathbf{0}_L$ are used in the transmit blocks the block length is N. Then, the first L samples of the response to transmit block $k + 1$ overlap with the last L samples of the response to the previous transmit block k, as demonstrated in Fig. 6.2b of Example 13.

(a)

		k	$=$	0				k	$=$	1				k	$=$	2	
$u_{nk}:$	u_{00}	u_{10}	u_{20}	u_{30}	0	0	u_{01}	u_{11}	u_{21}	u_{31}	0	0	u_{02}	u_{12}	u_{22}	$u_{32} \cdots$	
$v_{n0}:$	v_{00}	v_{10}	v_{20}	v_{30}	v_{40}	v_{50}											
$v_{n1}:$							v_{01}	v_{11}	v_{21}	v_{31}	v_{41}	v_{51}					
$v_{n2}:$													v_{02}	v_{12}	v_{22}	$v_{32} \cdots$	

(b)

		k	$=$	0		k	$=$	1		k	$=$	2		k	$=$	3
$u_{nk}:$	u_{00}	u_{10}	u_{20}	u_{30}	u_{01}	u_{11}	u_{21}	u_{31}	u_{02}	u_{12}	u_{22}	u_{32}	u_{03}	u_{13}	u_{23}	$u_{33} \cdots$
$v_{n0}:$	v_{00}	v_{10}	v_{20}	v_{30}	v_{40}	v_{50}										
$v_{n1}:$					v_{01}	v_{11}	v_{21}	v_{31}	v_{41}	v_{51}						
$v_{n2}:$									v_{02}	v_{12}	v_{22}	v_{32}	v_{42}	v_{52}		
$v_{n3}:$													v_{03}	v_{13}	v_{23}	$v_{33} \cdots$

Fig. 6.2 Block-wise transmission of $\mathbf{u}_0(k)$ and $\mathbf{u}(k)$, $k = 0, 1, 2$ over FIR channel with $L = 2$. **a** Input $\mathbf{u}_0(k)$ with length $N = 4$ and zero padding, block length 6, no inter-block interference. **b** Input $\mathbf{u}(k)$ with length $N = 4$, no zero padding, block length 4, inter-block interference indicated with color, only output samples v_{2k}, v_{3k} without interference

Example 13

A channel impulse response $h(n)$ with $L = 2$ and an input signal $u(n)$ with length $N = 4$ shall be given, as in Example 12. Define a sequence of input and output blocks all with the same length. Demonstrate the block structure with blocks $k = 0, 1, 2$ for the following cases: (a) Zero padding of $\mathbf{u}(k)$, block length 6, (b) no padding of $\mathbf{u}(k)$ and block length 4. The timing interval corresponding to n shall be T_S.

Solution:

(a) From (6.7 right) we conclude that each block $\mathbf{u}_0(k)$ and $\mathbf{v}(k)$ is composed of 6 samples and the block rate is $\frac{1}{6T_s}$. In Fig. 6.2a, the input block $\left(u_{00}\ u_{10}\ u_{20}\ u_{30}\ 0\ 0\right)^T =$ $\mathbf{u}_0(0)$ generates the output $\left(v_{00}\ v_{10}\ u_{20}\ v_{30}\ v_{40}\ v_{50}\right)^T = \mathbf{v}(0)$ listed beneath. The next input $\left(u_{01}\ u_{11}\ u_{21}\ u_{31}\ 0\ 0\right)^T = \mathbf{u}_0(1)$ results in the output $\left(v_{01}\ \cdots\ v_{51}\right)^T = \mathbf{v}(1)$ etc. As the channel is linear, the channel output sequence is obtained by superposition yielding $v_{00}, v_{10}, \ldots, v_{50}, v_{01}, v_{11}, \ldots, v_{51}, \ldots$ with no overlap, thus, no inter-block interference. The same holds for all following blocks. Please note that inside a receive block the samples normally have suffered intersymbol interference, because the channel impulse response is not ideal, $h(n) = \delta_n$.

(b) Figure 6.2b shows the input sequences, which are structured into blocks of length 4. The zero padding part is not transmitted to avoid unuseful samples. The dedicated responses v_{nk} obtained by convolution are shown beneath. The superposition of all responses results in the channel output sequence

$v_{00}, v_{10}, \ldots, v_{30}, (v_{40}+v_{01}), (v_{50}+v_{11}), v_{21}, v_{31}, (v_{41}+v_{02}), (v_{51}+v_{12}), v_{22}, v_{32}, \ldots,$
which exhibits inter-block interference indicated by brackets and at positions illustrated with color in Fig. 6.2b. If the transmit block length 4 is used at the receiver, the first two samples of the response to the actual transmit block (e.g. $k = 2$) overlap with the last two samples of the response to the previous transmit block (e.g. $k = 1$). In general, the first L samples of response $k + 1$ overlap with the last L samples of the previous response k, etc. Consequently, in Fig. 6.2b only the receive samples v_{2k}, v_{3k} are free of inter-block interference.

6.5   Block-Wise Time-Variant Channel

Parameters of strictly time-variant channels change from time instants n to $n + 1$, e.g., on wireless communication links between base stations and fast trains and airplanes. For a large number of other applications, e.g., cars, pedestrians or indoor users, block-wise time-variance can serve as an acceptable approximation, when the channel parameters do not change significantly during a block. Practically, this is fulfilled, if the coherence time of the fading channel is larger than the block length. Hence, the channel matrix changes just with k and not with n resulting approximately in

$$\mathbf{v}(k) \approx \mathbf{H}_0(k)\mathbf{u}_0(k). \tag{6.14}$$

6.6 Equalization of Block-Wise Receive Signals

Now, we are interested in getting rid of intersymbol interference for block-wise transmission with zero padding and a transmit block length $N + L$. To reconstruct $\mathbf{u}_0(k)$ a receive filter, called zero-forcing equalizer with matrix

$$\mathbf{E} = \mathbf{H}_0^{-1} \tag{6.15}$$

can be accommodated, as illustrated in Fig. 6.1b. Using the identity matrix $\mathbf{I} = \mathbf{H}_0^{-1}\mathbf{H}_0$ the equalizer output

$$\mathbf{y}(k) = \mathbf{E}\left(\mathbf{v}(k) + \mathbf{n}(k)\right) = \mathbf{u}_0(k) + \mathbf{H}_0^{-1}\mathbf{n}(k) \tag{6.16}$$

is composed of the original transmit block $\mathbf{u}_0(k)$ corrupted by noise $\mathbf{H}_0^{-1}\mathbf{n}(k)$. Please note that the resulting noise can be enhanced by $\mathbf{H}_0^{-1}$. For low noise, $\mathbf{n}(k) \approx \mathbf{0}$, we get $\mathbf{y}(k) \approx \mathbf{u}_0(k)$. An equalizer matrix $\mathbf{E}$ exists for $h_0 \neq 0$, owing to (6.10).

6.7 Block-Wise Channel Input with Prefix

Separation of consecutive blocks of the channel input sequence can be accomplished by insertion of prefix samples at the beginning of each block. In the following, we introduce the concept of a prefix with fixed length G per block, extend it later to a cyclic prefix with $G = L$, and derive the channel input-output relation. To simplify notation the block number k is dropped. In principle, an input block with prefix contains

$$\underbrace{u_{-G}, \ldots, u_{-1}}_{\text{prefix}}, \underbrace{u_0, u_1, \ldots, u_{N-1}}_{\text{useful samples}} \tag{6.17}$$

We obtain with convolution (6.2) the channel output signal

$$v(n) = \sum_{m=-G}^{-1} u(m)h(n-m) + \sum_{m=0}^{N-1} u(m)h(n-m) \ ; \ n = -G, \ldots, -1, 0, 1, \ldots, N-1+L \tag{6.18}$$

representing a block with elements

$$v_{-G}, \ldots, v_{-1}, v_0, v_1, \ldots, v_{N-1+L} \tag{6.19}$$

As will be shown, $v_{-G}, \ldots, v_{-1}$ depend only on the prefix $u_{-G}, \ldots, u_{-1}$ and not on the remaining input samples u_n, thus, determine the response of the channel to the prefix. The principle is explained with the next example.

The prefix provides a kind of guard interval between blocks. However, it wastes G samples, which do not carry useful information. Thus, the relative number of useful samples per input block

$$\eta = \frac{N}{N + G} \tag{6.20}$$

falls below one. If N is chosen much larger than G, the quotient η tends to one and becomes feasible for practical system design. Insertion of a guard interval is the task of a multiplexer. Let T_S be the interval between samples u_n. A memoryless multiplexer generates an output block with the same duration as the input block, $N T_S = (N + G) T_S'$, where T_S' is the time interval between output samples. Thus, the rate $1/T_S'$ at the output is increased by factor $1 + G/N$.

Example 14

Consider a channel impulse response $h(n)$ with $L = 2$, as in Example 13. To the useful input signal $u(n)$ with length $N = 4$ a prefix with $G = 2$ is allocated generating the sequence $u_{-2}, u_{-1}, u_0, u_1, u_2, u_3$, where u_{-2}, u_{-1} defines the given prefix. Apparently, a guard interval with length $G = L$ is introduced between successive input blocks. Find the output signal (a) with convolution and (b) with matrix notation. (c) Demonstrate that the useful receive samples $v_0, \ldots, v_3$ are not corrupted by inter-block interference in case of a transmit block length $N + G$.

Solution:

(a) From (6.18) follows

$v(n) = u_{-2}h_{n+2} + u_{-1}h_{n+1} + u_0 h_n + u_1 h_{n-1} + u_2 h_{n-2} + u_3 h_{n-3};$
$n = -2, -1, 0, 1, \ldots, 5.$

(b) We easily construct the channel matrix applying lower triangular Toeplitz form resulting in (6.21 left).

$$
\begin{array}{ccccccc}
u_{-2} & u_{-1} & u_0 & u_1 & u_2 & u_3 & 0 & 0 \\
\end{array} =
$$

$$
\begin{array}{|cccccccc|c}
h_0 & 0 & & & & & & & v_{-2} \\
h_1 & h_0 & 0 & & & 0 & & & v_{-1} \\
h_2 & h_1 & h_0 & 0 & & & & & v_0 \\
0 & h_2 & h_1 & h_0 & 0 & & & & v_1 \\
& 0 & h_2 & h_1 & h_0 & 0 & & & v_2 \\
& & 0 & h_2 & h_1 & h_0 & 0 & & v_3 \\
\mathbf{0} & & & 0 & h_2 & h_1 & h_0 & 0 & v_4 \\
& & & & 0 & h_2 & h_1 & h_0 & v_5 \\
\end{array}
$$

$$
\Rightarrow
\begin{array}{|cccc|}
u_0 & u_1 & u_2 & u_3 \\
h_0 & 0 & 0 & 0 \\
h_1 & h_0 & 0 & 0 \\
h_2 & h_1 & h_0 & 0 \\
0 & h_2 & h_1 & h_0 \\
\end{array}
\begin{array}{c}
= \\
v_0 - h_2 u_{-2} - h_1 u_{-1} \\
v_1 - h_2 u_{-1} \\
v_2 \\
v_3 \\
\end{array}
\tag{6.21}
$$

The vector $\mathbf{u}$ is zero padded to ensure the same length as $\mathbf{v}$. We use a matrix notation in tabular form. Obviously, the first two receive samples, v_{-2} and v_{-1}, solely depend on the prefix. v_6 and v_7 are automatically zero, owing to zero padding of the input block and thus not shown. There are various ways to recover $u_0, \ldots, u_3$, e.g., for a given prefix the linear equations for $v_0, \ldots, v_3$ will do, resulting in (6.21 right). Their solution is straightforward.

(c) We transmit the input block k as $\left(u_{-2k} \; u_{-1k} \; u_{0k} \; u_{1k} \; u_{2k} \; u_{3k} \right)^T$ without zero padding, as zeros carry no useful samples. The received block still is

		$k = 0$					$k = 1$				$k = 2$	
u_{nk}:	u_{-20}	u_{-10}	u_{00}	u_{10}	u_{20}	u_{30}	u_{-21}	u_{-11}	u_{01}	u_{11}	u_{21}	u_{31}	u_{-22}	u_{-12}...
v_{n0}:	v_{-20}	v_{-10}	v_{00}	v_{10}	v_{20}	v_{30}	v_{40}	v_{50}						
v_{n1}:							v_{-21}	v_{-11}	v_{01}	v_{11}	v_{21}	v_{31}	v_{41}	v_{51}
v_{n2}:													v_{-22}	v_{-12}...

Fig. 6.3 Block-wise transmission of u_{nk} and reception of v_{nk}, $k = 0, 1, 2$. FIR channel impulse response with $L = 2$, prefix with $G = L = 2$, input block length $N + G = 6$, block rate $\frac{1}{6T_S}$. Receive samples $v_{0k}, \ldots, v_{3k}$, $k = 0, 1, \ldots$ without block overlap

$\left(v_{-2k} \; v_{-1k} \; v_{0k} \; v_{1k} \; v_{2k} \; v_{3k} \; v_{4k} \; v_{5k} \right)^T$. For $k = 0, 1, \ldots$ we obtain the sequence of blocks in Fig. 6.3. Apparently, v_{-21} overlaps with v_{40} and v_{-11} with v_{50}. In other words, the prefix responses of output block $k + 1$ overlap with the last two elements of output block k, but the useful samples $v_{0k}, \ldots, v_{3k}$ do not suffer from interference.

Guided by this example we generalize the vector notation, define the prefix vector,

$$\mathbf{u}_{pre}(k) = \left(u_{-G,k} \; \cdots \; u_{-1,k} \right)^T \tag{6.22}$$

and get the compound input vector

$$\mathbf{u}_c(k) = \begin{pmatrix} \mathbf{u}_{pre}(k) \\ \mathbf{u}_0(k) \end{pmatrix} = \left(u_{-G,k} \; \cdots \; u_{-1,k} \mid u_{0k} \; \cdots \; u_{N-1,k} \mid \mathbf{0}_L^T \right)^T \tag{6.23}$$

with zero padding $\mathbf{0}_L$ conforming to convolution and to achieve a square matrix. The vectors of output and noise are also structured into a main and a prefix part

$$\mathbf{v}_c(k) = \begin{pmatrix} \mathbf{v}_{pre}(k) \\ \mathbf{v}(k) \end{pmatrix} = \left(v_{-G,k} \; \cdots \; v_{-1,k} \mid v_{0k} \; \cdots \; v_{N-1+L,k} \right)^T \tag{6.24}$$

$$\mathbf{n}_c(k) = \begin{pmatrix} \mathbf{n}_{pre}(k) \\ \mathbf{n}(k) \end{pmatrix} = \left(n_{-G,k} \; \cdots \; n_{-1,k} \mid n_{0k} \; \cdots \; n_{N-1+L,k} \right)^T \tag{6.25}$$

The channel input-output relation (without noise) then follows as

$$\mathbf{v}_c(k) = \mathbf{H}_c \mathbf{u}_c(k) \tag{6.26}$$

and $\mathbf{v}_c(k) + \mathbf{n}_c(k)$ is the noisy receive signal. $\mathbf{H}_c$ is a lower triangular $(G + N + L) \times (G + N + L)$ Toeplitz matrix similar to $\mathbf{H}_0$ in (6.9) with h_0 along the main diagonal. The block diagram in Fig. 6.4 shows the processing steps starting with the attachment of $\mathbf{u}_{pre}(k)$. At the receiver the prefix part in $\mathbf{v}_c(k) + \mathbf{n}_c(k)$ is finally detached resulting in $\mathbf{v}(k) + \mathbf{n}(k)$ and the input-output relation between $\mathbf{u}_0(k)$ and $\mathbf{v}(k)$ is given again by (6.13).

Fig. 6.4 Block-wise transmission with prefix $\mathbf{u}_{pre}(k)$ and channel matrix $\mathbf{H}_c$

6.8 Block-Wise Transmission with Cyclic Prefix

6.8.1 Input Signal with Cyclic Prefix

Now, a special prefix called *cyclic prefix* will be attached, which is the copy of the last L useful samples of the input block, yielding

$$\underbrace{u_{N-L}, \ldots, u_{N-1}}_{\text{cyclic prefix}}, \underbrace{u_0, \ldots, u_{N-L-1}, u_{N-L}, \ldots, u_{N-1}}_{\text{useful samples}} \qquad (6.27)$$

where $G = L$. Block index k is dropped to simplify notation. Before we look at the general solution we study the principle with the following example.

Example 15

Consider a channel impulse response $h(n)$ with $L = 2$, as in Example 14. To the useful input signal $u(n)$ with length $N = 4$ a cyclic prefix with $G = L = 2$ is allocated. Noise shall be neglected.
(a) Determine the input sequence and (b) the output signal using matrix notation.
(c) Find the set of equations to recover the useful samples and show that the channel matrix has become *circulant*.

Solution:
(a) The input sequence is $u_{-2}, u_{-1}, u_0, u_1, u_2, u_3$ and with the cyclic prefix $u_{-2} = u_2$ and $u_{-1} = u_3$ we get $u_2, u_3, u_0, u_1, u_2, u_3$.
(b) We refer to (6.21 left), which is copied in (6.28 left) ignoring the new entries marked in red color. The new entries arise owing to the following operations: According to linear algebra, setting $u_{-2} = u_2$ eliminates u_{-2} and the first column of the matrix is added to the fifth one (below u_2). In the same way, condition $u_{-1} = u_3$ results in an addition of the second to the sixth column (below u_3). In a next step, first and second column of the matrix as well as first and second equation can be dropped, because they are now redundant. The final result is given in (6.28 right).

$$
\begin{array}{cc|cccccc|c}
u_{-2} & u_{-1} & u_0 & u_1 & u_2 & u_3 & 0 & 0 & = \\
\hline
h_0 & 0 & 0 & 0 & h_0 & 0 & 0 & 0 & v_{-2} \\
h_1 & h_0 & 0 & 0 & h_1 & h_0 & 0 & 0 & v_{-1} \\
h_2 & h_1 & h_0 & 0 & h_2 & h_1 & 0 & 0 & v_0 \\
0 & h_2 & h_1 & h_0 & 0 & h_2 & 0 & 0 & v_1 \\
 & 0 & h_2 & h_1 & h_0 & 0 & 0 & 0 & v_2 \\
 & 0 & h_2 & h_1 & h_0 & 0 & 0 & & v_3 \\
 & 0 & 0 & 0 & h_2 & h_1 & h_0 & 0 & v_4 \\
 & & 0 & 0 & 0 & h_2 & h_1 & h_0 & v_5 \\
\end{array}
\quad\Rightarrow\quad
\begin{array}{cccccc|c}
u_0 & u_1 & u_2 & u_3 & 0 & 0 & = \\
\hline
h_0 & 0 & h_2 & h_1 & 0 & 0 & v_0 \\
h_1 & h_0 & 0 & h_2 & 0 & 0 & v_1 \\
h_2 & h_1 & h_0 & 0 & 0 & 0 & v_2 \\
0 & h_2 & h_1 & h_0 & 0 & 0 & v_3 \\
0 & 0 & h_2 & h_1 & h_0 & 0 & v_4 \\
0 & 0 & 0 & h_2 & h_1 & h_0 & v_5 \\
\end{array}
\tag{6.28}
$$

(c) Only the equations for $v_0, \ldots, v_3$ are required to calculate the useful samples $u_0, \ldots, u_3$. Samples v_4 and v_5 are of no interest. Thus, the last two equations and the last two columns in (6.28 right) can be dropped yielding

$$
\begin{array}{cccc|c}
u_0 & u_1 & u_2 & u_3 & = \\
\hline
h_0 & 0 & h_2 & h_1 & v_0 \\
h_1 & h_0 & 0 & h_2 & v_1 \\
h_2 & h_1 & h_0 & 0 & v_2 \\
0 & h_2 & h_1 & h_0 & v_3 \\
\end{array}
\tag{6.29}
$$

Compared to the matrix in (6.21 right), the cyclic prefix has caused new entries in the upper right corner marked in red color. Apparently, a *circulant matrix* occurs in (6.29) with the property that the rows from top to bottom can be generated by a shift to the right and the matrix element dropping out becomes the first one in the next row. Circulant matrices own interesting properties useful for digital transmission and are discussed next.

6.8.2 Circulant Channel Matrix and Cyclic Convolution

Circulant Channel Matrix

Consider again a channel impulse response $h(n)$ with length $L + 1$ and the input signal in (6.27) with cyclic prefix of length L. The determination of the useful output samples $v_0, \ldots, v_{N-1}$ as a function of the useful input samples $u_0, \ldots, u_{N-1}$ can be generalized from (6.29) with the following matrix notation

$$
\begin{array}{cccccccccccc}
u_0 & u_1 & \cdots & u_{L-1} & u_L & u_{L+1} & \cdots & u_{N-L} & u_{N-L+1} & \cdots & u_{N-2} & u_{N-1} \\
\hline
h_0 & 0 & \cdots & 0 & 0 & 0 & \cdots & h_L & h_{L-1} & \cdots & h_2 & h_1 \\
h_1 & h_0 & \cdots & 0 & 0 & 0 & \cdots & 0 & h_L & \cdots & h_3 & h_2 \\
\vdots & \vdots & \cdots & 0 & 0 & 0 & \cdots & 0 & 0 & \cdots & \vdots & \vdots \\
h_{L-1} & h_{L-2} & \cdots & h_0 & 0 & 0 & \cdots & & & \cdots & 0 & h_L \\
h_L & h_{L-1} & \cdots & h_1 & h_0 & 0 & \cdots & & & \cdots & 0 & 0 \\
0 & h_L & \cdots & h_2 & h_1 & h_0 & \cdots & 0 & 0 & \cdots & 0 & 0 \\
\vdots & \ddots & \ddots & & \ddots & \ddots & & & & & \vdots & \vdots \\
\vdots & & & & & \ddots & & & \ddots & & \vdots & \vdots \\
0 & 0 & \cdots & & & & \ddots & h_{L-2} & h_{L-3} & \ddots & h_0 & 0 \\
0 & 0 & 0 & 0 & 0 & 0 & \cdots & h_{L-1} & h_{L-2} & \cdots & h_1 & h_0 \\
\end{array}
$$

$$
\mathbf{v}_N = \mathbf{H}_{circ}\mathbf{u} \tag{6.30}
$$

with

$$
\mathbf{v}_N = \begin{pmatrix} v_0 & v_1 & \cdots & v_{N-1} \end{pmatrix}^T \tag{6.31}
$$

where k was dropped to simplify notation. In general, $\mathbf{u} = \mathbf{u}(k)$ and $\mathbf{v}_N = \mathbf{v}_N(k)$ holds with a second index k in their entries. As expected, the $N \times N$ channel matrix $\mathbf{H}_{circ}$ consists of a lower triangular part with h_0 in the main diagonal and in the upper right corner a sub-matrix of upper triangular form with h_L in the main diagonal displayed in color. $\mathbf{H}_{circ}$ is a *circulant* Toeplitz matrix. Circulant matrices show interesting properties [1], which are summarized shortly and applied in the following sections. Finally, please note that an extension of the derivation for $G > L$ is straightforward by "virtually" increasing the length of $h(n)$ with zeros and notional replacement of L by G.

Cyclic Convolution

As is well known from signal processing, (6.30) can be expressed by the *cyclic (or circulant) convolution*

$$
v(n) = u(n) \circledast h(n) = \sum_{m=0}^{N-1} u(m)h\left((n - m) \bmod N\right) \tag{6.32}
$$

where $h\left((n - m) \bmod N\right)$ defines the periodic (cyclic) repetitions of $h(n)$ with period N. Cyclic convolution is an operation defined for periodic signals. Applying the DFT on (6.32) results in the product of the DFTs of $u(n)$ and $h(n)$. This holds for cyclic convolution only and not in general for standard convolution.

Properties of Circulant Matrices

- A circulant (or cyclic) matrix is a Toeplitz matrix. From top to bottom each row can be generated from the previous row by a right shift and the matrix entry dropping out becomes the first one in the next row. After arriving at the last row the cycle continues with the first row.
- The eigenvalues λ_μ and eigenvectors $\mathbf{p}_\mu$ of an $N \times N$ circulant matrix with entries $h_0, \ldots, h_L$ are given by

$$\lambda_\mu = \sum_{n=0}^{N-1} h_n w^{\mu n} , \quad \mathbf{p}_\mu = \frac{1}{\sqrt{N}} \left(1 \ w^{-\mu} \ w^{-2\mu} \cdots w^{-(N-1)\mu}\right)^T$$

$$w = e^{-j2\pi/N} ; \qquad\qquad \mu = 0, 1, \ldots, N - 1 \tag{6.33}$$

see [1]. A general introduction on eigenvalues and eigenvectors is summarized in Appendix B. Apparently, the eigenvectors $\mathbf{p}_\mu$ do not depend on $h(n)$, thus are valid for any channel impulse response of length $L + 1$.

- λ_μ has an interesting property. The frequency response of a channel with impulse response $h(n)$ of length $L + 1 \leq N$ is $H\left(e^{j2\pi f/f_s}\right) = \sum_{n=0}^{N-1} h_n e^{-j2\pi n f/f_s}$. By comparison with $\lambda_\mu = \sum_{n=0}^{N-1} h_n e^{-j\frac{2\pi}{N}\mu n}$ in (6.33) and using $f_s = N/T$ we get

$$\lambda_\mu = H\left(e^{j2\pi f/f_s}\right)\big|_{f=\mu\frac{1}{T}} = H\left(e^{j2\pi\mu\frac{1}{N}}\right) ; \quad \mu = 0, 1, \ldots, N - 1 \tag{6.34}$$

which is the channel transfer function at multiples μ of the QAM symbol rate $1/T$.

- As is well known from signal processing, the $N \times N$ matrix of the *Discrete Fourier Transform* (DFT) is defined as

$$\mathbf{F} = \frac{1}{\sqrt{N}} \begin{pmatrix} 1 & 1 & 1 & \cdots & 1 \\ 1 & w^1 & w^2 & \cdots & w^{(N-1)} \\ 1 & w^2 & w^4 & \cdots & w^{2(N-1)} \\ \vdots & \vdots & \vdots & \cdots & \vdots \\ 1 & w^{(N-1)} & w^{2(N-1)} & \cdots & w^{(N-1)^2} \end{pmatrix} = \left(\mathbf{p}_0^* \ \mathbf{p}_1^* \ \mathbf{p}_2^* \cdots \mathbf{p}_{N-1}^*\right) \tag{6.35}$$

where $*$ indicates the complex conjugation. The *Inverse Discrete Fourier Transform* (IDFT) is given by the matrix

$$\mathbf{F}^{-1} = \mathbf{F}^* = \left(\mathbf{p}_0 \ \mathbf{p}_1 \ \mathbf{p}_2 \cdots \mathbf{p}_{N-1}\right) \tag{6.36}$$

because $\mathbf{F}$ is unitary and symmetric (for details of unitary matrices, please see Appendix B). Let $\mathbf{x} = \left(x_0 \ x_1 \cdots x_{N-1}\right)^T \in \mathbb{C}^{N \times 1}$, then the DFT results in

$$\underline{\mathbf{x}} = \left(X_0 \ X_1 \cdots X_{N-1}\right)^T = \mathbf{F}\mathbf{x} \tag{6.37}$$

by executing

$$x_n \rightarrowtail X_\nu = \frac{1}{\sqrt{N}} \sum_{n=0}^{N-1} x_n w^{\nu n} \; ; \; \nu = 0, 1, \ldots, N - 1 \tag{6.38}$$

This set represents the discrete Fourier spectrum of the finite length signal $x_0, x_1, \ldots, x_{N-1}$. Apparently, the DFT maps N samples x_n to N spectral coefficients X_ν. Multiplying (6.37) from the left by $\mathbf{F}^{-1}$ yields with $\mathbf{F}^{-1}\mathbf{F} = \mathbf{I}$

$$\mathbf{x} = \mathbf{F}^* \underline{\mathbf{x}} \tag{6.39}$$

which is accomplished by

$$X_\nu \leftrightarrowtail x_n = \frac{1}{\sqrt{N}} \sum_{\nu=0}^{N-1} X_\nu w^{-\nu n} \; ; \; n = 0, 1, \ldots, N - 1 \tag{6.40}$$

DFT and IDFT are also applicable to the primitive period of sequences x_n and X_ν, which are periodic with N yielding the same results (6.38) and (6.40), respectively.

6.8.3 Transmission Scheme with Cyclic Prefix

The various steps to achieve transmission with a cyclic prefix are summarized as follows. At the transmitter in Fig. 6.5a the cyclic prefix with length $G = L$

$$\mathbf{u}_{pre}(k) = \left(u_{N-L,k} \cdots u_{N-1,k} \right)^T \tag{6.41}$$

is attached to each input block $\mathbf{u}_0(k) = \begin{pmatrix} \mathbf{u}(k) \\ \mathbf{0}_L \end{pmatrix}$, where $\mathbf{u}(k) = \left(u_{0k}\ u_{1k} \cdots u_{N-1,k} \right)^T$ contains the N useful samples of transmit block k. The result is $\mathbf{u}_c(k) = \begin{pmatrix} \mathbf{u}_{pre}(k) \\ \mathbf{u}_0(k) \end{pmatrix}$. The output $\mathbf{v}_c(k) = \mathbf{H}_c \mathbf{u}_c(k)$ of the channel $\mathbf{H}_c$ is corrupted by noise $\mathbf{n}_c(k)$ yielding

$$\mathbf{v}_c(k) + \mathbf{n}_c(k) = \begin{pmatrix} \mathbf{v}_{pre}(k) \\ \mathbf{v}(k) \end{pmatrix} + \begin{pmatrix} \mathbf{n}_{pre}(k) \\ \mathbf{n}(k) \end{pmatrix} \tag{6.42}$$

with $\mathbf{v}(k) = \left(v_{0k}\ v_{1k} \cdots v_{N-1,k}\ v_{Nk} \cdots v_{N-1+L,k} \right)^T$ in (6.11). The receiver detaches the prefix part $\mathbf{v}_{pre}(k) + \mathbf{n}_{pre}(k)$ as well as the last L samples from $\mathbf{v}(k) + \mathbf{n}(k)$ resulting in the output

$$\mathbf{v}_N(k) + \mathbf{n}_N(k) = \left(v_{0k} \cdots v_{N-1,k} \right)^T + \left(n_{0k} \cdots n_{N-1,k} \right)^T \tag{6.43}$$

Fig. 6.5 Block-wise transmission with *cyclic prefix* $\mathbf{u}_{pre}(k)$ of length $G = L$. **a** Principal scheme with channel matrix $\mathbf{H}_c$. **b** Equivalent model with circulant channel matrix $\mathbf{H}_{circ}$. **c** Equivalent model with eigenmode channel matrix $\Lambda = \mathbf{F}\mathbf{H}_{circ}\mathbf{F}^{-1}$ and optional DFT-domain equalizer $\mathbf{E} = \Lambda^{-1}$

Then the system can be modeled as in Fig. 6.4b with input $\mathbf{u}(k)$, circulant matrix $\mathbf{H}_{circ}$ in (6.30), noise $\mathbf{n}_N(k)$, and output of $\mathbf{H}_{circ}$

$$\mathbf{v}_N(k) = \mathbf{H}_{circ}\mathbf{u}(k) \tag{6.44}$$

All vectors own the dimension $N \times 1$.

6.9 Transmission with Cyclic Prefix and DFT

6.9.1 DFT Domain System Model

The diagonalization of a matrix is defined in Appendix B and requires the matrix of eigenvalues and eigenvectors. For the circulant matrix $\mathbf{H}_{circ}$ the diagonal form then is

$$\mathbf{H}_{circ} = \mathbf{F}^{-1}\Lambda\mathbf{F} \ , \quad \Lambda = \text{diag}\,(\lambda_0, \lambda_1, \ldots, \lambda_{N-1}) \tag{6.45}$$

where the λ_μ are given by (6.33). The corresponding block diagram is portrait in Fig. 6.5b. Plugging (6.45) into (6.44) yields with additive noise $\mathbf{v}_N(k) + \mathbf{n}_N(k) = \mathbf{F}^{-1}\Lambda\mathbf{F}\mathbf{u}(k) + \mathbf{n}_N(k)$. Applying the DFT on both sides results in $\mathbf{F}\,(\mathbf{v}_N(k) + \mathbf{n}_N(k)) = \Lambda\mathbf{F}\mathbf{u}(k) + \mathbf{F}\mathbf{n}_N(k)$ and with

$$\underline{\mathbf{u}}(k) = \mathbf{F}\mathbf{u}(k) = \left(\underline{u}_{0k} \ \underline{u}_{1k} \ \cdots \ \underline{u}_{N-1,k}\right)^T \ , \quad \underline{\mathbf{v}}_N(k) = \mathbf{F}\mathbf{v}_N(k) = \left(\underline{v}_{0k} \ \underline{v}_{1k} \ \cdots \ \underline{v}_{N-1,k}\right)^T$$
$$\underline{\mathbf{n}}_N(k) = \mathbf{F}\mathbf{n}_N(k) \tag{6.46}$$

follows

$$\underline{\mathbf{v}}_N(k) + \underline{\mathbf{n}}_N(k) = \Lambda \underline{\mathbf{u}}(k) + \underline{\mathbf{n}}_N(k) \tag{6.47}$$

which is the output including noise in the DFT domain. Hence, a channel model based on eigenmodes with a diagonal matrix Λ depicted in Fig. 6.5c is obtained. For the νth component we write without noise

$$\underline{v}_{\nu k} = \lambda_\nu \underline{u}_{\nu k} \tag{6.48}$$

Thus, the spectral component $\underline{v}_{\nu k}$ of the output signal is just the input $\underline{u}_{\nu k}$ multiplied by factor λ_ν.

6.9.2   DFT Domain Equalizer

Equation (6.48) gives rise to a simple equalizer, which multiplies the receive spectral component by factor $1/\lambda_\mu = 1/H\left(e^{j2\pi\mu\frac{1}{N}}\right)$ to achieve $\underline{v}_{\nu k} = \underline{u}_{\nu k}$. For all spectral components an equalizer with matrix

$$\mathbf{E} = \Lambda^{-1} = \operatorname{diag}\left(\frac{1}{\lambda_0}, \frac{1}{\lambda_1}, \dots, \frac{1}{\lambda_{N-1}}\right) \tag{6.49}$$

has to be employed, as shown in Fig. 6.5c for transmit signals with cyclic prefix. The output block $\mathbf{y}(k)$ is readily determined as $\mathbf{y}(k) = \mathbf{F}^{-1}(\mathbf{E}\Lambda\mathbf{F}\mathbf{u}(k) + \mathbf{E}\underline{\mathbf{n}}_N(k))$ yielding with (6.49)

$$\mathbf{y}(k) = \mathbf{u}(k) + \mathbf{H}_{circ}^{-1}\mathbf{n}_N(k) \tag{6.50}$$

with $\mathbf{F}^{-1}\Lambda^{-1}\mathbf{F} = \mathbf{H}_{circ}^{-1}$. Apparently, intersymbol interference caused by the FIR channel is completely removed by this zero-forcing equalizer and the receiver output is just corrupted by noise, which may be enhanced by $\mathbf{H}_{circ}^{-1}$.

Reference

1. Gray, R.M.: Toeplitz and Circulant Matrices: A Review. NOW Publisher Inc., Boston (2006)

Chapter 7
Multicarrier Modulation and OFDM

7.1 Introduction

The division of a frequency band into dedicated channels by modulation, called frequency-division multiplexing (FDM), has a long history. Prominent examples are the analog telephone networks until the 1970s. They were replaced by digital telephony, operating with time-division multiplex (TDM), in which dedicated time slot are allocated to each application. Moreover, the analog and later the digital television broadcasting networks—terrestrial, satellite and cable based—operate with FDM. Most mobile and cellular networks today utilize combinations of FDM and TDM. Conventionally, the channels in FDM schemes are strictly separated by filters to avoid cross-talk of information. To this end guard spaces are required between the different frequency bands, however, they reduce spectral efficiency. Furthermore, the analog filters and modulators in the past were bulky, costly and a hurdle for micro-electronic integration. Therefore, research work started already in the late 1960s to allow overlapping signal spectra and the development towards digital circuits. Weinstein [1] gives a comprehensive survey on the history. Fundamental ideas on multicarrier FDM with orthogonal signals have been published by Chang [2] and Saltzberg [3] already at the end of the 1960s. Zimmermann and Kirsch [4] designed a high frequency transceiver (KATHRYN) using a discrete Fourier transform (DFT), which was implemented with analog circuits. Weinstein and Ebert [5] showed how Inverse DFT and DFT can be used for Orthogonal Frequency-Division Multiplexing (OFDM) modulation and demodulation, respectively. The efficient Cooley–Tukey algorithm and others known as Fast Fourier Transform (FFT) reduced the number of operations from $O(N^2)$ to $O(N \log N)$ and opened the path for widespread applications [6]. But it took quite some time until the 1990s for first industrial hardware prototypes [7], e.g., for Digital Subscriber Line (DSL) [8], Digital Audio Broadcasting (DAB), and Digital Video Broadcasting (DVB) [9–11]. Several wireless local area and metropolitan area network standards, WLAN IEEE 802.11 and WiMAX IEEE 802.16 followed with OFDM at the end of the 1990 and the standardization of

© Springer Nature Switzerland AG 2021
J. Speidel, *Introduction to Digital Communications*,
Signals and Communication Technology,
https://doi.org/10.1007/978-3-030-67357-4_7

new versions is still ongoing [12, 13]. Also the main developments of Ultra Wide-band (UWB) systems adopt OFDM [14–16]. Furthermore, systems for information transmission over power line [17] and several solutions for fiber optic transmission and visible light communications utilize OFDM today [18, 19]. Starting in the early 2000s with releases for Third Generation 3G/UMTS/IMT2000, later Fourth Generation 4G/LTE and around 2017 followed by Fifth Generation 5G, the multicarrier scheme OFDM and some of its variants entered into the technology of wireless cellular networks [20, 21].

7.2 Discrete-Time Multicarrier Modulation

7.2.1 System Overview

Modern transmitters and receivers are implemented with discrete-time digital circuits or by software programs running on processors. A multicarrier modulation scheme is shown in Fig. 7.1. The inputs of the transmitter are parallel sequences of QAM symbols $a_\nu(k)$, $\nu = 0, \ldots, N - 1$, each with symbol rate $v_S = \frac{1}{T}$ coming from QAM modulators, as described in Sect. 1.2. The sequences are transmitted in separate frequency bands divided by the carrier frequencies $0 \leqslant f_0 < f_1 \cdots < f_{N-1}$ over a single channel and the method is therefore also called *Frequency-Division Multiplexing* (FDM). Each of the N parallel branches is equipped with a transmit lowpass filter and a modulator. An adder yields the transmit signal $u(n)$. The channel is modeled as an equivalent baseband channel with lowpass characteristic and a cut-off frequency larger than the highest carrier frequency f_{N-1}. Time-invariant

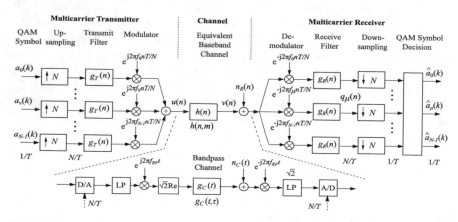

Fig. 7.1 Discrete-time transmitter and receiver with multicarrier modulation (upper part). Details of front ends with high frequency modulation f_{RF} as interface to a physical bandpass channel (lower part)

channels are characterized by the discrete-time impulse response $h(n)$. Most wireless channels are time-variant and thus modeled by the discrete-time delay spread function $h(n, m)$ as in Sect. 4.5, where n is discrete time and m discrete delay. To simplify the mathematical notations throughout the following sections we use $h(n)$. It will be clear from the context or by notice whether it has to be replaces by $h(n, m)$ in case of time-variance. The receive signal $v(n)$ corrupted by additive noise $n_R(n)$ is entering N branches with demodulator, receive lowpass filter, and down-sampling. Finally, the output signals undergo QAM symbol decision. Synchronous demodulation is applied. Thus, the demodulator frequencies $f_0, f_1, \ldots, f_{N-1}$ are exactly the same as those of the modulators.

Details of the physical link are depicted in the lower part of Fig. 7.1. Several physical channels provide transmission only at high frequencies around f_{RF}, such as the UHF bands for digital television, the satellite frequency bands or the frequency regions 2.4, 5, 30 or 60 GHz for cellular and wireless local area networks. An optical fiber even operates around 190 THz. For an economic digital implementation of the multicarrier scheme, the carriers must have much lower frequencies $f_0, \ldots, f_{N-1}$ than f_{RF}. For that reason the spectrum of the output signal $u(n)$ has to be shifted to the higher transmission band around f_{RF} by a second modulator stage mostly implemented by analog circuits, called radio frequency (RF) modulator or up-converter. To this end the signal $u(n)$ first undergoes digital-to-analog (D/A) conversion with a clock frequency N/T and after lowpass filtering (LP) modulation by a carrier $e^{j2\pi f_{RF}t} = \cos(2\pi f_{RF}t) + j\sin(2\pi f_{RF}t)$ follows. The real part of the modulator output feeds the RF bandpass channel with impulse response $g_C(t)$ for time-invariant or $g_C(t, \tau)$ for time-variant channels. At the receiver side the reverse operations take place in principle and after analog-to-digital (A/D) conversion the signal $v(n) + n_R(n)$ is available. In case of an optical fiber a Laser accomplishes the up-conversion. In the following the digital multicarrier transmission scheme (upper part of Fig. 7.1) is described in more detail.

7.2.2 Single Input Single Output

Transmitter

Firstly, only one sequence of QAM symbols $a_\nu(k)$ shall be sent with symbol rate $v_S = 1/T$ and solely branch ν is active at the transmitter. Apparently, together with the receiver branch $\mu = \nu$ a single input single output (SISO) transmission scheme with QAM results, similar as in Fig. 1.1 of Sect. 1.2, but here with a discrete-time implementation. The sequence $a_\nu(k)$ must first undergo pulse shaping and lowpass filtering using a transmit filter with impulse response $g_T(n)$. According to the theory of digital signal processing, these operations have to be executed at a higher sampling rate compared to the symbol rate v_S. To this end an up-sampler is installed, which introduces $N - 1$ zero samples between two symbols $a_\nu(k)$ and $a_\nu(k + 1)$, $k \in \mathbb{Z}$. In addition to the discrete-time variable k we introduce

Fig. 7.2 **a** QAM symbol $a_\nu(k) = a_\nu \delta_k$, **b** $a_\nu \delta_n$ after up-sampling with N, **c** Example of transmit filter impulse response $g_T(n)$, **d** Continuous time axis t

$$n = kN \tag{7.1}$$

representing the new sampling interval

$$T_S = \frac{T}{N} = \frac{1}{f_S} \tag{7.2}$$

after up-sampling, where f_S is the sampling rate. We select the up-sampling factor identical to the number of parallel branches, N, and will witness later that this definition helps to simplify the scheme, in particular when we implement the modulators by the inverse discrete Fourier transform for OFDM. As an example, in Fig. 7.2a one real symbol a_ν shall be sent, thus $a_\nu(k) = a_\nu \delta_k$, where δ_k is the unit impulse equal to one for $k = 0$ and zero for $k = \pm 1, \pm 2, \ldots$. Then we obtain $a_\nu \delta_n$ after up-sampling in Fig. 7.2b and a response $g_T(n)$ in Fig. 7.2c, as an example. The relation (7.1) is evident from the Figs. 7.2a, b. The continuous time axis t in Fig. 7.2d reveals the QAM symbol interval T and the sampling interval $\frac{T}{N}$.

For a symbol sequence $a_\nu(k)$ at the input of branch ν in Fig. 7.1 the up-converter outputs $\sum_{k=-\infty}^{\infty} a_\nu(k)\delta_{n-kN}$ and the transmit filter output is $\sum_{k=-\infty}^{\infty} a_\nu(k)g_T(n-kN)$. This signal is multiplied by the discrete-time carrier $e^{j2\pi f_\nu nT/N}$ yielding the transmitter output

Table 7.1 z- transform and Fourier transform of discrete-time signals

Signal	z- Transform	Fourier transform $\left(z = e^{j2\pi f/f_s}\right)$
$x(n)$	$X(z) = \sum_{n=-\infty}^{\infty} x(n)z^{-n}$	$X\left(e^{j2\pi f/f_s}\right) =$ $\sum_{n=-\infty}^{\infty} x(n)e^{-j2\pi nf/f_s}$
δ_n	1	1
$x(n)e^{j2\pi nf_0/f_s}$	$X\left(ze^{-j2\pi f_0/f_s}\right)$	$X\left(e^{j2\pi(f-f_0)/f_s}\right)$
$x(n-p)$	$X(z)z^{-p}$	$X\left(e^{j2\pi f/f_s}\right)e^{-j2\pi pf/f_s}$
$x(n) * y(n)$	$X(z)Y(z)$	$X\left(e^{j2\pi f/f_s}\right)Y\left(e^{j2\pi f/f_s}\right)$

$$u(n) = e^{j2\pi f_\nu nT/N} \sum_{k=-\infty}^{\infty} a_\nu(k)g_T(n-kN) \tag{7.3}$$

$u(n)$ is composed of successive impulses $g_T(n)$ as in Fig. 7.2c, shifted by kN and equipped with complex amplitude $a_\nu(k)e^{j2\pi f_\nu nT/N}$. For the following analysis we use the frequency domain and summarize beforehand some helpful relations from digital signal processing in Table 7.1.

Let $G_T\left(e^{j2\pi f/f_s}\right)$ be the Fourier transform (also called discrete-time Fourier transform) of $g_T(n)$ and thus the filter transfer function. For the Fourier transform of (7.3) follows $U\left(e^{j2\pi f/f_s}\right) = G_T\left(e^{j2\pi(f-f_\nu)/f_s}\right) \sum_{k=-\infty}^{\infty} a_\nu(k)e^{-j2\pi k(f-f_\nu)N/f_s}$ and the Fourier spectrum of $a(k)$ is

$$A_\nu(e^{j2\pi fN/f_s}) = \sum_{k=-\infty}^{\infty} a_\nu(k)e^{-j2\pi kfN/f_s} \tag{7.4}$$

noting the sampling rate f_s/N of $a(k)$. Then we get

$$U\left(e^{j2\pi f/f_s}\right) = A_\nu(e^{j2\pi(f-f_\nu)N/f_s})G_T\left(e^{j2\pi(f-f_\nu)/f_s}\right) \tag{7.5}$$

As an example, Fig. 7.3a shows the transfer function of the transmit filter, which shall be real, with ideal lowpass shape of the baseband, and cut-off frequency f_g. In general, any discrete-time signal owns a spectrum, which is periodic with the sampling frequency f_s, as indicated by the argument $e^{j2\pi f/f_s}$. Figure 7.3b depicts $U\left(e^{j2\pi f/f_s}\right)$ indicated by filled color, if only branch $\nu = 0$ is active with input $a_0(k) = a_0\delta_k$, $A_0(e^{j2\pi fN/f_s}) = a_0$, and real a_0 yielding from (7.5) $U\left(e^{j2\pi f/f_s}\right) = a_0G_T\left(e^{j2\pi(f-f_0)/f_s}\right)$. Apparently, $G_T\left(e^{j2\pi f/f_s}\right)$ is shifted by f_0 and weighted by a_0. Please ignore the other spectral components for the moment.

Fig. 7.3 a Example of a transfer function $G_T\left(e^{j2\pi f/f_s}\right)$ of the transmit lowpass filter. **b** Output spectrum $U\left(e^{j2\pi f/f_s}\right)$ in (7.13) of the transmitter for real-valued input signals $a_\nu(k) = a_\nu\delta_k$, $\nu = 0, 1, \ldots, N-1$

Receiver

Branch μ of the receiver in Fig. 7.1 is composed of a demodulator with carrier $e^{-j2\pi f_\mu nT/N}$, which shifts the input spectrum by f_μ to the left on the frequency axis. After the receive filter with impulse response $g_R(n)$ and transfer function $G_R\left(e^{j2\pi f/f_s}\right)$ the down-sampling unit decimates the sampling rate to the symbol rate $1/T$. The resulting samples undergo decision yielding estimates $\hat{a}_\mu(k)$ of the transmit symbols. We prefer calculation in the frequency domain and neglect the additive noise $n_R(n)$ for the moment. The spectrum of the receive filter output $q_\mu(n)$ can be obtained easily as

$$Q_\mu\left(e^{j2\pi f/f_s}\right) = U\left(e^{j2\pi(f+f_\mu)/f_s}\right) H\left(e^{j2\pi(f+f_\mu)/f_s}\right) G_R\left(e^{j2\pi f/f_s}\right) \qquad (7.6)$$

where $H\left(e^{j2\pi f/f_s}\right)$ is the transfer function of the time-invariant equivalent baseband channel. Inserting (7.5) yields

$$Q_\mu\left(e^{j2\pi f/f_s}\right) = A_\nu(e^{j2\pi(f-f_\nu+f_\mu)N/f_s})G_T\left(e^{j2\pi(f-f_\nu+f_\mu)/f_s}\right) G_R\left(e^{j2\pi f/f_s}\right) H\left(e^{j2\pi(f+f_\mu)/f_s}\right) \qquad (7.7)$$

If the receive filters own a similar lowpass characteristic as the transmit filters with cut-off frequencies $\pm f_g$ and if the carrier frequencies are selected in such a way that the passbands of $G_T\left(e^{j2\pi(f-f_\nu+f_\mu)/f_s}\right)$ and $G_R\left(e^{j2\pi f/f_s}\right)$ do not overlap for $f_\mu \neq f_\nu$, then

$$G_T\left(e^{j2\pi(f-f_\nu+f_\mu)/f_s}\right) G_R\left(e^{j2\pi f/f_s}\right) = \begin{cases} G_T\left(e^{j2\pi f/f_s}\right) G_R\left(e^{j2\pi f/f_s}\right) \; ; \; \mu = \nu \\ \qquad\qquad 0 \qquad\qquad\quad ; \; \mu \neq \nu \end{cases}$$

$$(7.8)$$

holds, which is a strict form of orthogonality also called condition for removal of intercarrier interference. Then follows from (7.7) for the receiver output $\mu = \nu$

$$Q_\nu\left(e^{j2\pi f/f_s}\right) = A_\nu(e^{j2\pi k f N/f_s})G_T\left(e^{j2\pi f/f_s}\right) G_R\left(e^{j2\pi f/f_s}\right) H\left(e^{j2\pi(f+f_\nu)/f_s}\right) \quad (7.9)$$

and $Q_\mu\left(e^{j2\pi f/f_s}\right) = 0$ for $\mu \neq \nu$.

From (7.9) the time-domain signal is readily determined using Table 7.1

$$q_\nu(n) = \left(\sum_{k=-\infty}^{\infty} a_\nu(k)\delta_{n-kN}\right) * g_T(n) * g_R(n) * \left(h(n)e^{-j2\pi n f_\nu/f_s}\right) + g_R(n) * \left(n_R(n)e^{-j2\pi n f_\nu/f_s}\right)$$

$$(7.10)$$

for $\mu = \nu$ and $q_\mu(n) = 0$ for $\mu \neq \nu$. The second term in (7.10) characterizes the additive noise.

7.2.3 Multiple Input Multiple Output

Transmitter Output Signal
If all N branches of the transmitter are active with input sequences $a_\nu(k)$, $\nu = 0, 1, \ldots, N-1$, the output signal follows with (7.3)

$$u(n) = \sum_{\nu=0}^{N-1} e^{j2\pi f_\nu nT/N} \sum_{k=-\infty}^{\infty} a_\nu(k)g_T(n - kN) \quad (7.11)$$

and the Fourier spectrum is obtained with (7.5)

$$U\left(e^{j2\pi f/f_s}\right) = \sum_{\nu=0}^{N-1} A_\nu(e^{j2\pi(f-f_\nu)N/f_s})G_T\left(e^{j2\pi(f-f_\nu)/f_s}\right) \quad (7.12)$$

using the superposition principle for the linear transmitter. In Fig. 7.3b an example is depicted, where the input to each branch ν is a single real QAM symbol $a_\nu(k) = a_\nu\delta_k$. Then follows for the output

$$U\left(e^{j2\pi f/f_s}\right) = \sum_{\nu=0}^{N-1} a_\nu G_T\left(e^{j2\pi(f-f_\nu)/f_s}\right) \quad (7.13)$$

Because the baseband of $G_T\left(e^{j2\pi f/f_s}\right)$ in Fig. 7.3a is strictly bandlimited, the partial spectra of $U\left(e^{j2\pi f/f_s}\right)$ in Fig. 7.3b do not overlap for the given carrier frequencies $f_0, \ldots, f_{N-1}$. In contrast, for OFDM discussed in Sect. 7.3 and Fig. 7.5 overlapping and even not bandlimited spectra will be employed. Therefore, in that case recovery of the transmit signals can not be done just by lowpass filtering with $G_R\left(e^{j2\pi f/fs}\right)$ after demodulation. An additional orthogonality condition has to be imposed on the carriers.

Receiver Output Signal

Plugging (7.12) into (7.6) yields the spectrum of $q_\mu(n)$ in branch μ

$$Q_\mu\left(e^{j2\pi f/fs}\right) = \sum_{\nu=0}^{N-1} A_\nu(e^{j2\pi(f-f_\nu+f_\mu)N/fs})G_T\left(e^{j2\pi(f-f_\nu+f_\mu)/fs}\right)G_R\left(e^{j2\pi f/fs}\right)H\left(e^{j2\pi(f+f_\mu)/fs}\right)$$

(7.14)

With condition (7.8) follows again $Q_\nu\left(e^{j2\pi f/fs}\right)$ in (7.9) and $q_\nu(n)$ in (7.10), which hold for $\nu = 0, 1, \ldots, N-1$.

7.3 Orthogonal Frequency Division Multiplexing (OFDM)

7.3.1 OFDM Transmitter with IDFT

We build upon the multicarrier scheme in Fig. 7.1, however, the carriers shall have equidistant frequency spacing Δf coupled with the symbol rate according to

$$\Delta f = \frac{1}{T}$$

(7.15)

The carrier frequencies are

$$f_\nu = \nu\Delta f = \nu\frac{1}{T} \; ; \; \nu = 0, 1, \ldots, N-1$$

(7.16)

and the N carriers $1, e^{j2\pi n/N}, \ldots, e^{j2\pi \mu n/N}, \ldots, e^{j2\pi(N-1)n/N}$ are pairwise orthogonal. The first carrier is constant, hence the corresponding signal remains unmodulated.

Consider in Fig. 7.1 QAM sequences $a_\nu(k)$, $\nu = 0, 1, \ldots, N-1$ at all transmitter inputs. Then follows from (7.11) with (7.16) the transmitter output

$$u(n) = \sum_{k=-\infty}^{\infty}\left(\sum_{\nu=0}^{N-1} e^{j2\pi\nu n/N} a_\nu(k)\right) g_T(n-kN) \; ; \; n\in\mathbb{Z}$$

(7.17)

Apparently, $u(n)$ is composed of successive impulses $g_T(n)$ shifted by kN and with amplitudes indicated by brackets $(\ldots)$. Let $g_T(n)$ be a rectangular impulse of length N

$$g_T(n) = \begin{cases} \frac{1}{\sqrt{N}} \; ; & n = 0, 1, \ldots, N-1 \\ 0 \; ; & else \end{cases} \tag{7.18}$$

resulting in the block-wise signal

$$u(n) = \frac{1}{\sqrt{N}} \sum_{\nu=0}^{N-1} e^{j2\pi\nu n/N} a_\nu(k) \; ; \; kN \leq n < (k+1)N \; ; \; k \in \mathbb{Z} \tag{7.19}$$

where k is the block number. Comparing with (6.40) we recognize that $u(n)$ resembles in each block k the IDFT of the set $a_\nu(k)$, $\nu = 0, 1, \ldots, N-1$

$$\frac{1}{\sqrt{N}} \sum_{\nu=0}^{N-1} a_\nu(k) e^{j2\pi\nu n/N} = \tilde{a}_n(k) \; , \; n = 0, 1, \ldots, N-1 \; ; \; k \text{ fixed} \tag{7.20}$$

where n can run just from 0 to $N-1$, because $e^{j2\pi\nu n/N}$ is periodic with N. With vector notation we write

$$\mathbf{a}(k) = \left(a_0(k) \; a_1(k) \; \cdots \; a_{N-1}(k) \right)^T \quad \text{IDFT} \quad \tilde{\mathbf{a}}(k) = \left(\tilde{a}_0(k) \; \tilde{a}_1(k) \; \cdots \; \tilde{a}_{N-1}(k) \right)^T \tag{7.21}$$

$\tilde{\mathbf{a}}(k)$ is called *OFDM symbol* or *OFDM block*. Alternatively, we can formulate the IDFT as

$$\tilde{\mathbf{a}}(k) = \mathbf{F}^{-1}\mathbf{a}(k) \tag{7.22}$$

using the inverse DFT matrix $\mathbf{F}^{-1}$ in (6.36). Inserting (7.20) into (7.19) yields the block-wise transmitter output

$$u(n) = \tilde{a}_n(k) \; ; \; kN \leq n < (k+1)N \; ; \; k \in \mathbb{Z} \tag{7.23}$$

Artificially, we have to interpret the QAM symbols $a_\nu(k)$ as elements in the DFT domain, whereas the $\tilde{a}_n(k)$ are in the original domain. The multicarrier transmitter in Fig. 7.1 can be modified as shown in Fig. 7.4. The parallel branches and the summing node are replaced by a processor executing an IDFT and a parallel-to-serial (P/S) converter, which translates the column vector $\tilde{\mathbf{a}}(k)$ into a row vector $\mathbf{u}^T(k) = \tilde{\mathbf{a}}^T(k)$ for serial transmission. For the moment we do not consider the insertion of a cyclic prefix. Hence, the P/S converter ignores the prefix. Block-wise transmission with cyclic prefix and the IDFT are treated in detail in Sects. 6.7–6.9. Here, we just give references and the reader not familiar with block-wise processing is recommended to study details in Chap. 6.

Fig. 7.4 OFDM transmitter and receiver for block-wise transmission. Cyclic prefix $\tilde{a}_{N-G}(k), \ldots, \tilde{a}_{N-1}(k)$ (upper part). Details of front ends with high frequency modulation f_{RF} as interfaces to a physical bandpass channel (lower part)

7.3.2 OFDM Output Spectrum

The z-transform of $g_T(n)$ in (7.18) is $G_T(z) = \sum_{n=0}^{N-1} g_T(n)z^{-n} = \frac{1}{\sqrt{N}}\frac{z^{-N}-1}{z^{-1}-1}$, using the sum formula of a finite geometric series. Then follows the frequency response

$$G_T\left(e^{j2\pi f/f_S}\right) = \frac{1}{\sqrt{N}}\frac{sin(\pi Nf/f_S)}{sin(\pi f/f_S)}e^{-j\pi(N-1)f/f_S} \tag{7.24}$$

with $G\left(e^{j2\pi f/f_S}\right) = 0$ for $f = f_\mu = \mu\frac{f_S}{N}$, $\mu \in \mathbb{Z} \setminus \mu \neq 0, \pm N, \pm 2N, \ldots$ and $\left|G_T\left(e^{j2\pi f/f_S}\right)\right| = \frac{1}{\sqrt{N}}$ for $f = \mu f_S$, $\mu \in \mathbb{Z}$. $f_S = \frac{N}{T}$ was defined in (7.2). Apparently, $\left|G_T\left(e^{j2\pi f/f_S}\right)\right|$ is periodic with f_S, as expected. For an input signal $a_\nu(k) = \delta_k$, $\nu = 0, 1, \ldots, N-1$ the output spectrum of the multicarrier transmitter, now an OFDM transmitter, follows from (7.13) as

$$U\left(e^{j2\pi f/f_S}\right) = \sum_{\nu=0}^{N-1} G_T\left(e^{j2\pi(f-f_\nu)/f_S}\right) \tag{7.25}$$

The magnitudes of the spectral parts are depicted in Fig. 7.5. Owing to the zeros of $G_T\left(e^{j2\pi f/f_S}\right)$, there is no overlap with other spectral components at positions of the maxima, which is typical for OFDM. Hence, sampling in the frequency domain at $f_\mu = \mu\frac{f_S}{N}$, $\mu = 0, 1, \ldots, N-1$, what is inherently done by a DFT, does not suffer from spectral aliasing. The D/A converter in Fig. 7.5 owns the principal transfer

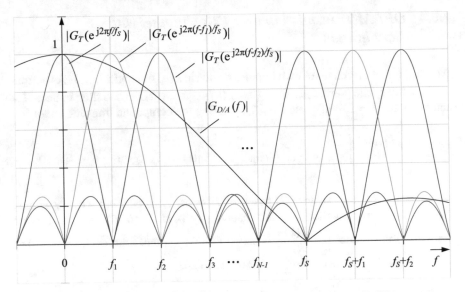

Fig. 7.5 Spectral parts $\left|G_T\left(\mathrm{e}^{\mathrm{j}2\pi(f-f_\nu)/f_s}\right)\right|$ of output Fourier spectrum $U\left(\mathrm{e}^{\mathrm{j}2\pi f/f_s}\right)$ in (7.25) for multicarrier (OFDM) transmitter with $f_\nu = \nu\frac{f_s}{N}$ and magnitude response $\left|G_{D/A}\left(f\right)\right|$ of the D/A converter. Curves normalized with respect to maxima

function $G_{D/A}\left(f\right) = sinc(f/f_s)$ with zeros at integer multiples of f_s, except $f = 0$, and provides a smooth lowpass filtering.

7.3.3 OFDM Transmission over Ideal Channel

To show the principle of OFDM transmission, assume an ideal channel, $h(n) = \delta_n$, with no noise. Furthermore, no prefix shall be used. Thus, the channel input vector is $\mathbf{u}(k) = \tilde{\mathbf{a}}(k)$ and the receiver gets the vector $\mathbf{v}_N(k) = \mathbf{u}(k) = \tilde{\mathbf{a}}(k)$. To recover $\mathbf{a}(k)$ an OFDM receiver has to be furnished with a DFT that outputs

$$\mathbf{a}'(k) = \mathbf{F}\mathbf{v}_N(k) = \mathbf{F}\tilde{\mathbf{a}}(k) = \mathbf{a}(k) \tag{7.26}$$

and the original QAM symbol vector $\mathbf{a}(k)$ is perfectly reconstructed in this ideal case.

7.3.4 OFDM Transmission over FIR Channel with Equalization

We consider an FIR channel with impulse response $h(n)$ of length $L + 1$. As portrait in Fig. 7.4, a cyclic prefix $\tilde{\mathbf{a}}_{pre}(k) = \left(\tilde{a}_{N-L}(k) \cdots \tilde{a}_{N-1}(k)\right)^T$ according to Sects. 6.8 and 6.9 with $G = L$ is allocated to $\tilde{\mathbf{a}}(k)$ yielding the compound transmit vector

$$\mathbf{u}_c(k) = \begin{pmatrix} \tilde{\mathbf{a}}_{pre}(k) \\ \tilde{\mathbf{a}}(k) \\ \mathbf{0}_L \end{pmatrix} = \left(\tilde{a}_{N-L}(k) \cdots \tilde{a}_{N-1}(k) \mid \tilde{a}_0(k) \cdots \tilde{a}_{N-L}(k) \cdots \tilde{a}_{N-1}(k) \mid \mathbf{0}_L^T\right)^T$$

(7.27)

where $\mathbf{0}_L$ is a column vector composed of L zeros. Zero padding with $\mathbf{0}_L$ is only required for proper notation with the channel matrix $\mathbf{H}_c$ in (6.26) and the zeros are not really transmitted. $\begin{pmatrix} \tilde{\mathbf{a}}_{pre}(k) \\ \tilde{\mathbf{a}}(k) \end{pmatrix}$ is called OFDM symbol with cyclic prefix. The receive block follows according to (6.24) and (6.25)

$$\mathbf{v}_c(k) + \mathbf{n}_c(k) = \begin{pmatrix} \mathbf{v}_{pre}(k) \\ \mathbf{v}(k) \end{pmatrix} + \begin{pmatrix} \mathbf{n}_{pre}(k) \\ \mathbf{n}(k) \end{pmatrix}$$

(7.28)

where signal $\mathbf{v}_c(k)$ and noise $\mathbf{n}_c(k)$ are separated into a pefix and a main part. Furthermore

$$\mathbf{v}_c(k) = \mathbf{H}_c \mathbf{u}_c(k)$$

(7.29)

holds as in (6.26). $\mathbf{H}_c$ is a lower triangular $(N + 2L)\text{x}(N + 2L)$ Toeplitz matrix similar to $\mathbf{H}_0$ in (6.9) with h_0 along the main diagonal. The S/P converter at the receiver drops the prefix part as well as the last L samples from the receive block $\mathbf{v}_c(k) + \mathbf{n}_c(k)$ resulting in

$$\mathbf{v}_N(k) + \mathbf{n}_N(k) = \left(v_{0k} \cdots v_{N-1,k}\right)^T + \left(n_{0k} \cdots n_{N-1,k}\right)^T$$

(7.30)

given by (6.43) with signal part

$$\mathbf{v}_N(k) = \mathbf{H}_{circ}\mathbf{u}(k)$$

(7.31)

from (6.44). $\mathbf{u}(k) = \tilde{\mathbf{a}}(k)$ is the transmit vector and $\mathbf{n}_N(k)$ a residual $N\text{x}1$ noise vector. As outlined in Sect. 6.8.2 in quite detail, the transmission scheme with cyclic prefix can be described equivalently by the input-output relation $\mathbf{v}_N(k) = \mathbf{H}_{circ}\mathbf{u}(k)$ in the noise free case, where $\mathbf{H}_{circ}$ is an $N\text{x}N$ cyclic matrix defined in (6.30).

The DFT device with $N\text{x}N$ matrix $\mathbf{F}$ in Fig. 7.4 outputs

$$\mathbf{a}'(k) = \mathbf{F}\left(\mathbf{v}_N(k) + \mathbf{n}_N(k)\right) = \mathbf{F}\mathbf{H}_{circ}\tilde{\mathbf{a}}(k) + \mathbf{F}\mathbf{n}_N(k)$$

(7.32)

With $\mathbf{H}_{circ} = \mathbf{F}^{-1}\mathbf{\Lambda}\mathbf{F}$ from (6.45) and $\tilde{\mathbf{a}}(k) = \mathbf{F}^{-1}\mathbf{a}(k)$ of (7.22) follows

$$\mathbf{a}'(k) = \Lambda \mathbf{a}(k) + \mathbf{Fn}_N(k) \tag{7.33}$$

where $\Lambda = \text{diag}\,(\lambda_0, \lambda_1, \ldots, \lambda_{N-1})$ represents the matrix of eigenvalues of $\mathbf{H}_{circ}$. The DFT domain equalizer $\mathbf{E} = \Lambda^{-1} = \text{diag}\left(\frac{1}{\lambda_0}, \frac{1}{\lambda_1}, \ldots, \frac{1}{\lambda_{N-1}}\right)$ in (6.49) provides the output

$$\mathbf{Ea}'(k) = \mathbf{a}(k) + \Lambda^{-1}\mathbf{Fn}_N(k) \tag{7.34}$$

Apparently, the output of the equalizer $\mathbf{Ea}'(k)$ in Fig. 7.4 is composed of the transmit QAM symbol vector $\mathbf{a}(k)$ corrupted by the DFT spectrum of the noise, filtered and possibly enhanced by the equalizer $\mathbf{E} = \Lambda^{-1}$. Finally, the equalizer output is subject to QAM symbol detection.

The eigenvalues are derived in Sect. 6.8.2 and own the interesting property, $\lambda_\mu = \sum_{n=0}^{N-1} h(n)e^{-j\frac{2\pi}{N}\mu n} = H\left(e^{j2\pi f/fs}\right)|_{f=\mu\frac{1}{T}}$ in (6.34), where $H\left(e^{j2\pi f/fs}\right)$ is the transfer function of the FIR channel with impulse response $h(n)$. In the noise free case, the vector components of the equalizer output $\mathbf{Ea}'(k)$ are simply $a'_\mu(k)/\lambda_\mu$, as indicated in Fig. 7.4. This low complexity equalizer is a major advantage of OFDM transmission with cyclic prefix. However, it comes with the penalty that a prefix with length G in the transmit block reduces the number of useful samples by $\eta = \frac{N}{N+G}$, as discussed earlier in (6.20). Finally, it should be mentioned that the clock rate at channel input and output in Fig. 7.4 is $\frac{N}{T}(1 + \frac{G}{N})$. Hence, the relation (7.1) between n and k changes to $n = (N + G)k$.

7.3.5 Summary on OFDM

OFDM is a powerful transmission scheme for N parallel input sequences of QAM symbols. Modulation and demodulation is done with processors executing an $N{\times}N$ IDFT and DFT, respectively. Fast and efficient computational methods are present, such as the Cooley–Turkey algorithm [6]. The available transmission frequency band is separated by N "carrier" frequencies of the IDFT, which outputs a vector, called OFDM symbol, with N components. A cyclic prefix of length G samples can be attached to the OFDM symbols before transmission. Equalization at the receiver after DFT is enabled by simply dividing each output μ by $\lambda_\mu = H\left(e^{j2\pi f/fs}\right)|_{f=\mu\frac{1}{T}}$, where $H\left(e^{j2\pi f/fs}\right)$ is the transfer function of the equivalent baseband channel and $1/T$ the QAM symbol rate. Given an FIR channel impulse response with length $L + 1$, the cyclic prefix must consist of $G \geqslant L$ samples to completely remove intersymbol interference in the receive signal. The prefix samples span a guard interval that reduces the number of useful samples by factor $\eta = N/(N + G)$, because the cyclic prefix just allocates redundancy. On the other hand, if N is made large, the loss is tolerable and, in addition, the transmission frequency band is separated into even smaller and thus almost flat sub-bands. This is the major reason why OFDM became the dominant method for mobile wireless communications allowing a frequency selective fading channel to be handled as a piece-wise frequency flat fading channel.

Some examples for the number N of carriers are: Digital Audio Broadcasting (DAB) 1576, Digital Video Broadcasting Terrestrial (DVB-T) about 8000, Wireless LAN 802.11a about 48, and 4G/LTE up to 1200 carrier frequencies.

Very High Speed Digital Subscriber Line (VHDSL) on twisted pair telephone lines is using Digital Multitone (DMT) modulation with 4096 carrier frequencies. DMT is the baseband version of OFDM, where no RF up-conversion is required. Details of DMT are considered, e.g., in [22]. There, also a method for dedicated power allocation to the N transmit signals is described for maximization of the signal-to-noise ratio at the receiver. The water-filling algorithm is used, which is described in detail for MIMO systems in Sect. 17.5. Besides many advantages, also some critical points in the development of OFDM and DMT had to be solved, in particular the Peak-to-Average Power Ratio (PAPR) of the transmit signal. Due to the superposition of a large number of modulated carriers, the instantaneous output peak signal can be rather high, thus, driving the D/A converter out of range or the transmit amplifiers into the non-linear regime. Various methods have been found to minimize PAPR, such as rough clipping, signal shaping, e.g., by statistical methods, linear and nonlinear pre-emphasis, and the use of a few dedicated QAM input sequences specially designed to "flatten" the output signal over time.

The transmitter in Fig. 7.1 can also be implemented as a filter bank without dedicated modulators, where the $g_T(n)e^{j2\pi f_\nu nT/N}$, $\nu = 1, 2, \ldots, N-1$ are interpreted as filters. This method is useful for the design of transmitter output spectra to achieve low interference with neighboring spectra and applied in principle for the 5G cellular system. Also combinations of filter banks and OFDM have become interesting solutions [23].

References

1. Weinstein, S.B.: The history of orthogonal frequency-division multiplexing. IEEE Commun. Mag. (2009)
2. Chang, R.W.: Synthesis of band-limited orthogonal signals for multichannel data transmission. Bell Syst. Tech. J. (1966)
3. Salzberg, B.R.: Performance of an efficient parallel data transmission system. IEEE Trans. Commun. Tech. (1967)
4. Zimmermann, M.S., Kirsch, A.L.: The AN/GSC-10 (KATHRYN) variable rate data modem for HF radio. IEEE Trans. Commun. Tech. (1967)
5. Weinstein, S.B., Ebert, P.M.: Data transmission for frequency-division multiplexing using the discrete Fourier transform. IEEE Trans. Commun. Tech. (1971)
6. Cooley, J.W., Tukey, J.W.: An algorithm for the machine calculation of complex Fourier series. Math. Comput. 19(90), 297–301 (1965)
7. Bingham, J.: Multicarrier modulation for data transmission: an idea whose time has come. IEEE Commun. Mag. (1990)
8. Sistanizadeh, K., Chow, P., Cioffi, J.: Multi-tone transmission for asymmetric digital subscriber lines. In: IEEE International Conference on Communication, ICC (1993)
9. ETS 300 401: Digital Audio Broadcasting (DAB). DAB to mobile, portable and fixed receivers (1997)
10. Reimers, U.: Digital video broadcasting. IEEE Commun. Mag. (1998)

11. Reimers, U.: DVB - The Family of International Standards for Digital Video Broadcasting. Springer, Berlin (2013)
12. 802.11: IEEE Standards for Information Technology - Telecommunications
13. 802.16: IEEE Standard for Air Interface for Broadband Wireless Access Systems
14. 802.15: IEEE Standard for Low-Rate Wireless Networks
15. Zhang, J., Orlik, P.V., Sahinoglu, Z., Molisch, A.F., Kinney, P.: UWB systems for wireless sensor networks. Proc. IEEE (2009)
16. Kaiser, T., Zheng, F.: Ultra Wideband Systems with MIMO. Wiley, New York (2010)
17. Lampe, L., Tonello, A.M., Swart, T.G. (eds.): Power Line Communications: Principles, Standards and Applications from Multimedia to Smart Grid. Wiley, New York (2016)
18. Shieh, W., Djordjevic, I.: OFDM for Optical Communications. Academic, New York (2010)
19. Fernando, X., Farahneh, H.: Visible Light Communications. IOP Publishing, Bristol (2019)
20. 3GPP: The mobile broadband standards
21. Ahmadi, S.: 5G NR: Architecture, technology, implementation, and operation of 3GGP new radio standards. Academic, New York (2017)
22. Schur, R., Pfletschinger, S., Speidel, J.: Digital Multitone (DMT) Modulation. In: Wiley Encyclopedia of Telecommunications. Wiley, New York (2003)
23. Pfletschinger, S.: Multicarrier modulation for broadband return channels in cable TV networks. Ph.D. Thesis, University of Stuttgart, Institute of Telecommunications, Shaker Publ. ISBN 3-8322-1437-2 (2003)

Part II
Theory of Linear Time-Variant Systems

Chapter 8
Introduction and Some History

Time-variant systems are of general interest, because they play an important role in communications due to the emerging wireless networks for in-house and outdoor applications. As a matter of fact a wireless channel can change its parameters with time depending on the position of the mobile transmitter, the receiver, and on the change of the surroundings. During the education of electrical engineers the main focus is on time-invariant systems and the topic of time-variant systems is not always strongly alluded. Thus, also from this perspective a general view on time variant systems and their mathematical description is favorable.

First contributions to the subject have been made by Carson [1], later by the seminal papers of Zadeh [2], Bello [3], Kailath [4, 5], and Gersho [6]. Zadeh considers electrical circuits which he calls "variable networks" if their circuit elements vary with time. He points out that the fundamental characteristic of "fixed networks", which have constant elements, is the fact that their impulse response $w(t - s)$ is dependent solely upon the "age variable" $t - s$ that is the difference between the instant t of the observation of the response at the output and the instant s of a Dirac impulse at the input. Furthermore, he argues that no such property is possessed by variable networks and he concludes that in those the response must be of the general form $w(t, s)$ characterizing a two-dimensional function of the independent variables t and s. For the first time he defines a time-variant transfer function $H(f, t) = \int_{-\infty}^{\infty} w(t, s) e^{-j2\pi f(t-s)} ds$, where t acts as a parameter and he interprets $H(f, t)$ as the natural extension of the transfer function $H(f) = \int_{-\infty}^{\infty} h(t - s) e^{-j2\pi f(t-s)} ds$, which is the Fourier transform of $h(t)$ of a fixed network. To find the relation between the input $x(t)$ and the output $y(t)$ of variable networks Zadeh [2] takes the model of a linear electrical circuitry with time-variant circuit elements and describes these networks by an ordinary linear differential equation of higher order with time-variant coefficients. He reports that the general solution of this equation has the form

© Springer Nature Switzerland AG 2021
J. Speidel, *Introduction to Digital Communications*,
Signals and Communication Technology,
https://doi.org/10.1007/978-3-030-67357-4_8

$$y(t) = \int_{-\infty}^{\infty} w(t, s)x(s)ds \tag{8.1}$$

where $x(t)$ and $y(t)$ are the input and the output signals, respectively. Furthermore, he shows that $w(t, s)$ is the response at observation instant t to an input Dirac impulse at instant s and calls it impulse response of the time-variant system. The integral in (8.1) is denoted as general superposition integral, generalized convolution integral, or time-variant convolution.

With the Fourier transform $X(f)$ of $x(t)$ he concludes that

$$y(t) = \int_{-\infty}^{\infty} H(f, t)X(f)e^{j2\pi ft}df \tag{8.2}$$

holds and that many operations can be performed with $H(f, t)$ just like $H(f)$. While the focus of Zadeh [2] is on the description of variable networks in the frequency domain Bello [3] starts his considerations with the time domain and he builds upon the work of [2, 4, 5]. Bello added three more input-output relations to (8.1), where two pairs of them turn out to represent Fourier transform pairs and he showed that all are equivalent for the description of linear time-variant systems [7]. In this chapter we follow only the first one which is

$$y(t) = \int_{-\infty}^{\infty} K_1(t, s)x(s)ds \tag{8.3}$$

where $K_1(t, s)$ is identical with $w(t, s)$ in (8.1). Bello denotes $K_1(t, s)$ as kernel system function or time-variant impulse response and as already stated by Zadeh he interprets $K_1(t, s)$ as the response of the system at observation time instant t to an input Dirac impulse $\delta(t - s)$ applied at time instant s. However, he also points out that $K_1(t, s)$ has some drawbacks for system modeling in electrical engineering and therefore takes over the transformation of variables proposed already by Kailath [4, 5]

$$s = t - \tau \tag{8.4}$$

yielding the "modified impulse response"

$$g(t, \tau) = K_1(t, t - \tau) \tag{8.5}$$

which Bello [3] calls *(input) delay spread function*.[1] Then (8.3) results in

$$y(t) = \int_{-\infty}^{\infty} K_1(t, t - \tau)x(t - \tau)d\tau \tag{8.6}$$

---

[1] Please note that several authors of later literature denote $g(t, \tau)$ bewildered as impulse response.

Bello [3] gives an interpretation using an analog, densely tapped delay line with an infinite number of taps and the input signal $x(t)$. The differential output signals $K_1(t, t - \tau)x(t - \tau)d\tau$ of the taps are "summed up" by integration yielding the output signal $y(t)$. Later we will show that the Fourier transform of $g(t, \tau)$ provides the time-variant transfer function $H(f, t)$ proposed by Zadeh [2]. Using the previous explanation for $K_1(t, s)$ we can interpret $g(t, \tau)$ as the response of the system at the observation instant t to an input Dirac impulse applied at the instant $t - \tau$, thus the quantity τ earlier, which has also elicited the name *age variable* or *delay time* for τ and *output time* or simply *time* for t . Please note that the time variation of the system parameters is also determined by t.

In the next chapter we present an alternative derivation for (8.3), which is different from [2, 3]. Firstly, we derive an input-output relation for time-variant systems using a system theoretic approach and next we present the input-output relation of a discrete-time, time-variant system with the help of a discrete-time delay-line filter. Furnished with the latter results we will construe the discrete-time system as the sampled version of a continuous time one. With these approaches we can prove (8.3) directly. Then we discuss several important properties of the time-variant convolution, such as linearity, associativity, commutativity, and apply them for cascading time-variant and time-invariant systems. An important subject will be the Fourier spectrum of the modified impulse response $g(t, \tau)$, which results in the time-variant transfer function and the Doppler spread function. Of great interest is also the Fourier spectrum of the output signal. Two examples will illustrate that the output spectrum of a time-variant system, e.g., a wireless fading channel, can have a larger bandwidth than the input spectrum. This fact is quite in contrast to linear time-invariant systems. Finally, we close with a thorough consideration of important correlation functions for randomly changing time-variant systems and address in particular also stationary random time-variant systems. These considerations allow us to formulate useful parameters for the characterization of time-varying channels, such as coherence time and coherence bandwidth. We are going to describe time-variant systems from the perspective of system theory for communications and do not consider electrical circuits with time-variant elements as in [2] and later in [8].

References

1. Carson, J.R.: Theory and calculation of variable systems. Phys. Rev. **17** (1921)
2. Zadeh, L.A.: Frequency analysis of variable networks. Proceed. IRE (1950)
3. Bello, P.A.: Characterization of randomly time-variant linear channels. IEEE Trans. Commun. Syst. (1963)
4. Kailath, T.: Sampling models for linear time-variant filters. Technical report, M.I.T. Research Laboratory of Electronics, Cambridge, Mass. Report 352 (1959)
5. Kailath, T.: Channel characterization: Time-variant dispersive channels. In: Lectures on Communications System Theory. McGraw Hill (1961)

6. Gersho, A.J.: Characterization of time-varying linear systems. Proceedings of the IEEE (Correspondence) (1963)
7. Bello, P.A.: Time-frequency duality. IEEE Trans. Inf. Theory (1964)
8. Shmaliy, Y.: Continuous-Time Signals. Springer, Berlin (2006)

Chapter 9
System Theoretic Approach for the Impulse Response of Linear Time-Variant Systems

9.1 Continuous Time, Time-Variant Impulse Response

Let $T[\ldots]$ be a linear system operator, which maps the input signal $x(t)$ of a linear dynamic system to an output signal $y(t)$ as

$$y(t) = T[x(t)] \tag{9.1}$$

We assume empty system memories for $t \to -\infty$. First consider the case of a linear time-invariant system with impulse response $w(t)$. The Dirac impulse $\delta(t-s)$ yields the output response

$$w(t-s) = T[\delta(t-s)] \tag{9.2}$$

with the property that the shape of w does not change if s is varied, because the response $w(t)$ is just shifted by s along the time axis t. No such relation is possessed by a linear time-variant system. The shape of its response w depends also on the time instant s. Thus, as outlined previously, the response of a time-variant system is more general and given by

$$T[\delta(t-s)] = w(t, s) \tag{9.3}$$

as a function of the two independent variables t and s.

To derive the input-output relation of a linear time-variant system we start with the identity using the sifting property of the Dirac impulse

$$x(t) = \int_{-\infty}^{\infty} x(\zeta)\delta(t-\zeta)\,d\zeta \tag{9.4}$$

In the next step we apply the linear system operator $T[\ldots]$ on both sides

© Springer Nature Switzerland AG 2021
J. Speidel, *Introduction to Digital Communications*,
Signals and Communication Technology,
https://doi.org/10.1007/978-3-030-67357-4_9

$$T[x(t)] = T[\int_{-\infty}^{\infty} x(\zeta)\delta(t-\zeta)\,d\zeta] = \int_{-\infty}^{\infty} T[x(\zeta)\delta(t-\zeta)]d\zeta = \int_{-\infty}^{\infty} x(\zeta)T[\delta(t-\zeta)]d\zeta \quad (9.5)$$

noting that the system operator acts only with respect to t. With (9.1) and (9.3) follows

$$y(t) = \int_{-\infty}^{\infty} x(\zeta)w(t,\zeta)d\zeta = \int_{-\infty}^{\infty} x(s)w(t,s)ds = x(t) \circledast w(t,s) \quad (9.6)$$

where we have just replaced ζ by s in the second step. Obviously, (9.6) with (9.3) prove (8.1) in a different way. We denote the operation in (9.6) as the time-variant convolution and indicate it by $\circledast$ to differentiate from the well known time-invariant convolution $*$.

In summary, a linear time-variant system is determined by an impulse response $w(t,s)$, which is a function of the independent variables t and s. Furthermore, $w(t,s)$ is the response observed at time instant t to an input Dirac impulse $\delta(t-s)$, which is active at time instant s. To fully describe a time-variant system an infinite number of responses $w(t,s)$ as functions of t with the parameter $s \in \mathbb{R}$ must be considered compared to a time-invariant system, where only one impulse response $w(t,0) = h(t)$ suffices.

Example 1
Show that $w(t,s)$ is the response to $x(t) = \delta(t-s)$.

Solution:
Using (9.6) and the sifting property of the Dirac impulse yields
$\int_{-\infty}^{\infty} \delta(\zeta-s)w(t,\zeta)d\zeta = w(t,s)$.

Example 2
We consider the time-variant impulse response of a fading transmission channel

$$w(t,s) = \begin{cases} \text{sinc}\left(f_1(t-s-\frac{1}{f_1})\right)\text{sinc}(f_2 s) \; ; \; t \geqslant s \geqslant 0 \\ \qquad\qquad 0 \qquad\qquad\quad ; \quad else \end{cases} \quad (9.7)$$

where $\text{sinc}(x) = \sin(\pi x)/\pi x$.
(a) Sketch $w(t,s)$ and $\delta(t-s)$.
(b) Show that $w(t,s)$ is zero for $s = t - \frac{2}{f_1}$, $t - \frac{3}{f_1}$, ...(lines parallel to diagonal $s = t$) and for $s = \frac{1}{f_2}, \frac{2}{f_2}, \ldots$ (lines orthogonal to s-axis).

Solution:
(a) $w(t,s)$ is shown in Fig. 9.1a. For the moment please ignore the three planes at s_1, s_2, and s_3. Apparently, the Dirac impulse $\delta(t-s)$ is a Dirac plane along the diagonal $s = t$ in the tree-dimensional space.

The impulse response is causal, because $w(t,s) = 0$ for $t < s$ holds. We also recognize $w(t,s) \to 0$ for $s, t \to \infty$.

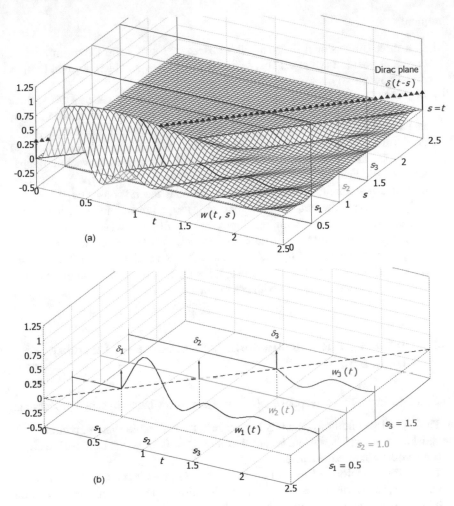

Fig. 9.1 **a** Time-variant impulse response $w(t, s)$ given by (9.7) with $f_1 = 4$, $f_2 = 1$ and stimulating Dirac plane $\delta(t - s)$ **b** Isolated impulse responses $w_i(t)$ as cutting curves of the surface $w(t, s)$ with the planes at $s = s_i$, $i = 1, 2, 3$, in **a**

(b) The zeros are obtained from the conditions $f_1(t - s - \frac{1}{f_1}) = 1, 2, 3, \ldots$, and $f_2 s = 1, 2, 3, \ldots$. e.g., $w(t, s) = 0$ for $s = s_2 = 1$ is clearly visible.

Example 3
The system with time-variant impulse response (9.7) shall be stimulated by an isolated Dirac impulse δ_1 at $t = s_1$ and $s = s_1$.

(a) Determine δ_1.
(b) Find the isolated response $w_1(t) = w(t, s_1)$ to δ_1 and sketch it as a function of t in a two-dimensional diagram.

Solution:

(a) In a three-dimensional diagram with abscissas t and s the Dirac impulse δ_1 occurs at position (s_1, s_1). Thus, $\delta_1 = \delta(t - s_1, s - s_1)$, which is equal to $\delta_1 = \delta(t - s_1)\delta(s - s_1)$.

(b) In principle, the isolated response $w_1(t)$ is obtained by cutting the surface $w(t, s)$ with the plane $s = s_1$, which is shown in Fig. 9.1a at location $s = s_1$ in parallel to the front plane given by the ordinate and the t-axis. $w_1(t)$ is indicated by a curve in bold. In the same way the cutting curves $w_2(t)$ and $w_3(t)$ are created and depicted in Fig. 9.1b, which shows three separate two-dimensional diagrams with parameters s_1, s_2, and s_3. As expected from a time-variant system, we clearly see that the responses $w_1(t)$, $w_2(t)$, and $w_3(t)$ are different, in particular $w_2(t) = 0$ holds. Mathematically, we can use (9.6) and get with the input $x(\zeta) = \delta(\zeta - s_i)\delta(s - s_i)$ the output $y_i(t) = \int_{-\infty}^{\infty} \delta(\zeta - s_i)\delta(s - s_i)w(t, \zeta)d\zeta = w(t, s_i)\delta(s - s_i)$. Thus, each cutting plane can be defined by a Dirac plane $\delta(s - s_i)$ and the cutting curves then are $w(t, s)\delta(s - s_i) = w(t, s_i)\delta(s - s_i)$, $i = 1, 2, 3$, if Fig. 9.1b shall represent a three dimensional diagram with continuous s.

9.2   Modified Time-Variant Impulse Response—The Delay Spread Function

The time-variant impulse response $w(t, s)$ has some drawbacks in signal processing.

- The causality condition, which is important for a realizable system, requires two variables, t and s. The system is causal, if $w(t, s) = 0$ for $t < s$, because the effect cannot occur before the cause.
- The Fourier transform of $w(t, s)$ does not directly provide meaningful frequency responses, such as the time-variant transfer function or the Doppler spread function of the system, which are introduced in Chap. 11.
- Furthermore, a minor point is as follows. If the system is time-invariant, the input-output relation (9.6) does not turn elegantly into the conventional convolution integral by just dropping one variable of $w(t, s)$, as will be the case for the delay spread function discusses next.

To overcome these drawbacks the transformation of variables (8.4) is proposed. With $s = t - \tau$ follows from (9.6)

$$y(t) = \int_{-\infty}^{\infty} x(t - \tau)w(t, t - \tau)d\tau \tag{9.8}$$

We introduce the *modified time-variant impulse response* also called *delay spread function* $g(t, \tau)$ as

$$w(t, t - \tau) = g(t, \tau) \tag{9.9}$$

and with $\tau = t - s$ follows on the other hand

$$w(t, s) = g(t, t - s) \tag{9.10}$$

With (9.9) we get from (9.8)

$$y(t) = \int_{-\infty}^{\infty} x(t - \tau)g(t, \tau)d\tau = \int_{-\infty}^{\infty} x(\zeta)g(t, t - \zeta)d\zeta = x(t) \circledast g(t, \tau) \tag{9.11}$$

where the substitution $\zeta = t - \tau$ was used for the second integral. (9.8) and (9.11) are equivalent, because they provide the same output signal $y(t)$.

Important Remark

Please note that the time-variant convolutions $x(t) \circledast w(t, s)$ in (9.6) and $x(t) \circledast g(t, \tau)$ in (9.11) are executed by slightly different integral operators. Nevertheless, we use the same symbol $\circledast$ just to simplify the notation. What operator has to be used will become clear from the context.

Finally, we see from (9.11) that the third drawback listed before can be overcome with $g(t, \tau)$. If the system is time-invariant, we can just skip the first argument t in $g(t, \tau)$. Then the delay spread function turns into the impulse response $g(\tau)$. We can rename τ by t and write $g(t)$. Hence, from (9.11) follows the well known input-output relation of the time-invariant convolution

$$y(t) = \int_{-\infty}^{\infty} x(t - \tau)g(\tau)d\tau = \int_{-\infty}^{\infty} x(\zeta)g(t - \zeta)d\zeta = x(t) * g(t) \tag{9.12}$$

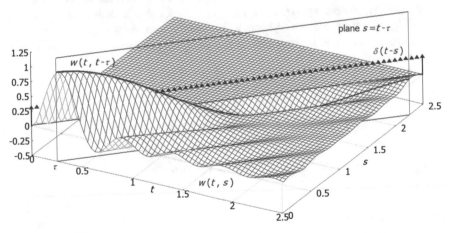

Fig. 9.2 Time-variant impulse response $w(t, s)$ for $f_1 = 4$ and $f_2 = 1$ cut by the plane $s = t - \tau$ to generate $w(t, t - \tau) = g(t, \tau)$

Example 4

We consider again the time-variant impulse response $w(t, s)$ in (9.7).

(a) Find the delay spread function $g(t, \tau)$.
(b) Show the interrelation between $w(t, s)$ and $g(t, \tau)$. Sketch the cutting curves.

Solution:

(a) Plugging $s = t - \tau$ into (9.7) yields

$$g(t, \tau) = \begin{cases} \text{sinc}\left(f_1(\tau - \frac{1}{f_1})\right) \text{sinc}\left(f_2(t - \tau)\right) & ; \ \tau \geqslant 0 \\ 0 & ; \ else \end{cases} \tag{9.13}$$

Figure 9.2 shows the surface of the time-variant impulse response $w(t, s)$, which is cut by the plane $s = t - \tau$. By varying τ the delay spread function $g(t, \tau)$ results and is depicted in Fig. 9.3a. Please ignore the diagonal planes there at the moment. As expected from (9.13), we recognize $g(t, \tau)$ approaching zero for t, $\tau \to \infty$. We also see that the diagonal Dirac plane $\delta(t - s)$ in Fig. 9.2 has turned into the Dirac plane $\delta(\tau)$ along the t-axis. Apparently, the distributed Dirac impulses along the Dirac plane are all active at the same delay $\tau = 0$ rather than at different delays along the diagonal $s = t - \tau$ as in Fig. 9.2. This is a major advantage of $g(t, \tau)$ compared to $w(t, s)$ for the modeling of time-variant systems.

(b) Now we consider the three diagonal planes along $\tau = t - s_i$ in Fig. 9.3a, which cut out the curves $g(t, t - s_i)$, $i = 1, 2, 3$, indicated in bold, from the surface $g(t, \tau)$. The isolated cutting curves are shown in Fig. 9.3b in detail. From (9.10) we conclude that $g(t, t - s_i) = w(t, s_i)$ must hold. The reader can convince oneself by comparing $g(t, t - s_i)$ in Fig. 9.3b for $\tau \geqslant 0$ with $w_i(t) = w(t, s_i)$ in Fig. 9.1b for $t \geqslant s_i$. In essence, this example again demonstrates that a time-variant system can be described by the delay spread function $g(t, \tau)$ owing to the direct relation (9.10).

9.3 Discrete-Time, Time-Variant System

9.3.1 Discrete-Time Delay Spread Function

We consider the discrete-time system in Fig. 9.4, which is a discrete-time tapped delay-line filter also called a finite impulse response (FIR) filter. The discrete-time input and output signals are denoted as $x_s(k)$ and $y_s(k)$, respectively. The index s stands for sampled. The $K + 1$ taps are the filter coefficients indicated by $g_s(k, m)$ with $k, m \in \mathbb{Z}$ and $g_s(k, m) = 0$ for $m > K$ and $m < 0$. Obviously, all parts in the filter are linear namely the delay elements with the delay Δ_τ, the multiplication of

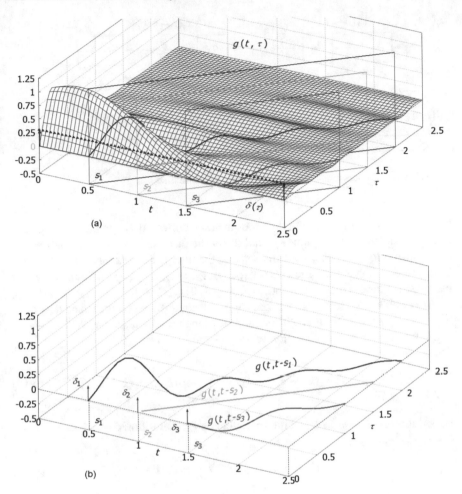

Fig. 9.3 a Delay spread function $g(t, \tau)$ given in (9.13) with $f_1 = 4$ and $f_2 = 1$ as well as stimulating Dirac plane $\delta(\tau)$ **b** Isolated delay spread functions $g(t, t - s_i)$ as cutting curves of the surface $g(t, \tau)$ with the diagonal planes $\tau = t - s_i, i = 1, 2, 3$, in **a**

the delayed inputs $x_s(k - m)$ with the filter coefficients yielding $g_s(k, m)x_s(k - m)$, and the addition of all component signals. Consequently, we get the output signal

$$y_s(k) = \sum_{m=0}^{K} g_s(k, m)x_s(k - m) = x_s(k) \circledast g_s(k, m) \qquad (9.14)$$

The reader is encouraged to convince oneself of the linearity by using the superposition of two input signals. We denote the relation (9.14) as the discrete-time, time-variant convolution between $g_s(k, m)$ and $x_s(k)$ and allocate the the symbol $\circledast$

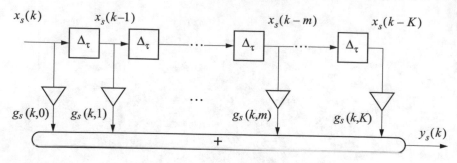

Fig. 9.4 Discrete-time, time-variant finite impulse response filter

to prevent from confusion with the time-invariant convolution operation $*$. For simplicity of the notation the same symbol $\circledast$ for the discrete-time and the continuous-time convolution is used. We clearly see that the filter is time-variant, because at the time instant k the set of coefficients is $g_s(k, 0), g_s(k, 1), \ldots, g_s(k, K)$ and at the next instant $k + 1$ the coefficients change to $g_s(k + 1, 0), g_s(k + 1, 1), \ldots, g_s(k + 1, K)$.

As is well known from signal processing, a discrete-time system can be described in general by a recursive filter with an infinitely long impulse response. With $K \to \infty$ and if we also allow non-causality, $m \to -\infty$, follows from (9.14)

$$y_s(k) = x_s(k) \circledast g_s(k, m) = \sum_{m=-\infty}^{\infty} g_s(k, m) x_s(k - m) \qquad (9.15)$$

By substituting $k - m = \zeta$ and finally replacing ζ by m follows

$$y_s(k) = x_s(k) \circledast g_s(k, m) = \sum_{m=-\infty}^{\infty} g_s(k, k - m) x_s(m) \qquad (9.16)$$

Now we define the unit impulse

$$\delta_k = \begin{cases} 1 ; & k = 0 \\ 0 ; & k = \pm 1, \pm 2, \ldots \end{cases} \qquad (9.17)$$

and apply at the input of the filter a unit impulse active at time instant $k_0 - m$, i.e. $x_s(k) = \delta_{k-(k_0-m)}$. The observed output $y_s(k)$ at time instant $k \geq k_0 - m$ follows from (9.15)

$$y_s(k) = \sum_{\nu=-\infty}^{\infty} g_s(k, \nu) x_s(k - \nu) = \sum_{\nu=-\infty}^{\infty} g_s(k, \nu) \delta_{k-(k_0-m)-\nu} = g_s(k, k - k_0 + m)$$

$$(9.18)$$

For $k_0 \rightarrow k$ the unit impulse at the input is active at time instant $k - m$ and the response observed at instant k follows from (9.18)

$$y_s(k) = g_s(k, m) \tag{9.19}$$

Hence, $g_s(k, m)$ can be considered as the response of the discrete-time, time-variant system at observation time k to a unit impulse applied at instant $k - m$, i.e. m time intervals earlier. According to our previous nomenclature for continuous-time systems we can call $g_s(k, m)$ as *discrete-time modified impulse response* or *discrete-time delay spread function*. A causal system cannot react before the instant of the unit impulse, i.e. at $k < k - m$, which is equivalent to $m < 0$. Thus

$$g_s(k, m) = 0 \, \forall \, m < 0 \tag{9.20}$$

must hold for a causal time-variant system.

We see that the discrete-time, time-variant system is fully defined by its delay spread function $g_s(k, m)$. The two variables $k \in \mathbb{Z}$ and $m \in \mathbb{Z}$ specify two different time coordinates. k characterizes the time variable of the input and output signal, in addition also the temporal change of the system parameters $g_s(k, m)$. The variable m is called delay time or just delay, because the observation instant is m time intervals later than the active unit impulse. Sometimes m is also referred to as integration time, because the summation in (9.15) and (9.16) is taken over m.

To grasp the mechanism of a time-variant system more closely assume that we would like to measure the coefficients $g_s(k, m)$ of the FIR filter in Fig. 9.4. In principle they can be found with a series of measurements using (9.19). To get $g_s(k, m)$ as a function of m, we can measure the output at a fixed instant k and change the instant $k - m$ of the unit impulse by varying $m = 0, 1, 2, \ldots, K$. The result are output samples $y_s(k) = g_s(k, 0)$ up to $g_s(k, K)$. In the next round the same procedure has to be executed for $k + 1$ with the measurement of $y_s(k + 1) = g_s(k + 1, 0)$ up to $g_s(k + 1, K)$ and so on.

This is quite in contrast to a time-invariant filter with coefficients $g_s(k, m) = g_s'(m)$, which are independent of k. As a consequence (9.14) turns into

$$y_s(k) = \sum_{m=0}^{K} g_s'(m) x_s(k - m) = x_s(k) * g_s'(k) \tag{9.21}$$

which is the well known convolution operation for linear, discrete-time, time-invariant systems. An example is a wire-line channel or a static wireless channel, where the transmitter and the receiver are non-moving and all scatterings and reflections of the electromagnetic waves do not change with time k. To find the time-invariant coefficients $g_s'(k)$ only one unit impulse $x_s(k) = \delta_k$ suffices, yielding from (9.21) the output sequence

$$y_s(k) = g_s'(k) \tag{9.22}$$

which is called the impulse response.

If we apply $x_s(k) = \delta_k$ to the time-variant system, we readily deduce from (9.15) that $y_s(k) = g_s(k, k)$, which does not completely determine the system, because $g_s(k, k)$ is just the delay spread function evaluated along the axis $m = k$. Therefore $g_s(k, m)$ should not be called impulse response.

9.3.2 Transition to Continuous-Time Delay Spread Function

In the following we extend our considerations to continuous-time, time-variant systems. For that purpose we interpret $g_s(k, m)$ as the result of a two-dimensional sampling of the continuous function $g(t, \tau)$ at $t = k\Delta_t$ and $\tau = m\Delta_\tau$ with $m, k \in \mathbb{Z}$. Δ_t and Δ_τ are the sampling intervals for the time t and delay-time τ, respectively. Hence

$$g(k\Delta_t, m\Delta_\tau) = g_s(k, m) \tag{9.23}$$

and $g(t, \tau)$ can be ideally reconstructed from $g_s(k, m)$, if the sampling theorem is fulfilled. The continuous-time input signal of the continuous-time system shall be $x(t)$ and the relation with the discrete-time version is

$$x(k\Delta_t) = x_s(k) \tag{9.24}$$

and similarly for the output signal $y(t)$

$$y(k\Delta_t) = y_s(k) \tag{9.25}$$

Roughly and without going into all mathematical details for $\Delta_t \to 0$ and $\Delta_\tau \to 0$ the discrete-time variables k and m are approaching the continuous-time variables t and τ, respectively. Moreover, the summations in (9.15) and (9.16) turn into integrals. Thus, the discrete-time, time-variant convolution becomes the continuous-time, time-variant convolution integral

$$y(t) = x(t) \circledast g(t, \tau) = \int_{-\infty}^{\infty} g(t, \tau)x(t - \tau)d\tau \tag{9.26}$$

where the same symbol $\circledast$ as in (9.15) is used to simplify notation. By changing the integration variable $\tau = t - \zeta$ we obtain alternatively

$$y(t) = x(t) \circledast g(t, \tau) = \int_{-\infty}^{\infty} g(t, t - \zeta)x(\zeta)d\zeta \tag{9.27}$$

From these equations it is quickly determined that $g(t, \tau)$ is the response at observation time instant t to an input Dirac impulse at time instant $t - \tau$.

For the proof we apply a Dirac impulse at $t_0 - \tau$, i.e. the input signal is $x(t) = \delta(t - (t_0 - \tau))$, and obtain from (9.26)

$$y(t) = \int_{-\infty}^{\infty} g(t, \eta) \delta(t - (t_0 - \tau) - \eta)\, d\eta = g(t, t - t_0 + \tau) \qquad (9.28)$$

For $t_0 \to t$ the Dirac impulse is active at $t - \tau$, which yields the output signal

$$y(t) = g(t, \tau) \qquad (9.29)$$

and the statement is proven.

Referring to the substitution $s = t - \tau$ of variables in (8.4) $g(t, \tau)$ is found by cutting the surface of the two-dimensional impulse response $w(t, s)$ with the plane $\tau = t - s$. Hence $g(t, \tau)$ is a two-dimensional function of t and the new axis $\tau = t - s$. Furthermore, we conclude from (9.20) for a causal delay spread function

$$g(t, \tau) = 0 \,\forall\, \tau < 0 \qquad (9.30)$$

as illustrated in Fig. 9.3.

We still owe some explanation how to find $w(t, s)$ practically. To this end, let us stimulate the system with a Dirac impulse at $t = s$, consider the response $w(t, s)$ as a function of t and continue this procedure for a large variety of parameters s, $-\infty < s < \infty$. Here again we recognize the main difference to an impulse response of a time-invariant system, for which an input Dirac impulse at a single s, in particular $s = 0$, suffices. Please note that according to (9.9) a Dirac impulse with $s = 0$ at the input of a time-variant system would just result in an output response $y(t) = w(t, 0) = g(t, t)$ and the determination of the system would not be complete.

Chapter 10
Properties of Time-Variant Convolution

In this chapter we proof some important properties of the time-variant convolution in detail. The results are also summarized in Tables 10.1 and 10.2.

10.1 Relation Between Time-Variant and Time-Invariant Convolution

It is interesting to check whether the time-variant convolution encompasses the time-invariant convolution. By definition, the impulse response $h(t)$ of a time-invariant system is the response to an input Dirac impulse $\delta(t)$. If the system is excited by the time-shifted Dirac impulse $\delta(t-s)$, the response $h(t-s)$ also incorporates a shift without changing its original shape. This holds for any s. Therefore $w(t, s)$ turns into a one-dimensional function of t

$$w(t, s) = w(t - s) = h(t - s) \tag{10.1}$$

where only one parameter value $s = 0$ suffices resulting in the impulse response $h(t)$. Consequently, the time-variant convolution (9.6) migrates to the time-invariant convolution as

$$
\begin{aligned}
y(t) &= x(t) \circledast w(t, s) = \int_{-\infty}^{\infty} x(s)w(t, s)ds \\
&= \int_{-\infty}^{\infty} x(s)w(t - s)ds = \int_{-\infty}^{\infty} x(s)h(t - s)ds = x(t) * h(t)
\end{aligned}
\tag{10.2}
$$

and we recognize that we can just replace $w(t, s)$ by $w(t - s)$ in (9.6), if the system is time-invariant. Thus, the name "general convolution" mentioned earlier for (8.1) and (9.6) is justified.

© Springer Nature Switzerland AG 2021
J. Speidel, *Introduction to Digital Communications*,
Signals and Communication Technology,
https://doi.org/10.1007/978-3-030-67357-4_10

The relation (10.2) can also be formulated with the delay spread function $g(t, \tau) = h(\tau)$ using the substitution $t - s = \tau$

$$x(t) \circledast g(t, \tau) = x(t) \circledast h(\tau) = x(t) * h(t) \tag{10.3}$$

Proof of (10.3)

The time-invariant system shall be described in general by the delay spread function $g(t, \tau) = h(t - s) = h(\tau)$. The time-variant convolution (9.11) yields $x(t) \circledast g(t, \tau) = \int_{-\infty}^{\infty} g(t, \tau)x(t - \tau)d\tau$. As is well known, the time-invariant convolution is defined as $x(t) * h(t) = \int_{-\infty}^{\infty} h(\tau)x(t - \tau)d\tau$. Both integrals are identical, if $g(t, \tau) = h(\tau)$. As any one-dimensional function $h(x)$ can be defined with any variable, $x = \tau$ or $x = t$, the proof of (10.3) is complete.

10.2   Properties

Linearity

The time-variant convolution (9.6) is *linear*. Let

$$x(t) = a_1 x_1(t) + a_2 x_2(t) \tag{10.4}$$

and

$$y_i(t) = x_i(t) \circledast w(t, s) \; ; \; i = 1, 2 \tag{10.5}$$

then

$$y(t) = a_1 y_1(t) + a_2 y_2(t) \tag{10.6}$$

The proof is straightforward, because the convolution integral represents a linear operation.

Convolution of Two Time-Variant Impulse Responses

As depicted in Fig. 10.1a, we can model the convolution of two time-variant impulse responses $w_1(t, s)$ and $w_2(t, s)$ by a cascade of two time-variant systems. The overall impulse response shall be $w_{12}(t, s)$. In the upper line we have indicated the input signal $x_0(t)$ as well as the output signals $x_1(t)$ and $x_2(t)$. Using (9.6) we get

$$x_i(t) = x_{i-1}(t) \circledast w_i(t, s) \; ; \; i = 1, 2 \tag{10.7}$$

In the lower line of Fig. 10.1a we indicate the input $x_0(t) = \delta(t - s)$ resulting in the impulse responses $w_1(t, s)$ and $w_{12}(t, s)$. We proof that the system possesses the overall impulse response

Fig. 10.1 Cascade of two time-variant systems **a** with impulse responses $w_1(t, s)$ and $w_2(t, s)$, **b** with delay spread functions $g_1(t, \tau)$ and $g_2(t, \tau)$

$$w_{12}(t, s) = w_1(t, s) \circledast w_2(t, s) = \int_{-\infty}^{\infty} w_1(\zeta, s)w_2(t, \zeta)d\zeta \qquad (10.8)$$

which is the time-variant convolution of the individual impulse responses. It is readily appreciated that the convolution integral (9.6) can be directly applied by interpreting $w_1(t, s)$ as a function of t with a parameter s, thus taking over the role of $x(\zeta)$ in (9.6).

Proof of (10.8)

Let $w_{12}(s, t)$ denote the overall impulse response. Then the response to the input signal $x_0(t)$ is according to (9.6)

$$x_2(t) = \int_{-\infty}^{\infty} x_0(s)w_{12}(t, s)ds \qquad (10.9)$$

Similarly we obtain $x_1(t) = \int_{-\infty}^{\infty} x_0(s)w_1(t, s)ds$ and $x_2(t) = \int_{-\infty}^{\infty} x_1(\zeta)w_2(t, \zeta)d\zeta$. Inserting $x_1(t)$ yields $x_2(t) = \int_{-\infty}^{\infty} \int_{-\infty}^{\infty} x_0(s)w_1(\zeta, s)w_2(t, \zeta)dsd\zeta$. Assuming that the integrals converge absolutely we can interchange the order of integration and get $x_2(t) = \int_{-\infty}^{\infty} x_0(s) \left[\int_{-\infty}^{\infty} w_1(\zeta, s)w_2(t, \zeta)d\zeta \right] ds$. From the comparison with (10.9) we conclude that the inner integral is $w_{12}(t, s)$ yielding the result $w_{12}(t, s) = \int_{-\infty}^{\infty} w_1(\zeta, s)w_2(t, \zeta)d\zeta$ and the proof ends.

Convolution of Two Delay Spread Functions

In Fig. 10.1b the two time-variant systems are characterized by their delay spread functions

$$g_i(t, \tau) = w_i(t, t - \tau); \; i = 1, 2; \; g_{12}(t, \tau) = w_{12}(t, t - \tau) \qquad (10.10)$$

We proof that the overall delay spread function is

$$g_{12}(t, \tau) = g_1(t, \tau) \circledast g_2(t, \tau) = \int_{-\infty}^{\infty} g_1(t - \eta, \tau - \eta)g_2(t, \eta)d\eta \qquad (10.11)$$

Please note, to simplify notation we use the same symbol $\circledast$ for the convolution of the impulse responses in (10.8) and the delay spread functions in (10.11) although the integral notations are different.

Proof of (10.11)

We start from (10.8) and obtain with $s = t - \tau$ and (10.10) for the left-hand side $w_{12}(t, t - \tau) = g_{12}(t, \tau) = g_1(t, \tau) \circledast g_2(t, \tau)$. For the integral in (10.8) the substitution $\zeta = t - \eta$, $d\zeta = -d\eta$ yields
$g_{12}(t, \tau) = \int_{-\infty}^{\infty} w_1(t - \eta, t - \tau) w_2(t, t - \eta) d\eta$. The term $w_1(t - \eta, t - \tau)$ has to be changed to the form $w_1(x, x - u) = g_1(x, u)$ as follows,
$w_1(t - \eta, t - \tau) = w_1(t - \eta, t - \eta - \tau + \eta) = w_1(t - \eta, \ t - \eta - (\tau - \eta))$
resulting in

$$w_1(t - \eta, t - \tau) = g_1(t - \eta, \tau - \eta) \tag{10.12}$$

Then follows with $w_2(t, t - \eta) = g_2(t, \eta)$ the result
$g_{12}(t, \tau) = \int_{-\infty}^{\infty} g_1(t - \eta, \tau - \eta) g_2(t, \eta) d\eta$ and the proof is finished.

Commutativity

- The time-variant convolution of two time-variant impulse responses $w_1(t, s)$ and $w_2(t, s)$ is in general *non-commutative*.

$$w_1(t, s) \circledast w_2(t, s) \neq w_2(t, s) \circledast w_1(t, s) \tag{10.13}$$

- The time-variant convolution of two delay spread functions $g_1(t, \tau)$ and $g_2(t, \tau)$ is in general *non-commutative*.

$$g_1(t, \tau) \circledast g_2(t, \tau) \neq g_2(t, \tau) \circledast g_1(t, \tau) \tag{10.14}$$

- From an engineering point of view we conclude that a change of the sequential arrangement of the two time-variant systems in Figs. 10.1a, b will result in different overall responses in general.
- The time-variant convolution of two impulse responses $w_1(t, s) = h_1(t - s)$ and $w_2(t, s) = h_2(t - s)$ of *time-invariant* systems is *commutative*.

$$h_1(t - s) \circledast h_2(t - s) = h_2(t - s) \circledast h_1(t - s) \tag{10.15}$$

As can be seen, the time-variant convolution boils down to the time-invariant convolution, where $s = 0$ suffices.

$$h_1(t) * h_2(t) = h_2(t) * h_1(t) \tag{10.16}$$

With $s = t - \tau$ we can also conclude from (10.15)

$$h_1(\tau) \circledast h_2(\tau) = h_2(\tau) \circledast h_1(\tau) \tag{10.17}$$

- Please note, if we consider an arrangement with time-variant and time-invariant systems, $h(\tau)$ characterizes the delay spread function of a time-invariant system and not a signal such as $x(t)$. In case of solely time-invariant systems the impulse response $h(t)$ can be treated just as a signal and the convolution operation is commutative, $x(t) * h(t) = h(t) * x(t)$. Thus, for time-variant systems a differentiation between a signal and an impulse response or delay spread function is required before executing the convolution. The following statement is an illustration:

• The time-variant convolution of a time-variant impulse response $w(t, s)$ and a signal $x(t)$ in general is *non-commutative*.

$$x(t) \circledast w(t, s) \neq w(t, s) \circledast x(t) \tag{10.18}$$

or equivalently

$$x(t) \circledast g(t, \tau) \neq g(t, \tau) \circledast x(t) \tag{10.19}$$

The proof is quite straightforward, because the two integrals $x(t) \circledast w(t, s) = \int_{-\infty}^{\infty} x(\zeta) w(t, \zeta) d\zeta = y_1(t)$ and $w(t, s) \circledast x(t) = \int_{-\infty}^{\infty} w(\zeta, s) x(\zeta) d\zeta = y_2(s)$ are executed over different variables of $w(t, s)$. Anyhow, $w(t, s) \circledast x(t)$ makes not much sense technically, as a system will only be cascaded by another one and not by a signal.

Proof of (10.13)

For the left-hand side we obtain $w_{12}(t, s) = \int_{-\infty}^{\infty} w_1(\zeta, s) w_2(t, \zeta) d\zeta$ according to (10.8). For the right-hand side follows $w_{21}(t, s) = \int_{-\infty}^{\infty} w_2(\eta, s) w_1(t, \eta) d\eta$. We see that in general both integrals are different. e.g., in $w_{12}(t, s)$ the integration is done over the first variable of $w_1(t, s)$ whereas in $w_{21}(t, s)$ the integration is executed with respect to the second variable of $w_1(t, s)$.

Proof of (10.14)

The proof is similar as before. For the left-hand side of (10.14) we obtain $g_{12}(t, \tau) = \int_{-\infty}^{\infty} g_1(t - \eta, \tau - \eta) g_2(t, \eta) d\eta$ according to (10.11). For the right-hand side follows $g_{21}(t, \tau) = \int_{-\infty}^{\infty} g_2(t - \eta, \tau - \eta) g_1(t, \eta) d\zeta$. We see that in general both integrals are different.

Proof of (10.15) and (10.16)

From (10.8) follows for the left-hand side $h_{12}(t - s) = h_1(t - s) \circledast h_2(t - s) = \int_{-\infty}^{\infty} h_1(\zeta - s) h_2(t - \zeta) d\zeta = \int_{-\infty}^{\infty} h_1(\eta) h_2(t - s - \eta) d\eta$, where we have used the substitution $\zeta - s = \eta$. The right-hand side of (10.15) is given by $h_{21}(t - s) = h_2(t - s) \circledast h_1(t - s) = \int_{-\infty}^{\infty} h_2(\zeta - s) h_1(t - \zeta) d\zeta = \int_{-\infty}^{\infty} h_1(\eta) h_2(t - s - \eta) d\eta$, where the substitution $t - \zeta = \eta$ was used. Obviously, $h_{12}(t - s) = h_{21}(t - s)$ is true. We also see that we get an overall impulse response which is just delayed by s, which is not surprising, because at the input of the system the delayed Dirac impulse $\delta(t - s)$ is active. For $s = 0$ we directly obtain the time-invariant convolution (10.16). This finalizes the proof.

Associativity of Time-Variant Convolution

- The time-variant convolution of a signal $x(t)$ with a time-variant impulse response is *associative*,

$$[x(t) \circledast w_1(t, s)] \circledast w_2(t, s) = x(t) \circledast [w_1(t, s) \circledast w_2(t, s)] \qquad (10.20)$$

- The time-variant convolution of a signal $x(t)$ with a delay spread functions is *associative*,

$$[x(t) \circledast g_1(t, \tau)] \circledast g_2(t, \tau) = x(t) \circledast [g_1(t, \tau) \circledast g_2(t, \tau)] \qquad (10.21)$$

- The time-variant convolution of time-variant impulse responses is *associative*,

$$[w_0(t, s) \circledast w_1(t, s)] \circledast w_2(t, s) = w_0(t, s) \circledast [w_1(t, s) \circledast w_2(t, s)] \qquad (10.22)$$

- The time-variant convolution of delay spread functions is *associative*,

$$[g_0(t, \tau) \circledast g_1(t, \tau)] \circledast g_2(t, \tau) = g_0(t, \tau) \circledast [g_1(t, \tau) \circledast g_2(t, \tau)] \qquad (10.23)$$

- The associativity also holds, if one or more systems are time-invariant characterized by $w_i(t, s) = w_i(t - s)$ and $g_i(t, \tau) = g_i(\tau)$, $i = 1, 2$, respectively.

Proof of (10.20) and (10.21)

For the left-hand-side of (10.20) we get in a first step
$x(t) \circledast w_1(t, s) = \int_{-\infty}^{\infty} x(\zeta) w_1(t, \zeta) d\zeta = y_1(t)$ and in a second step
$u_1(t) = y_1(t) \circledast w_2(t, s) = \int_{-\infty}^{\infty} y_1(\eta) w_2(t, \eta) d\eta =$
$\iint_{-\infty}^{\infty} x(\zeta) w_1(\eta, \zeta) w_2(t, \eta) d\eta d\zeta$. Similarly for the right-hand side we obtain
$w_1(t, s) \circledast w_2(t, s) = \int_{-\infty}^{\infty} w_1(\eta, s) w_2(t, \eta) d\eta = w_{12}(t, s)$ and
$u_2(t) = x(t) \circledast w_{12}(t, s) = \int_{-\infty}^{\infty} x(\zeta) w_{12}(t, \zeta) d\zeta =$
$\iint_{-\infty}^{\infty} x(\zeta) w_1(\eta, \zeta) w_2(t, \eta) d\eta d\zeta$. A comparison shows that $u_1(t) = u_2(t)$ and the proof of (10.20) is finished.
With $s = t - \tau$ we obtain $w_i(t, s) = w_i(t, t - \tau) = g_i(t, \tau)$; $i = 1, 2$. Then from (10.20) directly follows (10.21), which finalizes the proof of (10.21).

Proof of (10.22) and (10.23)

For the left-hand side of (10.22) we get
$w_0(t, s) \circledast w_1(t, s) = \int_{-\infty}^{\infty} w_0(\zeta, s) w_1(t, \zeta) d\zeta = y_1(t, s)$ and
$u_1(t, s) = y_1(t, s) \circledast w_2(t, s) = \int_{-\infty}^{\infty} y_1(\eta, s) w_2(t, \eta) d\eta =$
$\iint_{-\infty}^{\infty} w_0(\zeta, s) w_1(\eta, \zeta) w_2(t, \eta) d\eta d\zeta$.
Similarly the right-hand side yields with $w_{12}(t, s)$ from proof (10.20)
$u_2(t, s) = w_0(t, s) \circledast w_{12}(t, s) = \int_{-\infty}^{\infty} w_0(\zeta, s) w_{12}(t, \zeta) d\zeta =$
$\iint_{-\infty}^{\infty} w_0(\zeta, s) w_1(\eta, \zeta) w_2(t, \eta) d\eta d\zeta$. Obviously, $u_1(t, s) = u_2(t, s)$ holds and we conclude that the left and the right-hand side of (10.22) are identical, which finalizes the proof. The proof of (10.23) is similar to (10.21).

Distributivity of Time-Variant Convolution

The time-variant convolution is *distributive*,

$$w_0(t, s) \circledast [w_1(t, s) + w_2(t, s)] = [w_0(t, s) \circledast w_1(t, s)] + [w_0(t, s) \circledast w_2(t, s)] \tag{10.24}$$

$$g_0(t, \tau) \circledast [g_1(t, \tau) + g_2(t, \tau)] = [g_0(t, \tau) \circledast g_1(t, \tau)] + [g_0(t, \tau) \circledast g_2(t, \tau)] \tag{10.25}$$

The proof is straightforward, because $\circledast$ is a linear operator.

10.3 Summary

Table 10.1 summarizes important convolution operations and properties between impulse responses of time-variant and/or time-invariant systems. With the substitution $s = t - \tau$ of (8.4) the corresponding relations for the delay spread functions of the system are summarized in Table 10.2.

The convolution integrals in Table 10.2 for $g_1(\tau) \circledast g_2(t, \tau)$ and $g_1(t, \tau) \circledast g_2(\tau)$ are proven together with their Fourier transforms in Sect. 11.5. We recognize that we

Table 10.1 Summary of time-variant convolution and properties. Impulse response $w_i(t, s)$ of time-variant system, impulse response $w_i(t - s)$ of time-invariant system, input $x(t)$ and output signal $y(t)$, $i = 0, 1, 2$

$y(t) = x(t) \circledast w_i(t, s)$	$=$	$\int_{-\infty}^{\infty} x(s) w_i(t, s) ds$
$y(t) = x(t) \circledast w_i(t - s)$	$=$	$\int_{-\infty}^{\infty} x(s) w_i(t - s) ds$
$w_1(t, s) \circledast w_2(t, s)$	$=$	$\int_{-\infty}^{\infty} w_1(\zeta, s) w_2(t, \zeta) d\zeta$
$w_1(t - s) \circledast w_2(t, s)$	$= \int_{-\infty}^{\infty} w_1(\zeta - s) w_2(t, \zeta) d\zeta = \int_{-\infty}^{\infty} w_1(\eta) w_2(t, \eta + s) d\eta$	
$w_1(t, s) \circledast w_2(t - s)$	$= \int_{-\infty}^{\infty} w_1(\zeta, s) w_2(t - \zeta) d\zeta = \int_{-\infty}^{\infty} w_1(t - \eta, s) w_2(\eta) d\eta$	
$w_1(t - s) \circledast w_2(t - s)$	$=$	$\int_{-\infty}^{\infty} w_1(\zeta - s) w_2(t - \zeta) d\zeta$
$x(t) \circledast w_i(t, s)$	$\neq$	$w_i(t, s) \circledast x(t)$
$w_1(t, s) \circledast w_2(t, s)$	$\neq$	$w_2(t, s) \circledast w_1(t, s)$
$w_1(t - s) \circledast w_2(t - s)$	$=$	$w_2(t - s) \circledast w_1(t - s)$
$[x(t) \circledast w_1(t, s)] \circledast w_2(t, s)$	$=$	$x(t) \circledast [w_1(t, s) \circledast w_2(t, s)]$
$[w_0(t, s) \circledast w_1(t, s)] \circledast w_2(t, s)$	$=$	$w_0(t, s) \circledast [w_1(t, s) \circledast w_2(t, s)]$
$w_0(t, s) \circledast [w_1(t, s) + w_2(t, s)]$	$=$	$[w_0(t, s) \circledast w_1(t, s)] + [w_0(t, s) \circledast w_2(t, s)]$

Table 10.2 Summary of time-variant convolution and properties. Delay spread functions $g_i(t, \tau) = w_i(t, t - \tau)$ of time-variant system, delay spread function $g_i(\tau) = w_i(t - s)$ of time-invariant system, input $x(t)$ and output signal $y(t)$, $i = 0, 1, 2$

$$
\begin{array}{rcl}
y(t) = x(t) \circledast g_i(t, \tau) &=& \int_{-\infty}^{\infty} x(t - \eta) g_i(t, \eta) d\eta = \int_{-\infty}^{\infty} x(\zeta) g_i(t, t - \zeta) d\zeta \\
y(t) = x(t) \circledast g_i(\tau) &=& \int_{-\infty}^{\infty} x(t - \tau) g_i(\tau) d\tau
\end{array}
$$

$$
\begin{array}{rcl}
g_1(t, \tau) \circledast g_2(t, \tau) &=& \int_{-\infty}^{\infty} g_1(t - \zeta, \tau - \zeta) g_2(t, \zeta) d\zeta \\
g_1(\tau) \circledast g_2(t, \tau) &=& \int_{-\infty}^{\infty} g_1(\tau - \zeta) g_2(t, \zeta) d\zeta = \int_{-\infty}^{\infty} g_1(\eta) g_2(t, \tau - \eta) d\eta \\
g_1(t, \tau) \circledast g_2(\tau) &=& \int_{-\infty}^{\infty} g_1(t - \zeta, \tau - \zeta) g_2(\zeta) d\zeta \\
g_1(\tau) \circledast g_2(\tau) &=& \int_{-\infty}^{\infty} g_1(\tau - \zeta) g_2(\zeta) d\zeta
\end{array}
$$

$$
\begin{array}{rcl}
x(t) \circledast g_i(t, \tau) &\neq& g_i(t, \tau) \circledast x(t) \\
g_1(t, \tau) \circledast g_2(t, \tau) &\neq& g_2(t, \tau) \circledast g_1(t, \tau) \\
g_1(\tau) \circledast g_2(\tau) &=& g_2(\tau) \circledast g_1(\tau)
\end{array}
$$

$$
\begin{array}{rcl}
[x(t) \circledast g_1(t, \tau)] \circledast g_2(t, \tau) &=& x(t) \circledast [g_1(t, \tau) \circledast g_2(t, \tau)] \\
[g_0(t, \tau) \circledast g_1(t, \tau)] \circledast g_2(t, \tau) &=& g_0(t, \tau) \circledast [g_1(t, \tau) \circledast g_2(t, \tau)]
\end{array}
$$

$$
g_0(t, \tau) \circledast [g_1(t, \tau) + g_2(t, \tau)] = [g_0(t, \tau) \circledast g_1(t, \tau)] + [g_0(t, \tau) \circledast g_2(t, \tau)]
$$

can just drop t in the delay spread function, if the system is not time-variant. Please note that the same symbol $\circledast$ is used for the time-variant convolution of impulse responses w_i and of delay spread functions g_i, although the integral notations differ.

10.4   Examples

Furnished with the theory of time-variant systems we are now going to discuss some applications for signal transmission.

Example 5

A frequent application in communications is the cascade of a time-invariant transmit filter $w_1(t - s)$, a time-variant channel $w_2(t, s)$, and a time-invariant receive filter $w_3(t - s)$. We are interested in the overall impulse response $w_{13}(t, s)$, the corresponding delay spread function $g_{13}(t, \tau)$ and the response $y(t)$ to an input signal $x(t)$.

Solution:

Firstly please note, if time-variant and time-invariant systems are combined, the overall impulse response is the response to the Dirac impulse $\delta(t - s)$. Consequently,

Fig. 10.2 Example of a
time-variant channel with
impulse response $w_2(t, s)$
embedded between a
modulator with $w_1(t, s)$ and
a demodulator with $w_3(t, s)$

all the impulse responses should be defined with two variables t and s including
$w_i(t, s) = w_i(t - s)$, $i = 1, 3$ of the time-invariant systems, in which we just interpret the comma as the minus sign.

Regarding $w_2(t, s)$ in (10.8) as $w_2(t, s) \circledast w_3(t - s)$ and with
$w_1(t, s) = w_1(t - s)$ follows the overall impulse response of the cascade

$$w_{13}(t, s) = w_1(t - s) \circledast w_2(t, s) \circledast w_3(t - s) \tag{10.26}$$

which obviously characterizes an equivalent time-variant system.

We obtain the overall delay spread function from (10.26) by inserting $s = t - \tau$

$$g_{13}(t, \tau) = g_1(\tau) \circledast g_2(t, \tau) \circledast g_3(\tau) \tag{10.27}$$

Next we are interested in the output signal $y(t)$ for a given input signal $x(t)$. With
(9.6) and equivalently with (9.11) we obtain

$$y(t) = x(t) \circledast w_{13}(t, s) = x(t) \circledast g_{13}(t, \tau) \tag{10.28}$$

Example 6

Figure 10.2 shows a modulator, a time-variant channel, and a demodulator. Determine the time-variant impulse response $w_1(t, s)$ of the modulator, which multiplies
its input signal $x(t)$ with the complex carrier $e^{j2\pi f_0 t}$. Then find the overall impulse
response $w_{13}(t, s)$ of the cascade of the modulator, the channel with impulse response
$w_2(t, s)$, the demodulator with complex carrier $e^{-j2\pi f_1 t}$, and impulse response
$w_3(t, s)$. Discuss the important case that the demodulator is synchronous with the
modulator, $f_1 = f_0$, and determine the output signal $y(t)$.

Solution:
The output signal of the modulator is

$$x_1(t) = x(t)e^{j2\pi f_0 t} \tag{10.29}$$

With the input Dirac impulse $x(t) = \delta(t - s)$ the time-variant impulse response can
be determined as

$$w_1(t, s) = \delta(t - s)e^{j2\pi f_0 t} \tag{10.30}$$

Similarly, the impulse response of the demodulator follows as

$$w_3(t, s) = \delta(t - s)e^{-j2\pi f_1 t} \tag{10.31}$$

Then we get according to (10.8) the impulse response of the cascade

$$
\begin{aligned}
w_{13}(t, s) &= \quad w_1(t, s) \circledast w_2(t, s) \circledast w_3(t, s) \\
&= \left(\delta(t - s)e^{j2\pi f_0 t}\right) \circledast w_2(t, s) \circledast \left(\delta(t - s)e^{-j2\pi f_1 t}\right)
\end{aligned} \tag{10.32}
$$

Making use of the associativity (10.22) one can first calculate with (10.8) $w_{12}(t, s) = w_1(t, s) \circledast w_2(t, s) = \int_{-\infty}^{\infty} \delta(\zeta - s)e^{j2\pi f_0 \zeta} w_2(t, \zeta)d\zeta$, which represents the cascade of the modulator and the time-variant channel, resulting in

$$w_{12}(t, s) = e^{j2\pi f_0 s} w_2(t, s) \tag{10.33}$$

Next we determine
$w_{13}(t, s) = w_{12}(t, s) \circledast w_3(t, s) = \int_{-\infty}^{\infty} e^{j2\pi f_0 s} w_2(\eta, s)\delta(t - \eta)e^{-j2\pi f_1 t}d\eta$ and get the final result

$$w_{13}(t, s) = e^{-j2\pi f_1 t} e^{j2\pi f_0 s} w_2(t, s) \tag{10.34}$$

which is a time-variant impulse response, as expected. We can interpret this result as a modulation of $w_2(t, s)$ with respect to both coordinates s and t.

In the following we are going to write $w_{13}(t, s)$ in terms of the delay spread function $g_{13}(t, \tau)$. With $s = t - \tau$, $w_{13}(t, t - \tau) = g_{13}(t, \tau)$, and $w_2(t, t - \tau) = g_2(t, \tau)$ follows from (10.34)

$$g_{13}(t, \tau) = e^{-j2\pi(f_1 - f_0)t} e^{-j2\pi f_0 \tau} g_2(t, \tau) \tag{10.35}$$

The output signal of the cascade is obtained with (10.2) as $y(t) = x(t) \circledast w_{13}(t, s)$ and after some manipulation with the convolution integral follows

$$y(t) = \left[x(t) \circledast \left(w_2(t, s)e^{j2\pi f_0 s}\right)\right] e^{-j2\pi f_1 t} = [x_1(t) \circledast w_2(t, s)] e^{-j2\pi f_1 t} \tag{10.36}$$

with $x_1(t) = x(t)e^{j2\pi f_0 t}$. This result is plausible directly from Fig. 10.2, because $x_1(t)$ and $x_2(t) = x_1(t) \circledast w_2(t, s)$ are the output signals of the modulator and the channel, respectively.

Finally, for synchronous demodulation, $f_0 = f_1$, we obtain

$$w_{13}(t, s) = w_2(t, s)e^{-j2\pi f_0(t-s)} \tag{10.37}$$

With $s = t - \tau$ follows the overall delay spread function

$$g_{13}(t, \tau) = g_2(t, \tau)e^{-j2\pi f_0 \tau} \tag{10.38}$$

in which the modulation shows up only with respect to the delay variable τ.

Verification of Noncommutativity: This example is also a good opportunity to demonstrate that the time-variant convolution is noncommutative in general. Assume that we have a cascade of two systems, a modulator with $e^{j2\pi f_3 t}$ and $w_2(t, s)$ in Fig. 10.2. Then we obtain from (10.34) with $f_0 = f_3$ and $f_1 = 0$ the over-all impulse response $w_{13}(t, s) = e^{j2\pi f_3 s} w_2(t, s)$. If we interchange the sequential order, the overall impulse response is given by (10.34) with $f_0 = 0$ and $f_1 = -f_3$ yielding $w'_{13}(t, s) = w_2(t, s)e^{j2\pi f_3 t}$, which is clearly different. In a similar way we obtain the delay spread functions from (10.35) as $g_{13}(t, \tau) = g_2(t, \tau)e^{j2\pi f_3(t-\tau)}$ and $g'_{13}(t, \tau) = g_2(t, \tau)e^{j2\pi f_3 t}$.

Chapter 11
System Functions and Fourier Transform

We have already discussed the time-variant impulse response $w(t, s)$ and the delay spread function $g(t, \tau)$, which characterize a linear time-variant system completely and hence are called system functions. Now we will see that the latter provides meaningful Fourier transforms for applications in electrical engineering and thus system functions in the frequency domain. In the following sections we will apply the Fourier transform with respect to the variables t and/or τ. Therefore we define the corresponding variables as

$$t \leftrightarrow f_t \; ; \; \tau \leftrightarrow f_\tau$$

and use the symbol $\rightarrowtail$ for the transform. On top of the arrow we indicate in what respect the Fourier transform is executed. In wireless communications f_t is called Doppler frequency and f_τ is the "natural" frequency also used for the ordinary frequency response of a time-invariant system or the spectrum of a signal. The system functions as well as cascades of different systems are summarized in Table 10.2 of Sect. 10.3 and discussed in the following in quite some detail.

11.1 Time-Variant Transfer Function

The *time-variant transfer function* $G_t(t, f_\tau)$ is defined as the Fourier transform of the delay spread function $g(t, \tau)$ with respect to τ

$$g(t, \tau) \overset{\tau}{\rightarrowtail} G_t(t, f_\tau) = \int_{-\infty}^{\infty} g(t, \tau) e^{-j2\pi f_\tau \tau} d\tau \qquad (11.1)$$

© Springer Nature Switzerland AG 2021
J. Speidel, *Introduction to Digital Communications*,
Signals and Communication Technology,
https://doi.org/10.1007/978-3-030-67357-4_11

It represents the transfer function of a time-variant system at a fixed observation instant t, where t has to be regarded as a parameter. So $G_t(t, f_\tau)$ can help to understand the temporal change of the transfer function of a time-variant system. If the system parameters are slowly varying with time t, then $G_t(t, f_\tau)$ is approximately constant with respect to t during a short time interval Δt_{Coh}, which is called coherence time. In this time interval the transfer function $G_t(t, f_\tau)$ could be measured as a function of f_τ similar to a time-invariant system.

For the output signal $y(t)$ we proof that in general

$$y(t) = \int_{-\infty}^{\infty} G_t(t, \zeta) X(\zeta) e^{j2\pi\zeta t} d\zeta \tag{11.2}$$

holds, where

$$x(t) \overset{t}{\longmapsto} X(f_t) = \int_{-\infty}^{\infty} x(t) e^{-j2\pi f_t t} dt \tag{11.3}$$

is the Fourier transform of the input signal $x(t)$.[1] It should be pointed out that (11.2) does not represent the inverse Fourier transform of $G(t, \zeta) X(\zeta)$, although it looks similar on the first glance. Consequently, $G_t(t, f_t) X(f_t) \neq Y(f_t)$, where $Y(f_t)$ is the Fourier transform of $y(t)$.

Proof of (11.2)

From (11.3) follows by inverse Fourier transform

$$x(t) = \int_{-\infty}^{\infty} X(\zeta) e^{j2\pi\zeta t} d\zeta \tag{11.4}$$

which will be inserted into (9.26) yielding
$y(t) = \int_{-\infty}^{\infty} g(t, \tau) \int_{-\infty}^{\infty} X(\zeta) e^{j2\pi\zeta(t-\tau)} d\zeta d\tau$. For this proof and all the following we assume that the integrals are absolutely convergent and thus the interchange of the integration order is allowed. Then follows
$y(t) = \int_{-\infty}^{\infty} X(\zeta) \int_{-\infty}^{\infty} g(t, \tau) e^{-j2\pi\zeta\tau} d\tau e^{j2\pi\zeta t} d\zeta$. According to (11.1) the inner integral is $G_t(t, \zeta)$ resulting in $y(t) = \int_{-\infty}^{\infty} G_t(t, \zeta) X(\zeta) e^{j2\pi\zeta t} d\zeta$, which finalizes the proof.

11.2 Delay Doppler Spread Function

The *delay Doppler spread function* $G_\tau(f_t, \tau)$ also called *Doppler-variant impulse response* is determined by the Fourier transform of the delay spread function $g(t, \tau)$ with respect to t

---

[1] Please note, for the one-dimensional Fourier transform of a one-dimensional function it does not matter whether the pair t, f_t or τ, f_τ is used, because $x(t)$ and $x(\tau)$ are the same functions and also the spectra $X(f_t)$ and $X(f_\tau)$ are mathematically the same.

$$g(t, \tau) \overset{t}{\longmapsto} G_\tau(f_t, \tau) = \int_{-\infty}^{\infty} g(t, \tau) e^{-j2\pi f_t t} dt \qquad (11.5)$$

It represents the frequency response along the f_t-axis for a fixed delay τ as a parameter. In the field of wireless communications $G_\tau(f_t, \tau)$ is also called *spreading function*.

11.3 Doppler Spread Function

The *Doppler spread function* $G(f_t, f_\tau)$ also called *Doppler-variant transfer function* is defined as the two-dimensional Fourier transform of the delay spread function

$$g(t, \tau) \overset{t,\tau}{\longmapsto} G(f_t, f_\tau) = \iint\limits_{-\infty}^{\infty} g(t, \tau) e^{-j2\pi(f_t t + f_\tau \tau)} dt d\tau \qquad (11.6)$$

$G(f_t, f_\tau)$ can also be obtained by the Fourier transformation of the time-variant transfer function $G_t(t, f_\tau)$ with respect to t

$$G_t(t, f_\tau) \overset{t}{\longmapsto} G(f_t, f_\tau) = \int_{-\infty}^{\infty} G_t(t, f_\tau) e^{-j2\pi f_t t} dt \qquad (11.7)$$

or by the transformation of the delay Doppler spread function $G_\tau(f_t, \tau)$ with respect to τ

$$G_\tau(f_t, \tau) \overset{\tau}{\longmapsto} G(f_t, f_\tau) = \int_{-\infty}^{\infty} G_\tau(f_t, \tau) e^{-j2\pi f_\tau \tau} d\tau \qquad (11.8)$$

Example 7
Given the delay spread function

$$g(t, \tau) = \operatorname{sinc}\left(f_1 \left(\tau - \frac{1}{f_1} \right) \right) \operatorname{sinc}\left(f_2(t - \tau) \right) \, , \quad -\infty < t, \tau < \infty \qquad (11.9)$$

similar to (9.13) in Example 4, but with an extended range for t and τ.
(a) Calculate the delay Doppler spread function $G_\tau(f_t, \tau)$ of the system. (b) Find the Doppler spread function $G(f_t, f_\tau)$.

Solution:

(a) We use the Fourier transform pair $a \cdot \operatorname{sinc}(at) \overset{t}{\longmapsto} \operatorname{rect}\left(\frac{f}{a} \right) = \begin{cases} 1 \, ; & |f| \leq \frac{a}{2} \\ 0 \, ; & else \end{cases} \, , \, a \neq 0$

and obtain with the shifting property of the Fourier transform for the second term

Fig. 11.1 Top view of the magnitude $|G(f_t, f_\tau)|$ of the Doppler spread function for $f_1 = 4$ and $f_2 = 1$. Inside the parallelogram $|G(f_t, f_\tau)| = \frac{1}{f_1 f_2}$ and outside $|G(f_t, f_\tau)| = 0$

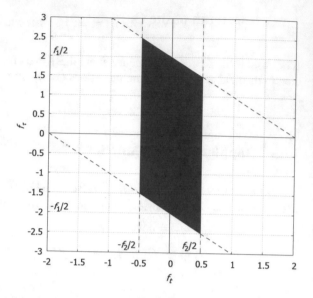

in (11.9) $\operatorname{sinc}(f_2(t - \tau)) \overset{t}{\longmapsto} \frac{1}{f_2} \operatorname{rect}\left(\frac{f_t}{f_2}\right) e^{-j2\pi f_t \tau}$ and thus the final result $g(t, \tau) \overset{t}{\longmapsto}$

$G_\tau(f_t, \tau) = \frac{1}{f_2} \operatorname{sinc}\left(f_1\left(\tau - \frac{1}{f_1}\right)\right) \operatorname{rect}\left(\frac{f_t}{f_2}\right) e^{-j2\pi f_t \tau}$

(b) We apply the Fourier transform on $G_\tau(f_t, \tau)$ with respect to τ. In a first step we get $\operatorname{sinc}\left(f_1\left(\tau - \frac{1}{f_1}\right)\right) \overset{\tau}{\longmapsto} \frac{1}{f_1} \operatorname{rect}\left(\frac{f_\tau}{f_1}\right) e^{-j2\pi f_\tau / f_1}$. Multiplication in the original domain with $e^{-j2\pi f_t \tau}$ results in a frequency shift to the left by f_t of the spectrum. Thus,

$$G(f_t, f_\tau) = \frac{1}{f_1 f_2} \operatorname{rect}\left(\frac{f_t}{f_2}\right) \operatorname{rect}\left(\frac{f_\tau + f_t}{f_1}\right) e^{-j2\pi(f_\tau + f_t)\tau} \qquad (11.10)$$

is the final result. For the magnitude we find $|G(f_t, f_\tau)| = \frac{1}{f_1 f_2} \operatorname{rect}\left(\frac{f_t}{f_2}\right) \operatorname{rect}\left(\frac{f_\tau + f_t}{f_1}\right)$. The first term is a constant factor. As depicted in Fig. 11.1 the second term describes a vertical stripe in the f_t-f_τ-plane and the third term represents a diagonal stripe from north-west to south-east. Consequently, the top view of $|G(f_t, f_\tau)|$ portraits a parallelogram. In reality, the transmission bandwidth $f_1/2$ of a broadband wireless channel is much larger than the bandwidth $f_2/2$ of the temporal fading and consequently, $|G(f_t, f_\tau)|$ is much more slim.

11.4 Spectrum of the Output Signal

Below we prove that the Fourier spectrum

$$Y(f_t) = \int_{-\infty}^{\infty} y(t)e^{-j2\pi f_t t}dt \tag{11.11}$$

of the output signal $y(t)$ of a time-variant system with Doppler spread function $G(f_t, f_\tau)$ is

$$y(t) \xrightarrow{t} Y(f_t) = \int_{-\infty}^{\infty} X(f_\tau)G(f_t - f_\tau, f_\tau)df_\tau = \int_{-\infty}^{\infty} X(\zeta)G(f_t - \zeta, \zeta)d\zeta \tag{11.12}$$

By substituting $f_t - f_\tau = \eta$ we get the alternative form

$$Y(f_t) = \int_{-\infty}^{\infty} X(f_t - \eta)G(\eta, f_t - \eta)d\eta \tag{11.13}$$

If we replace η by f_τ we can write (11.13) as

$$Y(f_t) = \int_{-\infty}^{\infty} X(f_t - f_\tau)G(f_\tau, f_t - f_\tau)df_\tau \tag{11.14}$$

The interesting Eqs. (11.12)–(11.14) show the relation between the input and the output spectrum of a time-variant system determined by its Doppler spread function. It is well known that $Y(f) = H(f)X(f)$ holds for time-invariant systems with the transfer function $H(f)$. In contrast, the relation (11.12) for time-variant systems contains an integral which shows some similarities to a convolution with respect to one frequency variable. From (11.12) and (11.13), where "neutral" variables ζ and η are used for integration, it is quite clear that the spectrum for any one-dimensional signal $y(t)$ can be given as a function of f_t or f_τ, as it makes no difference mathematically.

Proof of (11.12)

Inserting (11.2) into (11.11) yields
$Y(f_t) = \int_{-\infty}^{\infty} \int_{-\infty}^{\infty} G_t(t, \zeta)X(\zeta)e^{j2\pi\zeta t}e^{-j2\pi f_t t}d\zeta dt$. Provided that the integrals converge absolutely, the order of integration can be changed resulting in $Y(f_t) = \int_{-\infty}^{\infty} X(\zeta)\left(\int_{-\infty}^{\infty} G_t(t, \zeta)e^{-j2\pi(f_t-\zeta)t}dt\right)d\zeta$. The inner integral is related to the Doppler spread function given in (11.7) and recognized as $G(f_t - \zeta, \zeta)$. Then we obtain the result $Y(f_t) = \int_{-\infty}^{\infty} X(\zeta)G(f_t - \zeta, \zeta)d\zeta$ and the proof is finished.

11.5   Cascades of Time-Variant and Time-Invariant Systems

11.5.1   Cascade of Time-Invariant $g_1(\tau)$ and Time-Variant System $g_2(t, \tau)$

We consider the cascade in Fig. 10.1b of Sect. 10.2 with the prerequisite that the first system is time-invariant with $g_1(t, \tau) = g_1(\tau)$.

Delay Spread Function

$$g_{12}(t, \tau) = g_1(\tau) \circledast g_2(t, \tau) = \int_{-\infty}^{\infty} g_1(\tau - \zeta)g_2(t, \zeta)d\zeta = \int_{-\infty}^{\infty} g_1(\eta)g_2(t, \tau - \eta)d\eta$$

$$(11.15)$$

Fourier Transform with Respect to τ

$$g_{12}(t, \tau) = g_1(\tau) \circledast g_2(t, \tau) \overset{\tau}{\longmapsto} G_1(f_\tau)G_{2t}(t, f_\tau) \qquad (11.16)$$

where $G_1(f_\tau)$ is the transfer function of the time-invariant system $g_1(\tau)$ and $G_{2t}(t, f_\tau)$ represents the time-variant transfer function of the time-variant system $g_2(t, \tau)$ according to (11.1).

Fourier Transform with Respect to t and τ

$$g_{12}(t, \tau) = g_1(\tau) \circledast g_2(t, \tau) \overset{t,\tau}{\longmapsto} G_1(f_\tau)G_2(f_t, f_\tau) \qquad (11.17)$$

where $G_2(f_t, f_\tau)$ is the Doppler spread function of the time-variant system $g_2(t, \tau)$ defined in (11.6) and $G_1(f_\tau)$ is the transfer function of the time-invariant system $g_1(\tau)$.

Proof of (11.15)

In the following we are going to build upon the basic time-variant impulse responses. We start with (10.8), define $w_1(t, s) = w_1(t - s)$ and obtain $w_1(t - s) \circledast w_2(t, s) = \int_{-\infty}^{\infty} w_1(\eta - s)w_2(t, \eta)d\eta$. With the substitution $\eta = t - \zeta$ follows $w_1(t - s) \circledast w_2(t, s) = \int_{-\infty}^{\infty} w_1(t - \zeta - s)w_2(t, t - \zeta)d\zeta$. The transformation of variables (8.4), $s = t - \tau$, yields $w_1(t - s) = g_1(\tau), w_2(t, s){=}g_2(t, \tau)$, and $w_2(t, t - \zeta) = g_2(t, \zeta)$. Then we find $w_1(t - s) \circledast w_2(t, s) = g_1(\tau) \circledast g_2(t, \tau) = \int_{-\infty}^{\infty} g_1(\tau - \zeta)g_2(t, \zeta)d\zeta$. The last term in (11.15) follows directly with the substitution $\tau - \zeta = \eta$ and the proof is finished.

Proof of (11.16)

Let $g_1(\tau) \overset{\tau}{\longmapsto} G_1(f_\tau)$, then $g_1(\tau - \zeta) \overset{\tau}{\longmapsto} G_1(f_\tau)e^{-j2\pi f_\tau \zeta}$. From (11.15) follows $\int_{-\infty}^{\infty} g_1(\tau - \zeta)g_2(t, \zeta)d\zeta \overset{\tau}{\longmapsto} \int_{-\infty}^{\infty} G_1(f_\tau)e^{-j2\pi f_\tau \zeta}g_2(t, \zeta)d\zeta =$

$G_1(f_\tau) \int_{-\infty}^{\infty} g_2(t, \zeta) e^{-j2\pi f_\tau \zeta} d\zeta = G_1(f_\tau) G_{2t}(t, f_\tau)$, where
$G_{2t}(t, f_\tau) = \int_{-\infty}^{\infty} g_2(t, \zeta) e^{-j2\pi f_\tau \zeta} d\zeta$ is the Fourier transform of $g_2(t, \tau)$ with respect to τ and represents the time-variant transfer function of the time-variant system according to (11.1). This finalizes the proof of (11.16).

Proof of (11.17)
We take the Fourier spectrum in (11.16) and execute a transform with respect to t yielding with $G_{2t}(t, f_\tau) \overset{t}{\rightarrowtail} G_2(f_t, f_\tau)$ the result (11.17) and the proof is finished.

11.5.2 Cascade of Time-Variant $g_1(t, \tau)$ and Time-Invariant System $g_2(\tau)$

Again we consider the cascade in Fig. 10.1b of Sect. 10.2, however with the prerequisite that the second system is time-invariant with $g_2(t, \tau) = g_2(\tau)$.

Delay Spread Function

$$g_{12}(t, \tau) = g_1(t, \tau) \circledast g_2(\tau) = \int_{-\infty}^{\infty} g_1(t - \zeta, \tau - \zeta) g_2(\zeta) d\zeta \qquad (11.18)$$

Fourier Transform with Respect to t and τ

$$g_{12}(t, \tau) = g_1(t, \tau) \circledast g_2(\tau) \overset{t,\tau}{\rightarrowtail} G_1(f_t, f_\tau) G_2(f_t + f_\tau) \qquad (11.19)$$

where $G_1(f_t, f_\tau)$ represents the Doppler spread function of the time-variant system $g_1(t, \tau)$ defined in (11.6) and $G_2(f_\tau)$ is the transfer function of the time-invariant system $g_2(\tau)$.

Proof of (11.18)
A quick proof is obtained by just skipping the argument t in $g_2(t, \tau)$ and to directly apply the time-variant convolution integral (10.11). However, we will start with (10.8) and obtain $w_1(t, s) \circledast w_2(t - s) = \int_{-\infty}^{\infty} w_1(\eta, s) w_2(t - \eta) d\eta$. With the substitution $\eta = t - \zeta$ follows $w_1(t, s) \circledast w_2(t - s) = \int_{-\infty}^{\infty} w_1(t - \zeta, s) w_2(\zeta) d\zeta$. The transformation of variables (8.4), $s = t - \tau$, yields with (10.12)
$w_1(t - \zeta, t - \tau) = g_1(t - \zeta, \tau - \zeta)$ and $w_2(\zeta) = g_2(\zeta)$. Then we find
$w_1(t, s) \circledast w_2(t - s) = g_1(t, \tau) \circledast g_2(\tau) = \int_{-\infty}^{\infty} g_1(t - \zeta, \tau - \zeta) g_2(\zeta) d\zeta$.

Proof of (11.19)
Let

$$g_1(t - \zeta, \tau - \zeta) \overset{t,\tau}{\rightarrowtail} G_1(f_t, f_\tau) e^{-j2\pi(f_t + f_\tau)\zeta} \qquad (11.20)$$

then follows the transform of (11.18)
$g_{12}(t, \tau) \overset{t,\tau}{\rightarrowtail} G_1(f_t, f_\tau) \int_{-\infty}^{\infty} g_2(\zeta) e^{-j2\pi(f_t + f_\tau)\zeta} d\zeta = G_1(f_t, f_\tau) G_2(f_t + f_\tau)$,
where the integral represents $G_2(f_t + f_\tau)$ and the proof is finished.

11.5.3   Cascade of Two Time-Variant Systems $g_1(t, \tau)$ and $g_2(t, \tau)$

Finally, both systems in Fig. 10.1b of Sect. 10.2 shall be time-variant. We prove the following:

Fourier Transform with Respect to τ

$$g_{12}(t, \tau) = g_1(t, \tau) \circledast g_2(t, \tau) \overset{\tau}{\longmapsto} G_{12t}(t, f_\tau) = \int_{-\infty}^{\infty} G_{1t}(t - \eta, f_\tau) g_2(t, \eta) e^{-j2\pi f_\tau \eta} d\eta$$

(11.21)

where $G_{12t}(t, f_\tau)$ and $G_{1t}(t, f_\tau)$ are time-variant transfer functions.

Fourier Transform with Respect to t and τ

$$g_{12}(t, \tau) = g_1(t, \tau) \circledast g_2(t, \tau) \overset{t,\tau}{\longmapsto} G_{12}(f_t, f_\tau) = \int_{-\infty}^{\infty} G_1(\zeta, f_\tau) G_2(f_t - \zeta, f_\tau + \zeta) d\zeta$$

(11.22)

Here we can see the difference between the time-variant convolution (10.11) and the two-dimensional convolution between $g_1(t, \tau)$ and $g_2(t, \tau)$. For the latter the Fourier transform is $G_1(f_t, f_\tau) G_2(f_t, f_\tau)$.

Proof of (11.21) and (11.22)

$g_{12}(t, \tau)$ is defined in (10.11) as $g_{12}(t, \tau) = \int_{-\infty}^{\infty} g_1(t - \eta, \tau - \eta) g_2(t, \eta) d\eta$. We will execute the transform with respect to t and τ in two steps. First, we transform with respect to τ and obtain

$$g_{12}(t, \tau) \overset{\tau}{\longmapsto} G_{12t}(t, f_\tau) = \int_{-\infty}^{\infty} G_{1t}(t - \eta, f_\tau) e^{-j2\pi f_\tau \eta} g_2(t, \eta) d\eta, \text{ where}$$

$g_1(t, \tau) \overset{\tau}{\longmapsto} G_{1t}(t, f_\tau)$ and the proof of (11.21) is finished.

Next, the transformation is done with respect to t. As we recognize, the integrand in (11.21) is the product of two functions of t, namely $G_{1t}(t - \eta, f_\tau)$ and $g_2(t, \eta)$. Therefore we use the property that a product of functions in the time domain results in a convolution of their spectra after Fourier transform. Hence, with $G_{1t}(t - \eta, f_\tau) \overset{t}{\longmapsto}$ $G_1(f_t, f_\tau) e^{-j2\pi f_t \eta}$ and $g_2(t, \eta) \overset{t}{\longmapsto} G_{2\tau}(f_t, \eta)$ we obtain $G_{1t}(t - \eta, f_\tau) g_2(t, \eta) \overset{t}{\longmapsto}$ $G_1(f_t, f_\tau) e^{-j2\pi f_t \eta} *_{f_t} G_{2\tau}(f_t, \eta) = \int_{-\infty}^{\infty} G_1(\zeta, f_\tau) e^{-j2\pi \zeta \eta} G_{2\tau}(f_t - \zeta, \eta) d\zeta$. Then follows from (11.21)

$g_{12}(t, \tau) \overset{t,\tau}{\longmapsto} G_{12}(f_t, f_\tau) = \int_{-\infty}^{\infty} G_1(\zeta, f_\tau) \int_{-\infty}^{\infty} e^{-j2\pi(\zeta + f_\tau)\eta} G_{2\tau}(f_t - \zeta, \eta) d\eta d\zeta$. The inner integral is $G_2(f_t - \zeta, f_\tau + \zeta)$ and we get the final result $G_{12}(f_t, f_\tau) = \int_{-\infty}^{\infty} G_1(\zeta, f_\tau) G_2(f_t - \zeta, f_\tau + \zeta) d\zeta$, which finalizes the proof of (11.22).

11.6   Summary

Table 11.1 summarizes the Fourier transforms of delay spread functions and for some cascades. Please note that $G_1(f_\tau)$ and $G_2(f_\tau)$ represent transfer functions of time-invariant systems. If we consider these functions isolated, we can use any frequency variable, such as f_t and f_τ that is true for any one-dimensional function. However,

Table 11.1 Summary of Fourier transforms of delay spread functions $g(t, \tau) = w(t, t - \tau)$ of time-variant system, delay spread function $g(\tau) = w(t - s)$ of time-invariant system, signals $x(t)$ and $y(t)$

$x(t)$	$\overset{t}{\longmapsto}$	$X(f_t) = \int_{-\infty}^{\infty} x(t) e^{-j2\pi f_t t} dt$
$g(\tau)$	$\overset{\tau}{\longmapsto}$	$G(f_\tau) = \int_{-\infty}^{\infty} g(\tau) e^{-j2\pi f_\tau \tau} d\tau$
$g(t, \tau)$	$\overset{\tau}{\longmapsto}$	$G_t(t, f_\tau) = \int_{-\infty}^{\infty} g(t, \tau) e^{-j2\pi f_\tau \tau} d\tau$
$g(t, \tau)$	$\overset{t}{\longmapsto}$	$G_\tau(f_t, \tau) = \int_{-\infty}^{\infty} g(t, \tau) e^{-j2\pi f_t t} dt$
$g(t, \tau)$	$\overset{t,\tau}{\longmapsto}$	$G(f_t, f_\tau) = \iint_{-\infty}^{\infty} g(t, \tau) e^{-j2\pi (f_t t + f_\tau \tau)} dt d\tau$
$g(t - t_0, \tau - \tau_0)$	$\overset{t,\tau}{\longmapsto}$	$G(f_t, f_\tau) e^{-j2\pi (f_t t_0 + f_\tau \tau_0)}$
$g(t, \tau) e^{j2\pi (f_{t_0} t + f_{\tau_0} \tau)}$	$\overset{t,\tau}{\longmapsto}$	$G(f_t - f_{t_0}, f_\tau - f_{\tau_0})$
$g_1(\tau) \circledast g_2(t, \tau)$	$\overset{\tau}{\longmapsto}$	$G_1(f_\tau) G_{2t}(t, f_\tau)$
$g_1(\tau) \circledast g_2(t, \tau)$	$\overset{t,\tau}{\longmapsto}$	$G_1(f_\tau) G_2(f_t, f_\tau)$
$g_1(t, \tau) \circledast g_2(\tau)$	$\overset{t,\tau}{\longmapsto}$	$G_1(f_t, f_\tau) G_2(f_t + f_\tau)$
$g_1(t, \tau) \circledast g_2(t, \tau)$	$\overset{\tau}{\longmapsto}$	$\int_{-\infty}^{\infty} G_{1t}(t - \eta, f_\tau) g_2(t, \eta) e^{-j2\pi f_\tau \eta} d\eta$
$g_1(t, \tau) \circledast g_2(t, \tau)$	$\overset{t,\tau}{\longmapsto}$	$\int_{-\infty}^{\infty} G_1(\zeta, f_\tau) G_2(f_t - \zeta, f_\tau + \zeta) d\zeta$
$y(t) = x(t) \circledast g(t, \tau)$	$\overset{t}{\longmapsto}$	$Y(f_t) = \int_{-\infty}^{\infty} X(\zeta) G(f_t - \zeta, \zeta) d\zeta$ $= \int_{-\infty}^{\infty} X(f_t - \eta) G(\eta, f_t - \eta) d\eta$
$y(t) = \int_{-\infty}^{\infty} G_t(t, \zeta) X(\zeta) e^{j2\pi \zeta t} d\zeta$		Note : $Y(f_\tau) \neq G_t(t, f_\tau) X(f_\tau)$

in the context of cascading systems the frequency variable makes the difference and indicates in what direction the filtering takes place.

11.7 Applications

In the following we consider some examples to show the impact of the Doppler spread function $G(f_t, f_\tau)$ of a time-variant system on the output spectrum $Y(f_t)$ for a given input spectrum $X(f_t)$. As already pointed out, $G(f_t, f_\tau)$ is not just limiting the input spectrum in f_t- and f_τ-direction. Due to the integral in (11.12) we expect that the output spectrum $Y(f_t)$ can even have a larger bandwidth compared to the input signal, which does not occur for time-invariant systems. To show the details in principle we take simple rectangular shaped spectra as examples.

Example 8

In Fig. 11.2a the top view of the input spectrum $X(f_\tau)$ and the Doppler spread function

$$G(f_t, f_\tau) = \begin{cases} 1 \; ; \; |f_t| \le f_{t,G} \, , \, |f_\tau| \le f_{\tau,G} \\ 0 \; ; \qquad\qquad else \end{cases} \tag{11.23}$$

is given as a function of f_t and f_τ. Inside the shaded area the functions are equal to one and outside equal to zero. All spectra shall be real-valued. $G(f_t, f_\tau)$ is equipped with the cut-off frequencies $f_{\tau,G}$ and $f_{t,G}$. For a wireless channel $f_{t,G}$ represents the impact of the temporal fading and is normally much smaller than the transmission bandwidth $f_{\tau,G}$ of the channel. The spectrum

$$X(f_\tau) = \begin{cases} 1 \; ; \; |f_\tau| \le f_{\tau,X} \\ 0 \; ; \quad else \end{cases} \tag{11.24}$$

owns the cut-off frequency $f_{\tau,X}$. Please note that $X(f_\tau)$ is constant with respect to f_t. We also have assumed that $f_{\tau,G} > f_{\tau,X}$ so that the system is not limiting the input spectrum. Figure 11.2b shows $G(0, f_\tau)$ and $X(f_\tau)$ as a function of f_τ. In Fig. 11.2c the term $G(f_t - u, u)$ of the output spectrum $Y(f_t) = \int_{-\infty}^{\infty} X(u)G(f_t - u, u)du$ in (11.12) is illustrated. Similar to a convolution operation $G(f_t - u, u)$ "moves" along the u-axis for various f_t and covers $X(u)$ to compose the integrand $X(u)G(f_t - u, u)$. Please note the dashed line indicates that $G(f_t - u, u) = 0$ for $|u| > f_{\tau,G}$. For the indicated f_t the integrand starts to be unequal to zero and $Y(f_t)$ in Fig. 11.2d begins to rise. For $f_t = -(f_{\tau,X} - f_{t,G})$ both spectra completely overlap and consequently $Y(f_t)$ is maximal. Obviously, $Y(f_t)$ is an even function. The spectral parts of $G(f_t, f_\tau)$ in f_t-direction increase the cut-off frequency $f_{t,Y}$ of the output spectrum and make it larger than $f_{\tau,X}$ of the input signal, in our example $f_{t,Y} = f_{\tau,X} + f_{t,G}$. In wireless communications this effect is called Doppler spread. Thus, for a wireless, fading channel the larger the bandwidth $f_{t,G}$ of the temporal fading is, the more the cut-off frequency of the output spectrum is increased compared to the input. Note, we can use any frequency variable for a one-dimensional spectrum. Hence, the abscissa in Fig. 11.2d may alternatively be named as f_τ.

The reader can assure oneself that the solution is valid also, if the cut-off frequency $f_{\tau,G}$ of the system is equal to $f_{\tau,X}$.

Example 9

We are now considering the example in Fig. 11.3a where the cut-off frequency $f_{\tau,G}$ of the Doppler spread function $G(f_t, f_\tau)$ of the system limits the input spectrum and thus $f_{\tau,G} < f_{\tau,X}$ holds compared to Fig. 11.2a. The cut-off frequency $f_{t,G}$ of the temporal fading is unchanged. In Fig. 11.3b $X(f_\tau)$ and $G(0, f_\tau)$ are shown. Figure 11.3c depicts the shifted spectrum $G(f_t - u, u)$ as part of the integrand in

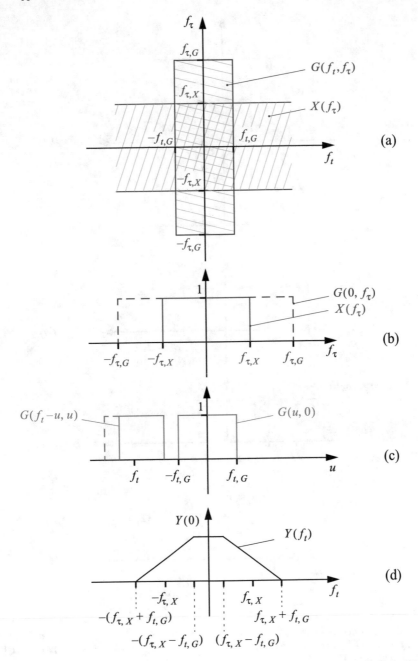

Fig. 11.2 **a** Top view of input $X(f_\tau)$, Doppler spread function $G(f_t, f_\tau)$, $f_{\tau,G} \geq f_{\tau,X}$ **b** Details of $G(f_t, f_\tau)$ and $X(f_\tau)$ for $f_t = 0$ **c** $G(f_t - u, u)$ of $Y(f_t) = \int_{-\infty}^{\infty} X(u)G(f_t - u, u)du$ in (11.12) **d** Output spectrum $Y(f_t)$

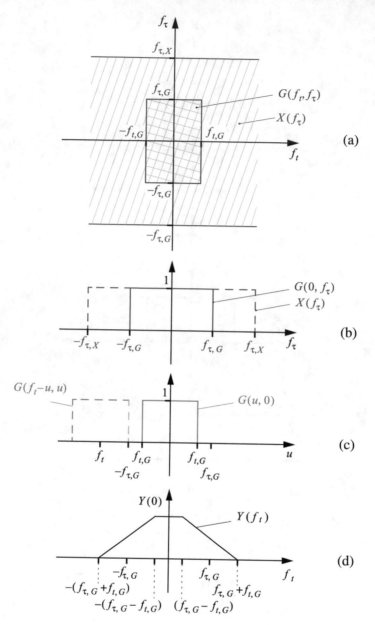

Fig. 11.3 **a** Top view of input $X(f_\tau)$, Doppler spread function $G(f_t, f_\tau)$, $f_{\tau,G} < f_{\tau,X}$ **b** Details of $G(f_t, f_\tau)$ and $X(f_\tau)$ for $f_t = 0$ **c** $G(f_t - u, u)$ of $Y(f_t) = \int_{-\infty}^{\infty} X(u)G(f_t - u, u)du$ in (11.12) **d** Output spectrum $Y(f_t)$

(11.12), where f_τ is renamed as u. Please note the dashed line indicates that $G(f_t - u, u) = 0$ for $|u| > f_{\tau,G}$. The resulting output spectrum $Y(f_t)$ is shown in Fig. 11.3d. Its cut-off frequency $f_{t,Y} = f_{\tau,G} + f_{t,G}$ is solely given by the parameters of the Doppler spread function of the system and exceeds the transmission bandwidth $f_{\tau,G}$ by the maximal Doppler frequency $f_{t,G}$. Again we recognize that the time-variance of the channel is the cause of the excess bandwidth of the output signal. As mentioned before, we can use any frequency variable for a one-dimensional spectrum Y, hence denote the abscissa in Fig. 11.3d also as f_τ.

Example 10
Now we consider a similar arrangement as in Fig. 11.2, however, the Doppler spread function $G(f_t, f_\tau)$ shall have the cut-off frequency $f_{t,G} = 0$. Hence, the system is time-invariant and in case of a transmission channel it would show no temporal fading. Thus an adequate model is

$$G(f_t, f_\tau) = \delta(f_t)G(0, f_\tau) \tag{11.25}$$

Then we obtain from (11.12)

$$Y(f_t) = \int_{-\infty}^{\infty} X(f_\tau)\delta(f_t - f_\tau)G(0, f_\tau)df_\tau = X(f_t)G(0, f_t) \tag{11.26}$$

As mentioned earlier for a one-dimensional signal, we can use any variable. To adapt to Fig. 11.2 we rename f_t as f_τ and get

$$Y(f_\tau) = X(f_\tau)G(0, f_\tau) \tag{11.27}$$

which is the well known frequency response of a time-invariant system. If $f_{\tau,G}$ and $f_{\tau,X}$ are the same as in Fig. 11.2, then $Y(f_\tau) = X(f_\tau)$, because the system is not limiting the input spectrum. Thus, no Doppler spread is present, as expected.

Example 11
Let us now discuss an artificial system, which has zero transmission bandwidth $f_{\tau,G} = 0$. As $G(f_t, f_\tau)$ still has spectral components in f_t direction with cut-off frequency $f_{t,G}$, it is interesting to know the output spectrum of the system. We model the Doppler spread function as

$$G(f_t, f_\tau) = G(f_t, 0)\delta(f_\tau) \tag{11.28}$$

which yields from (11.12)

$$Y(f_t) = \int_{-\infty}^{\infty} X(f_\tau)G(f_t - f_\tau, 0)\delta(f_\tau)df_\tau = X(0)G(f_t, 0) \tag{11.29}$$

Obviously, the output spectrum is non-zero and given by the frequency response of $G(f_t, f_\tau)$ along the f_t-axis multiplied by the mean value $X(0)$ of the input signal.

Hence, the output spectrum solely exhibits a Doppler spread. The system is not very practical for signal transmission, as only the spectral component $X(0)$ of the input is transferred to the output.

Example 12

Given the time-variant system in Fig. 10.2. Find the Doppler spread function $G_{13}(f_t, f_\tau)$ between input and output as well as the spectrum $Y(f_t)$ of the output signal $y(t)$.

Solution

From (10.35) follows $g_{13}(t, \tau) \overset{t,\tau}{\rightharpoonup} G_{13}(f_t, f_\tau) = G_2(f_t + f_1 - f_0, f_\tau + f_0)$. Table 11.1 provides the relation for output spectrum
$Y(f_t) = \int_{-\infty}^{\infty} X(\zeta) G_{13}(f_t - \zeta, \zeta) d\zeta$ and with $G_{13}(f_t, f_\tau)$ the final result
$Y(f_t) = \int_{-\infty}^{\infty} X(\zeta) G_2(f_t + f_1 - f_0 - \zeta, \zeta + f_0) d\zeta$. With the substitution
$\zeta = u - f_0$, $d\zeta = du$ follows alternatively
$Y(f_t) = \int_{-\infty}^{\infty} X(u - f_0) G_2(f_t + f_1 - u, u) du$.

In case, the system G_2 is time-invariant with delay spread function. $g_2(t, \tau) = h(\tau)$ we obtain $g_2(t, \tau) \overset{t,\tau}{\rightharpoonup} G_2(f_t, f_\tau) = \delta(f_t) H(f_\tau)$ and the output spectrum turns into $Y(f_t) = \int_{-\infty}^{\infty} X(u - f_0)\delta(f_t + f_1 - u) H(u) du = X(f_t + f_1 - f_0) H(f_t + f_1)$, which is a well known result.

11.8 Interrelation Between Time-Variant and Two-Dimensional Convolution

11.8.1 Input-Output Relation

We look back to the input-output relation of a time-variant system in (9.26) and (9.27), where we have recognized that a one-dimensional function $x(t)$ has to be convolved with a two-dimensional function $g(t, \tau)$ using a special form of the convolution operation. One may raise the question, whether we can apply the conventional two-dimensional convolution in some way, well bearing in mind that the input and the out signals finally must be interpreted as one-dimensional signals.

Following this idea we start to define the input signal as $x(t, \tau)$ and the output signal as $y(t, \tau)$. Then the two-dimensional convolution provides

$$y(t, \tau) = x(t, \tau) * g(t, \tau) = \iint\limits_{-\infty}^{\infty} g(u, v) x(t - u, \tau - v) du dv \qquad (11.30)$$

How can we extend $x(t)$ to a two-dimensional function? A similar question has already been touched upon in the Examples 8 and 9 in Sect. 11.7. There we had to plot in Figs. 11.2a and 11.3a the spectrum $X(f_\tau)$ of $x(t)$ in the two-dimensional frequency space f_t, f_τ together with the Doppler spread function $G(f_t, f_\tau)$. We have interpreted $X(f_\tau)$ as a constant with respect to f_t. Consequently we can extend $x(t)$ in the following way

$$x(t, \tau) = \delta(t)x(\tau) \tag{11.31}$$

because then its Fourier transform $X(f_t, f_\tau)$ is constant with respect to f_t. This is easily verified as

$$X(f_t, f_\tau) = \iint\limits_{-\infty}^{\infty} \delta(t)x(\tau)e^{-j2\pi(f_t t + f_\tau \tau)}\,dt\,d\tau = \int_{-\infty}^{\infty} x(\tau)e^{-j2\pi f_\tau \tau}\,d\tau = X(f_\tau) \tag{11.32}$$

From (11.30) results with (11.31)

$$y(t, \tau) = \iint\limits_{-\infty}^{\infty} g(u, v)\delta(t - u)x(\tau - v)\,du\,dv = \int_{-\infty}^{\infty} g(t, v)x(\tau - v)\,dv \tag{11.33}$$

which yields for $\tau = t$ the time-variant convolution integral

$$y(t, t) = y(t) = \int_{\infty}^{\infty} g(t, v)x(t - v)\,dv \tag{11.34}$$

where we use the short hand notation $y(t)$ for $y(t, t)$.

In summary we can operate with the two-dimensional input signal (11.31), apply the two-dimensional convolution integral (11.30) and set $\tau = t$ to get the final result (11.34).

11.8.2   *Fourier Spectrum of the Output Signal*

Now we are going to determine the Fourier spectrum $Y(f_t)$ of the output signal $y(t)$ of the time-variant system. To this end we consider (11.30) again. Knowing that the Fourier transform turns the two-dimensional convolution in the original domain into the product of the spectra in the frequency domain we get for the two-dimensional Fourier spectrum of $y(t, \tau)$

$$Y(f_t, f_\tau) = X(f_t, f_\tau)G(f_t, f_\tau) = X(f_\tau)G(f_t, f_\tau) \tag{11.35}$$

To introduce the condition $t = \tau$ used in (11.34) we consider the inverse transform

$$y(t, \tau) = \iint\limits_{-\infty}^{\infty} X(f_\tau)G(f_t, f_\tau)e^{j2\pi(f_t t + f_\tau \tau)}\,df_t\,df_\tau \tag{11.36}$$

With $t = \tau$ and the substitution $f_t = \eta - f_\tau$, $df_t = d\eta$ follows
$y(t, t) = y(t) = \iint_{-\infty}^{\infty} X(f_\tau)G(\eta - f_\tau, f_\tau)df_\tau e^{j2\pi\eta t}d\eta$. Next, we rename the integration variable η as f_t and get

$$y(t) = \int_{-\infty}^{\infty} \left(\int_{-\infty}^{\infty} X(f_\tau)G(f_t - f_\tau, f_\tau)df_\tau \right) e^{j2\pi f_t t}df_t \qquad (11.37)$$

The inner integral must be the Fourier transform $Y(f_t)$ of $y(t)$

$$Y(f_t) = \int_{-\infty}^{\infty} X(f_\tau)G(f_t - f_\tau, f_\tau)df_\tau \qquad (11.38)$$

Hence, with (11.38) we have confirmed our earlier result (11.12) in a different way.

Chapter 12
Randomly Changing Time-Variant Systems

12.1 Prerequisites

Hitherto we have considered signals and characteristic functions of time-variant system, in particular the delay spread function $g(t, \tau)$, as deterministic. With the Fourier transform different spectra or transfer functions, such as the Doppler spread function $G(f_t, f_\tau)$, have been defined. In many applications, e.g., wireless communications the time-variant channel can take on a fast of different characteristics depending on the environment, the speed of the transmitter or receiver, and other effects. Hence, there is a need for the introduction of a statistical description for the most important system parameters. The use of multivariate probability density functions could help in principle. However, it would be hard or even prohibitive to determine them exhaustively in many applications. For most practical cases second order statistics provide reasonable approximations, at least for the performance comparison of systems and can still be handled, as proposed in [1]. Several functions characterizing the time-variant system, such as the Doppler spread function, can be separated into a deterministic part, identified by its mean value and a pure stochastic part. In the following we are only interested in latter and therefore assume stochastic signals with zero mean. Consequently, we focus on the correlation rather than the covariance functions. For details of stochastic processes the reader is relegated to the survey given in the Appendix A.

In the following we start with the definitions of the autocorrelation functions with no restriction whether the processes are stationary or not. In a second step we focus on stationary processes in the wide sense. Finally, we consider the case that the stochastic processes under consideration are uncorrelated with respect to the delay time τ, which is also called uncorrelated scattering in case of a time-variant wireless channel.

© Springer Nature Switzerland AG 2021
J. Speidel, *Introduction to Digital Communications*,
Signals and Communication Technology,
https://doi.org/10.1007/978-3-030-67357-4_12

12.2   Correlation Functions of Randomly Changing Time-Variant Systems

We define the autocorrelation functions for various system functions of a randomly changing time-variant system.

Autocorrelation Function of the Delay Spread Function $g(t, \tau)$

$$R_{gg}(t, t'; \tau, \tau') = \mathbf{E}\left[g^*(t, \tau)g(t', \tau')\right] \tag{12.1}$$

If not otherwise noted, the expected value $\mathbf{E}[\ldots]$ is considered with respect to all time and delay instants t, t', τ, and τ', respectively. This holds similarly also for the following.

Autocorrelation Function of the Time-Variant Transfer Function $G_t(t, f_\tau)$

$$R_{G_t G_t}(t, t'; f_\tau, f_\tau') = \mathbf{E}\left[G_t^*(t, f_\tau)G_t(t', f_\tau')\right] \tag{12.2}$$

Autocorrelation Function of the Delay Doppler Spread Function $G_\tau(f_t, \tau)$

$$R_{G_\tau G_\tau}(f_t, f_t'; \tau, \tau') = \mathbf{E}\left[G_\tau^*(f_t, \tau)G_\tau(f_t', \tau')\right] \tag{12.3}$$

Autocorrelation Function of the Doppler Spread Function $G(f_t, f_\tau)$

$$R_{GG}(f_t, f_t'; f_\tau, f_\tau') = \mathbf{E}\left[G^*(f_t, f_\tau)G(f_t', f_\tau')\right] \tag{12.4}$$

With the autocorrelation function $R_{gg}(t, t'; \tau, \tau')$ of $g(t, \tau)$ in (12.1) we show at the end of this section that

$$R_{GG}(f_t, f_t'; f_\tau, f_\tau') = \iiiint\limits_{-\infty}^{\infty} R_{gg}(t, t'; \tau, \tau')e^{-j2\pi(-f_t t + f_t' t' - f_\tau \tau + f_\tau' \tau')}dt\,dt'\,d\tau\,d\tau'$$

$$\tag{12.5}$$

holds. Hence, $R_{GG}(f_t, f_t'; f_\tau, f_\tau')$ can be interpreted as the four-dimensional Fourier spectrum of $R_{gg}(t, t'; \tau, \tau')$, evaluated at the frequency positions $-f_t$, f_t', $-f_\tau$, and f_τ'.

Interrelation Between the Autocorrelation Functions

Because the frequency domain system functions $G_t(t, f_\tau)$, $G_\tau(f_t, \tau)$, and $G(f_t, f_\tau)$ are the result of the one- or two-dimensional Fourier transform of the delay spread function $g(t, \tau)$, it is evident that the autocorrelation functions will also show some interrelations through the Fourier transform. We just give the following two examples.

$$R_{gg}(t, t'; \tau, \tau') \overset{\tau, \tau'}{\longmapsto} R_{G_t G_t}(t, t'; -f_\tau, f_\tau') \quad \text{with}$$
$$R_{G_t G_t}(t, t'; -f_\tau, f_\tau') = \iint_{-\infty}^{\infty} R_{gg}(t, t'; \tau, \tau')e^{-j2\pi(f_\tau \tau + f_\tau' \tau')}d\tau\,d\tau' \tag{12.6}$$

$$R_{gg}(t, t'; \tau, \tau) \overset{t,t'}{\longmapsto} R_{G_\tau G_\tau}(-f_t, f'_t; \tau, \tau') \quad \text{with} \tag{12.7}$$
$$R_{G_\tau G_\tau}(-f_t, f'_t; \tau, \tau') = \iint_{-\infty}^{\infty} R_{gg}(t, t'; \tau, \tau') e^{-j2\pi(f_t t + f'_t t')} dt dt'$$

The proof of (12.6) and (12.7) is given at the end of this section.

Autocorrelation Function of the Output Signal $y(t)$

The autocorrelation function of the output signal $y(t)$ is given by

$$R_{yy}(t, t') = \mathbf{E}\left[y^*(t)y(t')\right] = \iint_{-\infty}^{\infty} R_{gg}(t, t'; \tau, \tau') R_{xx}(t - \tau, t' - \tau') d\tau d\tau' \tag{12.8}$$

with the autocorrelation function of $x(t)$

$$R_{xx}(t, t') = \mathbf{E}\left[x^*(t)x(t')\right] \tag{12.9}$$

We recognize from (12.8) that this relation has some similarities to the Wiener–Lee relation for time-invariant systems outlined in the Appendix A.

Proof of (12.5)
From (12.4) follows with (11.6) $R_{GG}(f_t, f'_t; f_\tau, f'_\tau) =$
$\mathbf{E}\left[\iint_{-\infty}^{\infty} g^*(t, \tau) \ e^{j2\pi(f_t t + f_\tau \tau)} dt d\tau \iint_{-\infty}^{\infty} g(t', \tau') e^{-j2\pi(f'_t t' + f'_\tau \tau')} dt' d\tau'\right] =$
$\iiiint_{-\infty}^{\infty} \mathbf{E}\left[g^*(t, \tau)g(t', \tau')\right] e^{-j2\pi(f'_t t' + f'_\tau \tau' - f_t t - f_\tau \tau)} dt d\tau dt' d\tau'$. Plugging in
$R_{gg}(t, t'; \tau, \tau')$ from (12.1) yields the final result
$R_{GG}(f_t, f'_t; f_\tau, f'_\tau) = \iiiint_{-\infty}^{\infty} R_{gg}(t, t'; \tau, \tau') e^{-j2\pi(-f_t t + f'_t t' - f_\tau \tau + f'_\tau \tau')} dt dt' d\tau d\tau'$ and
the proof is finished.

Proof of (12.6)
With (12.1) we obtain $R_{gg}(t, t'; \tau, \tau') \overset{\tau, \tau'}{\longmapsto}$
$\iint_{-\infty}^{\infty} \mathbf{E}\left[g^*(t, \tau)g(t', \tau')\right] e^{-j2\pi(f_\tau \tau + f'_\tau \tau')} d\tau d\tau'$. Due to the linearity we can exchange
the expectation operation and the integration and get
$\mathbf{E}\left[\int_{-\infty}^{\infty} g^*(t, \tau) e^{-j2\pi f_\tau \tau} d\tau \int_{-\infty}^{\infty} g(t', \tau') e^{-j2\pi f'_\tau \tau'} d\tau'\right]$. With (11.1) the first integral is
identified as $G_t^*(t, -f_\tau)$ and the second one as $G_t(t', f'_\tau)$. Hence, $R_{gg}(t, t'; \tau, \tau') \overset{\tau, \tau'}{\longmapsto}$
$\mathbf{E}\left[G_t^*(t, -f_\tau)G_t(t', f'_\tau)\right] = R_{G_t G_t}(t, t'; -f_\tau, f'_\tau)$ results and the proof is finished.

Proof of (12.7)
With (12.1) we obtain $R_{gg}(t, t'; \tau, \tau') \overset{t,t'}{\longmapsto}$
$\iint_{-\infty}^{\infty} \mathbf{E}\left[g^*(t, \tau)g(t', \tau')\right] e^{-j2\pi(f_t t + f'_t t')} dt dt'$. Due to the linearity we can exchange
the expectation operation and the integration and get
$\mathbf{E}\left[\int_{-\infty}^{\infty} g^*(t, \tau) e^{-j2\pi f_t t} dt \int_{-\infty}^{\infty} g(t', \tau') e^{-j2\pi f'_t t'} dt'\right]$. With (11.5) the first integral is
identified as $G_\tau^*(-f_t, \tau)$ and the second one as $G_\tau(f'_t, \tau')$. Hence, $R_{gg}(t, t'; \tau, \tau') \overset{t,t'}{\longmapsto}$
$\mathbf{E}\left[G_\tau^*(-f_t, \tau)G_\tau(f'_t, \tau')\right] = R_{G_\tau G_\tau}(-f_t, f'_t; \tau, \tau')$ results and the proof is finished.

Proof of (12.8)

With (9.26) follows

$R_{yy}(t, t') = \mathbf{E}\left[y^*(t)y(t')\right] = \mathbf{E}\left[\iint_{-\infty}^{\infty} g^*(t, \tau)x^*(t - \tau) \; g(t', \tau')x(t' - \tau')d\tau d\tau'\right] =$
$\iint_{-\infty}^{\infty} \mathbf{E}\left[g^*(t, \tau)g(t', \tau')x^*(t - \tau)x(t' - \tau')\right]d\tau d\tau'.$

We can assume that $g(t, \tau)$ and $x(t)$ are uncorrelated yielding

$\mathbf{E}\left[g^*(t, \tau)g(t', \tau')x^* \; (t - \tau)x(t' - \tau')\right] =$
$\mathbf{E}\left[g^*(t, \tau)g(t', \tau')\right]\mathbf{E}\left[x^*(t - \tau)x(t' - \tau')\right].$ With (12.1) and (12.9) follows

$R_{yy}(t, t') = \iint_{-\infty}^{\infty} R_{gg}(t, t'; \tau, \tau')R_{xx}(t - \tau, t' - \tau')d\tau d\tau'$ and the proof is finished.

12.3   Wide Sense Stationary Time-Variant Systems

12.3.1   Wide Sense Stationary

We are now specializing on the case that the system function $g(t, \tau)$ is a wide sense stationary (WSS) stochastic process with respect to the time t. The same shall hold for the input and output signal $x(t)$ and $y(t)$ of the system, respectively. In Appendix A a wide sense stationary process is characterized by the (joint) probability density function and the mean value, both independent of time. Furthermore, the autocorrelation functions depend on t and t' just through the time difference

$$\Delta t = t' - t \tag{12.10}$$

Please note, when the delay variables τ and τ' are present the autocorrelation functions still depend on τ and τ' separately and not automatically on the difference $\Delta \tau = \tau' - \tau$. The latter is only true, if the delay spread function $g(t, \tau)$ is also stationary with respect to the delay variable.

12.3.2   Autocorrelation Functions and Power Spectral Densities

With $t' = t + \Delta t$ from (12.10) we define the following functions for wide sense stationary processes

Autocorrelation Function of the Delay Spread Function $g(t, \tau)$
from (12.1)

$$R_{gg}(\Delta t; \tau, \tau') = \mathbf{E}\left[g^*(t, \tau)g(t + \Delta t, \tau')\right] \tag{12.11}$$

Autocorrelation Function of the Time-Variant Transfer Function $G_t(t, f_\tau)$
from (12.2)

$$R_{G_t G_t}(\Delta t; f_\tau, f_\tau') = \mathbf{E}\left[G_t^*(t, f_\tau)G_t(t + \Delta t, f_\tau')\right] \tag{12.12}$$

Power Spectral Density of the Delay Spread Function

According to the Wiener–Khintchine theorem in Appendix A the *cross-power spectral density* $S_{gc}(f_t'; \tau, \tau')$ between $g(t, \tau)$ and $g(t', \tau')$ can be defined by the Fourier transform of the autocorrelation function $R_{gg}(\Delta t; \tau, \tau')$ of $g(t, \tau)$ with respect to Δt

$$R_{gg}(\Delta t; \tau, \tau') \xrightarrow{\Delta t} S_{gc}(f_t'; \tau, \tau') \text{ with}$$
$$S_{gc}(f_t'; \tau, \tau') = \int_{-\infty}^{\infty} R_{gg}(\Delta t; \tau, \tau') e^{-j2\pi f_t' \Delta t} d(\Delta t) \tag{12.13}$$

For $\tau = \tau'$ this function just depends on two variables and we use the short hand notation $S_{gc}(f_t'; \tau, \tau) = S_{gc}(f_t'; \tau)$, which can be interpreted as a form of the *power spectral density* of the delay spread function $g(t, \tau)$. Please note that τ and τ' are parameters, which are not touches by the Fourier transform in (12.13).

Relation Between $R_{G_\tau G_\tau}(f_t, f_t'; \tau, \tau')$ and $S_{gc}(f_t'; \tau, \tau')$

We prove that the autocorrelation function of the delay Doppler spread function is given by

$$R_{G_\tau G_\tau}(f_t, f_t'; \tau, \tau') = S_{gc}(f_t'; \tau, \tau')\delta(f_t' - f_t) \tag{12.14}$$

According to Appendix A we can conclude from (12.14) that the delay Doppler spread functions $G_\tau(f_t, \tau)$ and $G_\tau(f_t', \tau')$ are uncorrelated with respect to any of the two Doppler frequencies $f_t \neq f_t'$.

Proof of (12.14)
From (12.7) follows with the WSS property (12.10), $t' = t + \Delta t$, $dt' = d(\Delta t)$, and by replacing $-f_t$ by f_t the autocorrelation function $R_{G_\tau G_\tau}(f_t, f_t'; \tau, \tau') = \int_{-\infty}^{\infty} e^{-j2\pi(f_t' - f_t)t} dt \int_{-\infty}^{\infty} R_{gg}(\Delta t; \tau, \tau') e^{-j2\pi f_t' \Delta t} d(\Delta t)$. Knowing that $\int_{-\infty}^{\infty} e^{-j2\pi(f_t' - f_t)t} dt = \delta(f_t' - f_t)$ and using (12.13) yields the result $R_{G_\tau G_\tau}(f_t, f_t'; \tau, \tau') = S_{gc}(f_t'; \tau, \tau')\delta(f_t' - f_t)$, which finalizes the proof.

Input-Output Relation of Correlation Functions and Power Spectral Densities

In the following, we assume that the input signal $x(t)$ and the time-variant system are WSS. Then the autocorrelation function (12.9)

$$R_{xx}(t, t') = R_{xx}(\Delta t) \tag{12.15}$$

is just a function of $t' - t = \Delta t$ and for the argument of R_{xx} in (12.8) follows $(t' - \tau') - (t - \tau) = \Delta t - (\tau' - \tau)$. Then we obtain from (12.8)

$$R_{yy}(\Delta t) = \iint_{-\infty}^{\infty} R_{gg}(\Delta t; \tau, \tau') R_{xx}\left(\Delta t - (\tau' - \tau)\right) d\tau d\tau' \tag{12.16}$$

which is the final relation between the input and the output correlation function of a time-variant system. It has some similarities with the Wiener–Lee theorem for time-invariant systems given in Appendix A.

Next we are going to consider the Fourier transforms with respect to Δt

$$R_{xx}(\Delta t) \overset{\Delta t}{\rightharpoonup} S_{xx}(f_t) \; ; \; R_{yy}(\Delta t) \overset{\Delta t}{\rightharpoonup} S_{yy}(f_t) \tag{12.17}$$

which are the power spectral density functions of $x(t)$ and $y(t)$ according to the Wiener–Khintchine theorem in the Appendix A, respectively. With (12.13) and $R_{xx}\left(\Delta t - (\tau' - \tau)\right) \overset{\Delta t}{\rightharpoonup} S_{xx}(f_t)e^{-j2\pi f_t(\tau'-\tau)}$ follows for the Fourier transform of the integrand in (12.16)

$$R_{gg}(\Delta t; \tau, \tau')R_{xx}\left(\Delta t - (\tau' - \tau)\right) \overset{\Delta t}{\rightharpoonup} S_{gc}(f_t; \tau, \tau') * \left(S_{xx}(f_t)e^{-j2\pi f_t(\tau'-\tau)}\right) \tag{12.18}$$

where $*$ indicates the time-invariant convolution with respect to f_t. Please note that we changed the argument of S_{gc} to f_t, because for a one-dimensional function we can take any variable, f_t or f_t'. Finally we obtain from (12.16) the power spectral density

$$S_{yy}(f_t) = \int\limits_{-\infty}^{\infty}\!\!\!\int S_{gc}(f_t; \tau, \tau') * \left(S_{xx}(f_t)e^{-j2\pi f_t(\tau'-\tau)}\right) d\tau d\tau' \tag{12.19}$$

We recognize that (12.19) is an extension of the Wiener–Khintchine theorem in Appendix A, whereby the time-variance of the system implies the additional integration with respect to τ and τ', as is also the case in (12.16).

12.4 Time-Variant Systems with Uncorrelated Scattering

The term uncorrelated scattering (US) stems from wireless channels and we apply it here for the general time-variant system. The condition defines a system with a delay spread function $g(t, \tau)$ and functions derived thereof, which are uncorrelated with respect to the delay variable τ. As shown in the Appendix A for white noise, the autocorrelation function must feature a Dirac impulse $\delta(\tau)$ in principle. In this section we review some important properties of uncorrelated scattering. Then in Sect. 12.5 we combine the characteristics of processes showing both, wide sense stationarity and uncorrelated scattering (WSSUS).

12.4.1   Delay Cross Power Spectral Density of $g(t, \tau)$

For a wireless channel with uncorrelated scattering, the delay spread functions $g(t, \tau)$ and $g(t, \tau')$ are by definition uncorrelated for any delay $\tau \neq \tau'$. Thus, the autocorrelation function (12.1) of $g(t, \tau)$ must possess the following form

$$R_{gg}(t, t'; \tau, \tau') = R_{gg;US}(t, t'; \tau, \tau') = P_g(t, t'; \tau)\delta(\tau' - \tau) \tag{12.20}$$

$P_g(t, t'; \tau)$ is called *delay cross power spectral density* of $g(t, \tau)$ and obviously can still be a function of time. Next we show that $P_g(t, t'; \tau)$ is represented by the Fourier transform of $R_{gg}(t, t'; \tau, \tau')$ in (12.1) with respect to τ' at $f_\tau' = 0$. This results in

$$\int_{-\infty}^{\infty} R_{gg;US}(t, t'; \tau, \tau')d\tau' = P_g(t, t'; \tau) \tag{12.21}$$

Proof of (12.21)
We apply the Fourier transform on (12.20) and obtain
$\int_{-\infty}^{\infty} R_{gg}(t, t'; \tau, \tau')e^{-j2\pi f_\tau' \tau'}d\tau' = P_g(t, t'; \tau)\int_{-\infty}^{\infty} \delta(\tau' - \tau)e^{-j2\pi f_\tau' \tau'}d\tau' = P_g(t, t'; \tau)e^{-j2\pi f_\tau' \tau}$. Then we set $f_\tau' = 0$ and get the final result, which finalizes the proof.

12.4.2   Autocorrelation Function of Time-Variant Transfer Function

Now we are going to consider the autocorrelation function $R_{G_t G_t}(t, t'; f_\tau, f_\tau')$ of the time-variant transfer function $G_t(t, f_\tau)$ in (12.2) for the case of uncorrelated scattering specified by (12.20). In the following we prove that the autocorrelation function of $G_t(t, f_\tau)$ then depends on the frequency difference

$$\Delta f_\tau = f_\tau' - f_\tau \tag{12.22}$$

and hence must exhibit the form

$$R_{G_t G_t}(t, t'; f_\tau, f_\tau') = R_{G_t G_t}(t, t'; \Delta f_\tau) \tag{12.23}$$

Proof of (12.23)
We first make use of (12.6) and obtain $R_{G_t G_t}(t, t'; f_\tau, f_\tau') = \iint_{-\infty}^{\infty} R_{gg}(t, t'; \tau, \tau')e^{-j2\pi(-f_\tau \tau + f_\tau' \tau')}d\tau d\tau'$. Plugging in (12.20) yields
$R_{G_t G_t}(t, t'; f_\tau, f_\tau') = \iint_{-\infty}^{\infty} P_g(t, t'; \tau)\delta(\tau' - \tau)e^{-j2\pi(-f_\tau \tau + f_\tau' \tau')}d\tau d\tau' = \int_{-\infty}^{\infty} P_g(t, t'; \tau)e^{-j2\pi \Delta f_\tau \tau}d\tau$. The last term is just a function of Δf rather than f_τ

and f'_τ. Consequently $R_{G_tG_t}(t, t'; f_\tau, f'_\tau) = R_{G_tG_t}(t, t'; \Delta f_\tau)$, which finalizes the proof.

12.5 Wide Sense Stationary Processes with Uncorrelated Scattering

12.5.1 Delay Cross Power Spectral Density of $g(t, \tau)$

Now we consider a time-variant system with a delay spread function $g(t, \tau)$, which represents a wide sense stationary process (with respect to the time t) and exhibits uncorrelated scattering (with respect to the delay time τ). To this end the conditions (12.11) for the wide sense stationarity and (12.20) for the uncorrelated scattering must be true simultaneously. Then follows from (12.20) with the replacement of t, t' by Δt the autocorrelation function of $g(t, \tau)$

$$R_{gg;WSSUS}(t, t'; \tau, \tau') = R_{gg}(\Delta t; \tau, \tau') = P_g(\Delta t; \tau)\delta(\tau' - \tau) \qquad (12.24)$$

$P_g(\Delta t; \tau)$ is denoted as the *delay cross power spectral density* of $g(t, \tau)$ that just depends on the time difference Δt and features a Dirac impulse at $\tau' = \tau$.

12.5.2 Doppler Power Spectrum

The Fourier transform of $P_g(\Delta t; \tau)$ with respect to Δt yields another power density spectrum of $g(t, \tau)$, which is called *Doppler power spectrum* $S_{gD}(f_t, \tau)$

$$P_g(\Delta t; \tau) \overset{\Delta t}{\rightharpoonup} S_{gD}(f_t, \tau) = \int_{-\infty}^{\infty} P_g(\Delta t; \tau)e^{-j2\pi f_t \tau \Delta t} d\Delta t \qquad (12.25)$$

12.5.3 Autocorrelation Function of Time-Variant Transfer Function

From (12.23) we can directly obtain the autocorrelation function of the time-variant transfer function $G_t(t, f_\tau)$ for the case of WWSUS, if the argument t, t' is replaced by Δt

$$R_{G_tG_t,WSSUS}(t, t'; f_\tau, f'_\tau) = R_{G_tG_t}(\Delta t; \Delta f_\tau) \qquad (12.26)$$

and is also called *time-frequency correlation function*.

12.6 Simplified Parameters for Time-Variant Systems

In several applications, such as wireless communications, different categories of the time-variance of a system can be distinguished. In case of a slow change of the environment and/or slow movement of the transmitter or receiver, the time-variant system can be considered as quasi static during time periods or blocks, which are much larger than the duration of a transmit symbol. The time-variance is then denoted as block fading. For other use cases, such as fast moving terminals in cars, trains or even airplanes, the channel characteristics are changing rapidly. Consequently, there is a need to differentiate between these applications and use dedicated, yet simple parameters. In many applications the *coherence bandwidth* $\Delta f_{\tau,coh}$ and the *coherence time* Δt_{coh} turn out to be useful indicators and therefore frequently used. In the following we discuss these parameters for deterministic and stochastic system functions.

12.6.1 Coherence Bandwidth

We define the coherence bandwidth as the frequency interval $\Delta f_{\tau,coh}$, in which the magnitude of the time-variant transfer function $G_t(t, f_\tau)$ does not change significantly. Thus, in case of a deterministic system we can just take the 3dB cut-off frequency and define

$$\left|G_t(t, \Delta f_{\tau,coh})\right|^2 = \frac{1}{2} \left|G_t(t, 0)\right|^2 \tag{12.27}$$

for a fixed time instant t, or a small time interval. Alternatively the coherence bandwidth can also be defined by means of the Doppler spread function $G(f_t, f_\tau)$ along the f_τ-axis.

For a randomly changing time-variant system we have to apply statistical measures. Presuming a channel with wide sense stationary uncorrelated scattering we can make use of the autocorrelation function $R_{G_t G_t}(\Delta t; \Delta f_\tau)$ of the time-variant transfer function given in (12.26) and define a 3dB cut-off frequency as coherence bandwidth $\Delta f_{\tau,coh}$

$$\left|R_{G_t G_t}(\Delta t; \Delta f_{\tau,coh})\right| = \frac{1}{2} \left|R_{G_t G_t}(\Delta t; 0)\right| \tag{12.28}$$

for a fixed time, e.g., $\Delta t = 0$. Please note that $\Delta t = 0$ reflects the time instant $t = t'$ according to (12.10). Often $R_{G_t G_t}(\Delta t; \Delta f_\tau)$ is frequency selective with many ripples rather than monotonic with respect to Δf_τ. In this case adequate root mean square values are useful [2]. Inside the coherence bandwidth the spectral components of $R_{G_t G_t}(\Delta t; \Delta f_\tau)$ are strongly correlated. Recall that f_τ characterizes the frequency for the transmission spectrum. Hence, the larger $f_{\tau,coh}$ is the more spectral components can be conveyed from the input to the output of the time-variant system.

12.6.2   *Coherence Time*

The coherence time Δt_{coh} can be defined as that time interval, in which the delay spread function $g(t, \tau)$ does not change significantly with respect to time t for any fixed delay time τ. For a deterministic system we may determine a 3dB cut-off frequency $f_{t,coh}$ using the Doppler spread function $G(f_t, f_\tau)$ as

$$\left| G(f_{t,coh}, f_\tau) \right|^2 = \frac{1}{2} |G(0, f_\tau)|^2 \tag{12.29}$$

for any fixed f_τ. For illustration let us consider the following example.

Example 12
Given a Doppler spread function $G(f_t, f_\tau)$, which shall exhibits the ideal lowpass spectrum

$$G(f_t, f_\tau) = \begin{cases} 1 : |f_t| \le f_{t,c} \ , \ |f_\tau| \le f_{\tau,c} \\ 0 ; \qquad\qquad else \end{cases} \tag{12.30}$$

with the cut-off frequencies $f_{t,c}$ and $f_{\tau,c}$. Of course due to the step-wise transition there is no 3dB point. Therefore we alternatively take $f_{t,c}$. If we apply the two-dimensional inverse Fourier transform we obtain

$$g(t, \tau) = 4 f_{t,c} f_{\tau,c} \ \text{sinc}(2 f_{t,c} t) \ \text{sinc}(2 f_{\tau,c} \tau) \tag{12.31}$$

Roughly speaking, we consider the function $\text{sinc}(2 f_{t,c} t)$ approximately as constant up to the first zero at $t_0 = \frac{1}{2 f_{t,c}}$ and can find the coherence time as

$$\Delta t_{coh} = t_0 = \frac{1}{2 f_{t,c}} \tag{12.32}$$

Apparently, Δt_{coh} is inversely proportional to the maximal Doppler frequency $f_{t,c}$.

As already mentioned for the coherence bandwidth, in case of a randomly time-variant system we have to work on the basis of the autocorrelation and the power spectral density functions. Then we define the coherence time as the time interval Δt_{coh}, in which the delay spread function $g(t, \tau)$ is strongly correlated. For a wide sense stationary system with uncorrelated scattering the delay cross power spectral density $P_g(\Delta t; \tau)$ given in (12.24) is of significance and should be large in the interval Δt_{coh}. Equivalently in the frequency domain we can consider the 3dB cut-off frequency $f_{t,c}$ of the Doppler power spectrum $S_{gD}(f_t, \tau)$ in (12.25) as a measure

$$\left| S_{gD}(f_{t,c}, \tau) \right| = \frac{1}{2} \left| S_{gD}(0, \tau) \right| \tag{12.33}$$

for any fixed delay τ, mostly taken at $\tau = 0$. Similar to the example above, if $S_{gD}(f_t, \tau)$ is a rectangular lowpass as a function of f_t and with cut-off frequency $f_{t,c}$, then the coherence time Δt_{coh} can be defined likewise as in (12.32).

References

1. Bello, P.A.: Characterization of randomly time-variant linear channels. IEEE Trans. Commun. Syst. (1963)
2. Molisch, A.F., Asplund, H., Heddergott, R., Steinbauer, M., Zwick, T.: The COST259 directional channel model—part i: overview and methodology. IEEE Trans. Wireless Commun. **5** (2006)

Part III
Multiple Input Multiple Output Wireless Transmission

Chapter 13
Principles of Multiple Input Multiple Output Transmission

13.1 Introduction and Background

After the first demonstrations of electromagnetic waves in the year 1887 by the physicist Heinrich Hertz at the Technical University of Karlsruhe in Germany wireless telegraphy transmission was demonstrated at the end of the 19th century by the radio pioneer and founder of the later company Guglielmo Marconi. Besides quite some important developments of different antenna technologies the early ideas for Multiple Input Multiple Output (MIMO) schemes using multiple antennas trace back to Kaye and George [1], Branderburg and Wyner [2], and van Etten (1975), [1–3]. A concise survey is given in [4]. Later in 1984 and 1986 Winters and Salz considered beamforming techniques at Bell Laboratories, [5]. In 1994 Paulraj and Kailath introduced a patent on the concept of spatial multiplexing using multiple antennas. Raleigh and Cioffi investigated the transmission of multiple data streams using spatial-temporal coding, [6]. In the same year Foschini introduced the concept of Bell Labs Layered Space-Time (BLAST), [7], which was refined and implemented later in 1999 by Golden et al. [8]. The digital cellular system GSM (Global System for Mobile Communications) put into operation around 1992 in Europe did not yet use the MIMO principle. However, later standards such as the 3.5 Generation (3.5G, UMTS advanced, IMT 2000), the 4G, and the 5G cellular systems adopt this key technology. Similar developments and standards prevailed for the wireless local area network WLAN and WIMAX IEEE 802.11.

Starting around the year 2000 ideas came up to introduce the MIMO principle not only in the area of wireless broadcasting but also in the field of wire-line digital transmission. There the multiple antennas are replaced in principle by wire-line multiple transceivers. Today applications and standards are present in the field of digital transmission on power-lines [9, 10], and digital subscriber lines (vectoring), [11]. A survey on advances in wireless MIMO research and technology is also found in [12]. In the following sections we introduce the principles of signal transmission over a Multiple Input Multiple Output (MIMO) channel and emphasize on wireless

© Springer Nature Switzerland AG 2021
J. Speidel, *Introduction to Digital Communications*,
Signals and Communication Technology,
https://doi.org/10.1007/978-3-030-67357-4_13

links. We will derive a block diagram of a MIMO system and compact it using the concept of the equivalent baseband channel, which is described in Part I for single input single output (SISO) channels. We characterize the physical channels between the various inputs and outputs of the MIMO system by the delay spread functions, which are functions of two variables t and τ to prepare for the case of time-variant wireless channels. For time-invariant MIMO channels, such as cables composed of many two-wire electrical lines or optical fibers, the delay spread function turns into the impulse response and just depends on the delay variable τ, which is then renamed as t.

13.2 MIMO Transmission System with Quadrature Amplitude Modulation

13.2.1 System Model

Figure 13.1 shows the principle block diagram of a MIMO transmission system. The transmitter and the receiver are equipped with M and N parallel branches, respectively. In principle, the branches on one side are composed of the same building blocks. Only transmit branch j and receive branch i are shown in more detail.

Fig. 13.1 Principle block diagram of a digital MIMO transmission system with single carrier QAM

Single Input Single Output Link

Assume for the moment that only the transmit branch j is active and that all other transmit signals are zero, $s_p(k) = 0 \; \forall \, p \neq j$. Then the link from transmit branch j to receive branch i represents a single input single output scheme, which is described in quite some detail in Part I. Briefly, its operation is as follows. The binary bit stream $b_j(l)$, which contains the information bits and in most cases redundancy bits for forward error correction (FEC), is fed into a mapper, which periodically allocates κ_j consecutive bits to a complex symbol $s_j(k)$. We denote $l \in \mathbb{Z}$ and $k \in \mathbb{Z}$ as the discrete-time of the bit and the symbol sequence, respectively. Thus, the symbol alphabet $\mathcal{B}$ consists of $L_j = 2^{\kappa_j}$ symbols, which can be portrait in the complex domain. Together with the complex modulation carrier $e^{j2\pi f_0 t}$ the scheme constitutes a quadrature amplitude modulation (QAM) transmitter. The pulse shaper, which is a lowpass filter, limits the infinite bandwidth of the symbol sequence $s_j(k)$ to at most half the symbol rate $v_S = \frac{1}{T}$. We call T the symbol interval, i.e. every T a symbol leaves the mapper. The modulator with the carrier frequency f_0 shifts the spectrum to the channel passband with the center frequency f_0. As all physical channels own real-valued impulse responses, only the real part of the complex modulator output signal can be transmitted. Consequently, the modulator in Fig. 13.1 also contains a unit, which selects this part to create the real-valued output signal $u_j(t)$. The channel will be discussed later in more detail. For the moment it shall be characterized by the delay spread function $g_{C,ij}(t, \tau)$.

In the receiver branch i a bandpass filter may be employed to limit the noise spectrum to the channel passband, yet leaving the signal part of the receive signal unchanged. After synchronous demodulation with the complex carrier $e^{-j2\pi f_0 t}$ the demodulated signal is lowpass filtered yielding the complex-valued signal $\tilde{r}_i(t)$, which is sampled with symbol rate $\frac{1}{T}$ at $t = t_0 + kT$, resulting in

$$r_i(k) = \tilde{r}_i(t_0 + kT) \; ; \quad i = 1, 2, ..., N \tag{13.1}$$

where the delay t_0 between the transmitter and the receiver has to be estimated at the receiver.

Physical Single Input Single Output Channel

Electrical cables or optical fibers are in general time-invariant and will be described by an impulse response $g_{C,ij}(t)$. A wireless and mobile channel is time-variant due to the temporal change of the environment and the movement of the transmitter or receiver. As is known from Part I and II, such a channel can be characterized by a time-variant impulse response $w_{C,ij}(t, s)$, which is the response at observation time t to a Dirac impulse active at time instant $s \leq t$, [13, 14]. The variables t and s are independent. The use of $w_{C,ij}(t, s)$ has some drawbacks for signal processing, in particular it does not provide a meaningful Fourier spectrum. Therefore the transformation of variables $s = t - \tau$ is applied yielding the *modified impulse response* or *delay spread function*

$$w_{C,ij}(t, t - \tau) = g_{C,ij}(t, \tau) \tag{13.2}$$

which can then be interpreted as the response at observation time t to an input Dirac impulse at $t - \tau \leq t$. We call t the observation time or just time and τ is referred to as delay time or age variable. The input-output relation between $u_j(t)$ and $v_i(t)$ in Fig. 13.1 is given by the time-variant convolution

$$v_i(t) = \int_{-\infty}^{\infty} u_j(s) w_{C,ij}(t, s) ds = \int_{-\infty}^{\infty} u_j(t - \tau) g_{C,ij}(t, \tau) d\tau \qquad (13.3)$$

where we have used $s = t - \tau$ and (13.2). In case of time-invariance the delay spread function turns into the impulse response, $g_{C,ij}(t, \tau) = g_{C,ij}(\tau)$. It just depends on the variable τ, which is then renamed as t yielding $g_{C,ij}(t)$. Then the time-variant convolution boils down to the well known convolution for time-invariant systems

$$v_i(t) = \int_{-\infty}^{\infty} u_j(t - \tau) g_{C,ij}(\tau) d\tau \qquad (13.4)$$

Equivalent Baseband System Model

To obtain a compact mathematical description without details of impulse shaping, modulation, demodulation etc., the described SISO link from transmitter j to receiver i in Fig. 13.1 can be simplified by allocating between the input symbol sequence $s_j(k)$ of the impulse shaper and the output signal $\tilde{r}_i(t)$ of the receive lowpass an equivalent baseband model with overall impulse response $\tilde{w}_{ij}(t, s)$ or with the delay spread function $\tilde{h}_{ij}(t, \tau) = \tilde{w}_{ij}(t, t - \tau)$ using $s = t - \tau$. Similarly, $\tilde{h}_{ij}(t, \tau)$ is the response observed at time instant t to a Dirac impulse at instant $t - \tau \leq t$. After sampling at $t = t_0 + kT$ and $\tau = \tau_0 + mT$ we define $\tilde{h}_{ij}(t_0 + kT, \tau_0 + mT) = h_{ij}(k, m)$, which we call in the following the *discrete-time modified impulse response of the equivalent baseband system* or *discrete-time delay spread function of the equivalent baseband system*. In Fig. 13.1 $h_{ij}(k, m)$ characterizes the SISO link from the input sequence $s_j(k)$ at the transmitter to the output sequence $r_i(k)$ of the sampling device at the receiver. Similarly, it can be interpreted as the response observed at discrete-time instant k to a unit impulse at instant $k - m \leq k$. This leads to the simplified block diagram in Fig. 13.2 of the MIMO system with discrete-time equivalent baseband channels.

In case of a time-invariant channel, such as a cable, the equivalent baseband system impulse response $\tilde{h}_{ij}(t)$ and its sampled version $\tilde{h}_{ij}(t_0 + kT) = h(k)$ are representing the SISO link. To simplify the notation in the remainder we always use the general term $h_{i,j}(k, m)$ and whether we mean $h_{ij}(m)$ of a time-invariant link will become clear from the context. Please note that due to modulation and demodulation with a complex carrier, $h_{ij}(k, m)$ is complex in general.

If all M transmit signals $s_j(k)$ in Fig. 13.2 are active, the receive signal $r_i(k)$ is the superposition. This holds for all receive signals $r_1(k), ..., r_N(k)$. They are "combined" by the building block detector and demapper in Fig. 13.1, which provides the estimates $\hat{b}_j(l)$ of the transmit bit sequences $b_j(l)$; $j = 1, 2, ..., M$. Various

Fig. 13.2 Principle block diagram of a MIMO system with discrete-time equivalent baseband channels characterized by the delay spread functions $h_{ij}(k, m)$

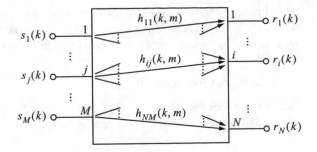

signal combiners and detection methods for MIMO receivers are described in the Chaps. 14 and 15 in detail.

13.2.2 Input-Output Relation of MIMO System with Time-Variant Channel

Single Input Single Output (SISO)

Let us assume that only the transmit branch j is active. As mentioned before, we characterize the SISO channel composed of the transmit branch j, the physical channel, and the receive branch i by the equivalent delay spread function $h_{ij}(k, m)$. Then we obtain at the output of the receive branch i

$$r_i(k) = s_j(k) \circledast h_{ij}(k, m) + n_i(k) \tag{13.5}$$

where $n_i(k)$ is the additive noise at the receive branch i and $\circledast$ denotes the discrete-time, time-variant convolution defined as

$$s_j(k) \circledast h_{ij}(k, m) = \sum_{m=-\infty}^{\infty} h_{ij}(k, m) s_j(k - m) \tag{13.6}$$

We have introduced the operator $\circledast$ to prevent from confusion with the time-invariant convolution, which we indicate by $*$. The prove of (13.6) is evident from (13.10), which is discussed later. The time-variant and the time-invariant convolution look similar, however, they differ in several properties. The two variables k and m specify two different time variables. m is called delay time, age variable, or integration time, as the summation in (13.6) is executed over m, and k characterizes the time variable of the output signal of the system and in addition also the temporal change of the system parameters. If the channel is time-invariant, then $h_{ij}(k, m)$ does not depend on k. Hence, we can just skip k in $h_{ij}(k, m)$ of (13.5) as well as (13.6) and write $h_{ij}(k, m) = h(m)$, as is shown in Part II. Consequently, $\circledast$ turns into the well known convolution operation ∗ for linear time-invariant systems. An example for $h_{ij}(k)$ is a wire-line channel or a static wireless channel, where the transmitter and the receiver are non-moving and all scatterers or reflectors of the electromagnetic waves do not vary with time k.

Multiple Input Single Output (MISO)

Now we consider that all transmit branches $j = 1, 2, ..., M$ and one receive branch i are active. Then we obtain for the receive branch i with (13.5)

$$r_i(k) = \sum_{j=1}^{M} s_j(k) \circledast h_{ij}(k, m) + n_i(k) \tag{13.7}$$

Multiple Input Multiple Output (MIMO)

Finally all transmit and all receive branches are active. Consequently, (13.7) holds for $i = 1, 2, ..., N$, which gives rise to a matrix notation, in which (13.7) represents the ith equation

$$\underbrace{\begin{pmatrix} r_1(k) \\ \vdots \\ r_i(k) \\ \vdots \\ r_N(k) \end{pmatrix}}_{\mathbf{r}(k)} = \underbrace{\begin{pmatrix} h_{11}(k,m) & \cdots & h_{1j}(k,m) & \cdots & h_{1M}(k,m) \\ & & \vdots & & \\ h_{i1}(k,m) & \cdots & h_{ij}(k,m) & \cdots & h_{1M}(k,m) \\ & & \vdots & & \\ h_{N1}(k,m) & \cdots & h_{Nj}(k,m) & \cdots & h_{NM}(k,m) \end{pmatrix}}_{\mathbf{H}(k,m)} \circledast \underbrace{\begin{pmatrix} s_1(k) \\ \vdots \\ s_j(k) \\ \vdots \\ s_M(k) \end{pmatrix}}_{\mathbf{s}(k)} + \underbrace{\begin{pmatrix} n_1(k) \\ \vdots \\ n_i(k) \\ \vdots \\ n_N(k) \end{pmatrix}}_{\mathbf{n}(k)}$$

$$\tag{13.8}$$

which can be written as

$$\mathbf{r}(k) = \mathbf{H}(k, m) \circledast \mathbf{s}(k) + \mathbf{n}(k) \tag{13.9}$$

$\mathbf{r}(k)$ is the receive vector, $\mathbf{H}(k, m)$ the matrix of the delay spread functions of the discrete-time equivalent MIMO channel model, $\mathbf{s}(k)$ the transmit signal vector, and $\mathbf{n}(k)$ the vector of the additive noise. Please note that $\circledast$ turns the matrix multiplication into the convolution operation defined in (13.6), where the order of operation is $s_j(k) \circledast h_{ij}(k, m)$ and has to be respected, because $\circledast$ is in general non-commutative.

13.3 Deterministic Models for Wireless MIMO Channels

In Part I on digital wireless communications various models for SISO channels and their properties are outlined in quite some detail. There are deterministic and stochastic models. All these findings can be applied for any link between the transmit branch j and the receive branch i in Figs. 13.1 and 13.2. The significance needs no detailed explanation and we restrict ourselves to a brief introduction of the finite impulse response channel model. Furthermore, we emphasize on the spatial interrelation of these SISO channels by considering their correlation properties.

13.3.1 Uniform Linear and Uniform Circular Antenna Arrays

Figure 13.3 shows in principle the configurations of MIMO antennas, which are frequently used, the uniform linear and the uniform circular array antenna. In both antennas the distance between adjacent elements is equal. These antennas can be applied for transmission and reception. From the theory of electromagnetic waves we know that the signals emitted by the antenna elements can interact and then are correlated, if approximately $l \leq \frac{\lambda_0}{2}$ holds, where $\lambda_0 = \frac{c}{f_0}$ is the operating wavelength, c is the speed of light in the air, and f_0 is the carrier frequency, [15]. The closer the antenna elements are, the higher the correlation of the signals can be, because in case of a transmit MIMO antenna the waves of neighboring elements overlap. Correlation also allows the shaping of the emitted electromagnetic beam into the direction of a hot spot of users by appropriate antenna feeding signals generated by a precoding vector. This technique is called transmit beamforming. In case of a receive MIMO antenna adjacent narrow elements are excited by signals, which are quite similar and thus receive correlation is introduced.

If the distance between adjacent antenna elements is increased, the correlation among the signals decreases. For $l \approx \lambda_0$ micro diversity starts and for larger distances such as $l > 3\lambda_0$ the correlation almost vanishes. This mode of operation is used for spatial diversity, which allows the transmission or reception of almost independent signals. Depending on the applications MIMO antennas with narrow and far spaced elements can also be combined.

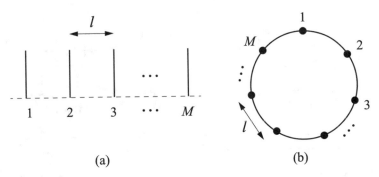

(a)   (b)

Fig. 13.3 Arrays with M antenna elements spaced by l, **a** uniform linear array, **b** uniform circular array

13.3.2 *Finite Impulse Response Channel Model*

As is well known from signal processing a linear discrete-time system is characterized by an impulse response with infinite duration (infinite impulse response, IIR). In addition, if the system is non-causal and time-invariant the input-output relation for the SISO link between transmitter j and receiver i is given by $r_i(k) = \sum_{m=-\infty}^{\infty} h_{ij}(m)s_j(k-m)$. Typically, the magnitude of the impulse response $h_{ij}(m)$ of a time-invariant equivalent baseband channel declines with increasing m. Therefore an adequate approximation can be obtained by pruning the response at $m = K'_{ij}$ resulting in a finite impulse response (FIR) filter model.

Similarly, the magnitude of the delay spread function $h_{ij}(k, m)$ of the equivalent baseband wireless channel approaches zero for increasing m and thus we can discard the samples $h_{ij}(k, m)$ for $m > K_{ij}$ to obtain an approximation with respect to m. This simplified channel model is depicted in Fig. 13.4 and shows the structure of a finite impulse response (FIR) filter, which is also called tapped delay line or transversal filter. The $K_{ij} + 1$ channel coefficients are given by $h_{ij}(k, m)$, where $m = 0, 1, ..., K_{ij}$. In contrast to a time-invariant filter the coefficients (taps) can change their values at every time instant k.

All operations in the filter are linear namely the delay Δ_τ of the input signal $s_j(k-m)$, the multiplication with the tap $h_{ij}(k, m)$ yielding $h_{ij}(k, m)s_j(k-m)$, and the addition of all component signals. Consequently, the output signal is

$$r_i(k) = \sum_{m=0}^{K_{ij}} h_{ij}(k, m)s_j(k - m) \ ; \ i = 1, 2, ..., N \ ; \ j = 1, 2, ..., M \qquad (13.10)$$

where we have imposed the requirement that the delay Δ_τ corresponds to the time base T of the input and output signal,

$$\Delta_\tau = T \qquad (13.11)$$

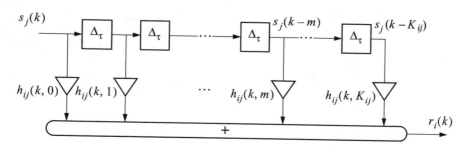

Fig. 13.4 Time-variant transversal filter for modeling the equivalent baseband SISO channel between transmitter j and receiver i

The reader is encouraged to convince oneself of the linearity by using the superposition of two input signals. Please note that the considered FIR filter is causal, because $h_{ij}(k, m) = 0 \, \forall m < 0$. A physical interpretation of (13.10) is a wireless channel where the transmit signal $s_j(k)$ is spread over $K_{ij} + 1$ different paths on its way to the receiver. Each path m introduces a dedicated delay mT and a path loss indicated by the complex channel coefficient $h_{ij}(k, m)$. Such a channel is called a time-variant multipath channel. It should be mentioned that the described modeling is done on the basis that the delay of each physical path is quantized by multiples of $\Delta_\tau = T$, as indicated by (13.11), where T is the symbol interval.

As portrait in Fig. 13.2, a wireless MIMO system is characterized by $N \cdot M$ of such multipath SISO channel models, where $j = 1, 2, ..., M$ and $i = 1, 2, ..., N$.

13.3.3 Spatial Channel Models

As outlined in Chap. 5 in most cases the transmit signal propagates along a multitude of paths with reflections and scattering. For MIMO systems several studies have been made to find models, which incorporate spatial parameters of the delay spread function. Among others each propagation path is divided into a large number of subpaths and their signals are characterized by the angle of depature from the transmit antenna, the angle of arrival at the receive antenna, the distances of the antenna elements, the phase difference of the waves, and the direction of the velocity of the moving receiver, [16]. Assuming that the angles change statistically, angular spread functions have been defined, similar to the statistical parameters discussed in Sect. 13.4.

13.3.4 Spectral Properties of the Channel Model

Time-Variant Transfer Function

We consider again the SISO channel model between the transmitter j and the receiver i with its delay spread function $h_{ij}(k, m)$. As described in Part II, a time-variant transfer function $\tilde{H}_{ij}(k, e^{j2\pi f_\tau T})$ can be defined by the Fourier transform of $h_{ij}(k, m)$ with respect to m. To this end we first apply the z-transform on $h_{ij}(k, m)$ with respect to m resulting in $\tilde{H}_{ij}(k, z) = \sum_{m=0}^{\infty} h_{ij}(k, m)z^{-m}$ and then substitute $z = e^{j2\pi f_\tau T}$ yielding

$$h_{ij}(k, m) \overset{m}{\longmapsto} \tilde{H}_{ij}(k, e^{j2\pi f_\tau T}) = \sum_{m=0}^{\infty} h_{ij}(k, m)e^{-j2\pi f_\tau T m} \qquad (13.12)$$

where the discrete delay variable mT corresponds to the frequency variable f_τ. Obviously, $\tilde{H}_{ij}(k, e^{j2\pi f_\tau T})$ is periodic with period $f_\tau = \frac{1}{T}$. Its baseband is located in $|f_\tau| \leq \frac{1}{2T}$. If $\left|\tilde{H}_{ij}(k, e^{j2\pi f_\tau T})\right|$ varies in this frequency interval significantly with f_τ for any fixed k, then we call this channel *frequency selective* and the effect as *frequency selective fading*. Given the fact that this property is owned by at least one SISO channel of the MIMO scheme in Fig. 13.2, the MIMO channel is referred to as frequency selective.

In contrast, if $\left|\tilde{H}_{ij}(k, e^{j2\pi f_\tau T})\right|$ is approximately constant with respect to f_τ in the aforesaid interval up to a certain cut-off frequency for any fixed k, then the SISO channel is referred to as non-frequency selective or *frequency flat*. Then we conclude from (13.12) that all channel coefficients must be zero except one, e.g.,

$$h_{ij}(k, 0) = h_{ij}(k) \tag{13.13}$$

where $h_{ij}(k)$ has to be understood as a short hand notation. Hence, the SISO channel is just time-variant and from (13.5) follows with (13.6) the input-output relation

$$r_i(k) = s_j(k)h_{ij}(k) + n_i(k) \tag{13.14}$$

Here we can study easily the effect of a time-variant channel. Assume there is no noise. Even if we sent a constant signal s_j, we get from (13.14) the response $r_i(k) = s_j h_{ij}(k)$ which varies with time.

If all SISO channels of the MIMO system are frequency flat, at least approximately, then the MIMO scheme is called frequency flat and from (13.9) follows

$$\mathbf{r}(k) = \mathbf{H}(k)\mathbf{s}(k) + \mathbf{n}(k) \tag{13.15}$$

with $\mathbf{H}(k, m) = \mathbf{H}(k, 0) = \mathbf{H}(k)$ as a short hand notation. Obviously, the time-variant convolution turns into the multiplication.

Delay Doppler Spread Function

As is known from the theory of time-variant systems in Part II, we obtain the delay Doppler spread function also called Doppler-variant impulse response, if we consider the Fourier spectrum of the delay spread function with respect to the time t or k. We start with the z-transform with respect to k yielding $\bar{H}_{ij}(z, m) = \sum_{k=0}^{\infty} h_{ij}(k, m)z^{-k}$. Substituting $z = e^{j2\pi f_t T}$ results in

$$h_{ij}(k, m) \overset{k}{\longrightarrow} \bar{H}_{ij}(e^{j2\pi f_t T}, m) = \sum_{k=0}^{\infty} h_{ij}(k, m)e^{-j2\pi f_t Tk} \tag{13.16}$$

where k of the original domain corresponds to the Doppler frequency variable f_t in the frequency domain. We see that $\bar{H}_{ij}(e^{j2\pi f_t T}, m)$ is periodic with period $f_t = \frac{1}{T}$ and its baseband is located in the interval $|f_t| \leq \frac{1}{2T}$. If the cut-off frequency of

Fig. 13.5 Dividing the magnitude of the time-variant transfer function $\left|\tilde{H}_{ij}(k,\mathrm{e}^{\mathrm{j}2\pi f_\tau T})\right|$ of the equivalent baseband channel into small regions of approximately flat segments using multi-carrier modulation with carrier frequencies $f_0, 2f_0,\dots$

$\left|\tilde{H}_{ij}(\mathrm{e}^{\mathrm{j}2\pi f_\tau T}, m)\right|$ in this interval is small, then the variations of $h_{ij}(k, m)$ with respect to k are slow. In this case the channel exhibits a weak temporal fading. The contrary is true, if the cut-off frequency is large resulting in fast changes.

Step-Wise Frequency Flat Channel Transfer Function

Frequency flat regions of the channel transfer function can be achieved, if multi-carrier modulation such as orthogonal frequency division multiplexing (OFDM) is applied discussed in Chap. 7. As illustrated in Fig. 13.5, this modulation technique subdivides the total transmission band into approximately flat segments between adjacent carriers $\mathrm{e}^{\mathrm{j}2\pi f_0 t}$, $\mathrm{e}^{\mathrm{j}2\pi 2 f_0 t}$, $\mathrm{e}^{\mathrm{j}2\pi 3 f_0 t}$, ..., which are in case of OFDM generated by an inverse discrete Fourier transform (IDFT) as a modulator and by a discrete Fourier transform (DFT) as a demodulator.

To this end each transmit and receive branch in Fig. 13.1 is composed of an IDFT and a DFT, respectively. In each flat frequency region an input-output relation similar to (13.14) holds per SISO channel. We will focus for all further considerations on MIMO systems with frequency flat fading channels described by (13.15).

13.4 Statistical Models for MIMO Channels

13.4.1 i.i.d. Gaussian MIMO Channel Model

As outlined in [15] and summarized in Part I on digital communications over single input single output links, multipath propagation, rich scattering, and reflections lead to a multitude of scenarios which in most cases cannot be described by detailed deterministic models. Suitable are statistical models, which will be reviewed in the following. As already stated, all SISO channels of the MIMO system shall exhibit a flat frequency response, $\left|\tilde{H}_{ij}(k, \mathrm{e}^{\mathrm{j}2\pi f_\tau T})\right| = const.$ with respect to f_τ, for any k. Therefore only the temporal fading of the complex-valued delay spread functions

$$h_{ij}(k) = \mathrm{Re}\left[h_{ij}(k)\right] + \mathrm{jIm}\left[h_{ij}(k)\right] \; ; \; j = 1, 2, ..., M \,; \, i = 1, 2, ..., N \quad (13.17)$$

has to be considered, which are now stochastic processes. A summary on the definitions and properties of random variables and stochastic processes is given in the Appendix A.

Assume that the output signal of an unidirectional transmit antenna element j undergoes rich scattering and reflections resulting in a multipath propagation with an infinite number of statistically independent paths carrying signals, which superimpose at the receive antenna element i resulting in the stochastic process $h_{ij}(k)$. Then the conditions of the central limit theorem are fulfilled, which states that $h_{ij}(k)$ exhibits a complex Gaussian probability density function. If there is no line of sight between the transmitter and the receiver, the receive signal has zero mean. In detail we can model each delay spread function in (13.17) as follows:

1. All $h_{ij}(k)$ exhibit zero mean,

$$\mathbf{E}\left[h_{ij}(k)\right] = 0 \tag{13.18}$$

from which follows that real and imaginary part of $h_{ij}(k)$ have zero mean. This prerequisite is fulfilled, if there is no line of sight between transmitter and receiver.
2. All $h_{ij}(k)$ are independent and identically distributed (i.i.d.) according to a Gaussian probability density function given by

$$p_x(x) = \frac{1}{\sqrt{2\pi}\sigma_x} e^{-\frac{x^2}{2\sigma_x^2}} \tag{13.19}$$

where x stands for Re $\left[h_{ij}(k)\right]$ and Im $\left[h_{ij}(k)\right]$ for any fixed k. The corresponding variances are $var(x) = \mathbf{E}\left[(x - \mathbf{E}[x])^2\right] = \sigma_x^2$.
3. All $h_{ij}(k)$ are circular symmetric Gaussian processes, i.e.

$$var\left(\text{Re}\left[h_{ij}(k)\right]\right) = var\left(\text{Im}\left[h_{ij}(k)\right]\right) \tag{13.20}$$

4. All $h_{ij}(k)$ are pairwise spatially uncorrelated,

$$\mathbf{E}\left[h_{pq}h_{rs}^*\right] = \begin{cases} 1 & ; \ p = r \ ; \ q = s \\ \mathbf{E}\left[h_{pq}(k)\right]\mathbf{E}\left[h_{rs}^*(k)\right] = 0 \ ; & else \end{cases} \tag{13.21}$$

The channel matrix $\mathbf{H}(k)$ is therefore denoted as $\mathbf{H}_w(k)$, where the index w stand for white and shall indicate the uncorrelated entries $h_{ij}(k)$. The second line in (13.21) is zero, because of (13.18). As the $h_{ij}(k)$ are Gaussian, they are even statistically independent.

The requirements above characterize a wide sense stationary (WSS) Gaussian process. The prerequisite (13.21) is often denoted as uncorrelated scattering. Therefore the considered channel model is also named *wide sense stationary uncorrelated scattering (WSSUS)* model.

With these prerequisites it can be shown that all $|h_{ij}|$ exhibit a Rayleigh probability density function and that $\arg[h_{ij}]$ is equally distributed in the interval $[-\pi, \pi]$. If there is a line of sight between transmitter and receiver, then $\mathbf{E}[h_{ij}(k)] \neq 0$ and $|h_{ij}|$ shows a Rice distribution.

13.4.2 Covariance Matrix of the MIMO Channel

If we use the term correlation, we always mean spatial correlation in the following. First we define the covariance matrix $\mathbf{R}_{HH}$ of our channel matrix $\mathbf{H}(k)$. As a covariance matrix is defined on vectors, we have to rearrange the matrix $\mathbf{H}(k)$ as a vector by stacking all column vectors of the matrix one over the other. In the following we drop k to simplify the notation. Let

$$\mathbf{H} = \begin{pmatrix} \mathbf{h}_1 & \mathbf{h}_2 & \cdots & \mathbf{h}_M \end{pmatrix} \tag{13.22}$$

be the channel matrix decomposed into column vectors $\mathbf{h}_j \in \mathbb{C}^{N \times 1}$; $j = 1, 2, ..., M$. Then the stacking operation is defined as

$$\text{vec}\,(\mathbf{H}) = \begin{pmatrix} \mathbf{h}_1 \\ \mathbf{h}_2 \\ \vdots \\ \mathbf{h}_M \end{pmatrix} \tag{13.23}$$

and the covariance matrix of $\mathbf{H}$ is

$$\mathbf{R}_{HH} = \mathbf{E}\left[\text{vec}\,(\mathbf{H})\, (\text{vec}\,(\mathbf{H}))^H \right] \tag{13.24}$$

where the superscript H denotes the Hermiteian[1] operation also called conjugate transposition. The reader can find a summary of useful theorems on matrix calculus in the Appendix B. Given the matrix $\mathbf{X}$ then $\mathbf{X}^H = (\mathbf{X}^T)^* = (\mathbf{X}^*)^T$, where the superscript T stand for the transposition. In a similar way the Hermiteian operation can also be applied on vectors $\mathbf{v}$. With a column vector $\mathbf{v}$ the product $\mathbf{v}\mathbf{v}^H$ defines a matrix whereas $\mathbf{v}^H\mathbf{v}$ is the scalar product of the two vectors resulting in a scalar. A matrix $\mathbf{X}$ is said to be Hermiteian, if $\mathbf{X}^H = \mathbf{X}$. One convinces easily that $\mathbf{R}_{HH}$ is a Hermiteian matrix.

Example 1

As an example we calculate the covariance matrix for a 2x2 channel matrix

$$\mathbf{H} = \begin{pmatrix} h_{11} & h_{12} \\ h_{21} & h_{22} \end{pmatrix} = \begin{pmatrix} \mathbf{h}_1 & \mathbf{h}_2 \end{pmatrix} \tag{13.25}$$

---

[1] Charles Hermite, French mathematician.

and get from (13.24)

$$\mathbf{R}_{HH} = \mathbf{E}\left[\begin{pmatrix}\mathbf{h}_1\\\mathbf{h}_2\end{pmatrix}\begin{pmatrix}\mathbf{h}_1^H\ \mathbf{h}_2^H\end{pmatrix}\right] = \mathbf{E}\left[\begin{pmatrix}h_{11}\\h_{21}\\h_{12}\\h_{22}\end{pmatrix}\begin{pmatrix}h_{11}^*\ h_{21}^*\ h_{12}^*\ h_{22}^*\end{pmatrix}\right] \qquad (13.26)$$

finally resulting in

$$\mathbf{R}_{HH} = \begin{pmatrix}\mathbf{E}\left[|h_{11}|^2\right] & \mathbf{E}\left[h_{11}h_{21}^*\right] & \mathbf{E}\left[h_{11}h_{12}^*\right] & \mathbf{E}\left[h_{11}h_{22}^*\right]\\[4pt] \mathbf{E}\left[h_{21}h_{11}^*\right] & \mathbf{E}\left[|h_{21}|^2\right] & \mathbf{E}\left[h_{21}h_{12}^*\right] & \mathbf{E}\left[h_{21}h_{22}^*\right]\\[4pt] \mathbf{E}\left[h_{12}h_{11}^*\right] & \mathbf{E}\left[h_{12}h_{21}^*\right] & \mathbf{E}\left[|h_{12}|^2\right] & \mathbf{E}\left[h_{12}h_{22}^*\right]\\[4pt] \mathbf{E}\left[h_{22}h_{11}^*\right] & \mathbf{E}\left[h_{22}h_{21}^*\right] & \mathbf{E}\left[h_{22}h_{12}^*\right] & \mathbf{E}\left[|h_{22}|^2\right]\end{pmatrix} \qquad (13.27)$$

A square matrix $\mathbf{X}$ is called Hermiteian, if $\mathbf{X}^H = \mathbf{X}$. Apparently, $\mathbf{R}_{HH}$ is a Hermiteian matrix. Moreover, we generalize from (13.27) that $\mathbf{R}_{HH} \in \mathbb{C}^{NM \times NM}$ holds, if $\mathbf{H} \in \mathbb{C}^{N \times M}$. For an uncorrelated MIMO channel with channel matrix $\mathbf{H} = \mathbf{H}_w$ the property (13.21) results in the covariance matrix

$$\mathbf{R}_{H_w H_w} = \begin{pmatrix}1&0&0&0\\0&1&0&0\\0&0&1&0\\0&0&0&1\end{pmatrix} = \mathbf{I}_4 \qquad (13.28)$$

which is the 4x4 identity matrix $\mathbf{I}_4$. Please note, if the stochastic process $h_{ij}(k)$ is at least wide sense stationary, then $\mathbf{R}_{HH}$ does not depend on time k.

13.4.3 MIMO Channel Model with Correlation

With $\mathbf{H}_w$ and its covariance matrix $\mathbf{R}_{H_w H_w} = \mathbf{I}_{NM}$ in (13.28) we have introduced a spatially uncorrelated MIMO channel. In the following we prove that a channel matrix $\mathbf{H} \in \mathbb{C}^{N \times M}$ defined with the stacked column vectors

$$\text{vec}\,(\mathbf{H}) = \mathbf{G}^H \text{vec}\,(\mathbf{H}_w) \qquad (13.29)$$

has the covariance matrix

$$\mathbf{R}_{HH} = \mathbf{G}^H \mathbf{G} \qquad (13.30)$$

where $\mathbf{G} = \mathbf{R}_{HH}^{\frac{1}{2}}$ is called the square root matrix of $\mathbf{R}_{HH}$.

Proof of (13.30)

From (13.24) follows with (13.29) $\mathbf{R}_{HH} = \mathbf{E}\left[\mathbf{G}^H \text{vec}\,(\mathbf{H}_w)\,(\text{vec}\,(\mathbf{H}_w))^H\,\mathbf{G}\right]$. As $\mathbf{G}$ shall be non-random, this matrix is not subject to the expected value. Consequently $\mathbf{R}_{HH} = \mathbf{G}^H \mathbf{E}\left[\text{vec}\,(\mathbf{H}_w)\,(\text{vec}\,(\mathbf{H}_w))^H\right]\mathbf{G}$ follows and $\mathbf{E}\left[\text{vec}\,(\mathbf{H}_w)\,(\text{vec}\,(\mathbf{H}_w))^H\right] = \mathbf{R}_{H_w H_w} = \mathbf{I}_{NM}$ results in $\mathbf{R}_{HH} = \mathbf{G}^H \mathbf{G}$, which finalizes the proof.

Example 2

Given a single input multiple output (SIMO) system with $M = 1$ transmit and $N = 2$ receive antennas. The channel matrix shall be $\mathbf{H} = \left(h_{11}\ h_{21}\right)^T = \left(1\ \frac{i}{2}\right)^T$ and approximately constant in a considered time interval. We are looking for the decomposition of the covariance matrix $\mathbf{R}_{HH} = \mathbf{G}^H \mathbf{G}$ with the square root matrix $\mathbf{G} = \mathbf{R}_{HH}^{\frac{1}{2}}$.

Solution:

First we determine the covariance matrix

$$\mathbf{R}_{HH} = \mathbf{E}\left[\begin{pmatrix} h_{11} \\ h_{21} \end{pmatrix}\left(h_{11}^*\ h_{21}^*\right)\right] = \mathbf{E}\left[\begin{pmatrix} 1 \\ \frac{i}{2} \end{pmatrix}\left(1\ -\frac{i}{2}\right)\right] = \begin{pmatrix} 1 & -\frac{i}{2} \\ \frac{i}{2} & \frac{1}{4} \end{pmatrix} \tag{13.31}$$

As $\mathbf{R}_{HH}$ is a Hermiteian matrix, we know from Appendix B that this matrix is unitarily diagonalizable with the eigenvalues $\lambda_i \geq 0$; $i = 1, 2$ and the unitary transform matrix $\mathbf{V}$ with $\mathbf{V}^{-1} - \mathbf{V}^H$

$$\mathbf{R}_{HH} = \mathbf{V}\Lambda\mathbf{V}^H = \mathbf{V}\Lambda^{\frac{1}{2}}\left(\Lambda^{\frac{1}{2}}\right)^H \mathbf{V}^H \tag{13.32}$$

From the comparison with (13.30) we conclude $\mathbf{V}\Lambda^{\frac{1}{2}} = \mathbf{G}^H$ and get the square root matrix $\mathbf{G} = \Lambda^{\frac{1}{2}}\mathbf{V}^H = \mathbf{R}_{HH}^{\frac{1}{2}}$ knowing that the entries of the diagonal matrix $\Lambda^{\frac{1}{2}}$ are real. Apparently, $\mathbf{R}_{HH}^{\frac{1}{2}}$ is not a Hermiteian matrix in general. The diagonal matrix $\Lambda = \text{diag}\,(\lambda_1, \lambda_2)$ contains the eigenvalues of $\mathbf{R}_{HH}$. The characteristic equation for the eigenvalues is

$$\begin{vmatrix} 1 - \lambda & -\frac{i}{2} \\ \frac{i}{2} & \frac{1}{4} - \lambda \end{vmatrix} = 0 \tag{13.33}$$

which yields $\lambda_1 = 0$, $\lambda_2 = \frac{5}{4}$ and thus

$$\Lambda = \begin{pmatrix} 0 & 0 \\ 0 & \frac{5}{4} \end{pmatrix} ;\ \Lambda^{\frac{1}{2}} = \begin{pmatrix} 0 & 0 \\ 0 & \frac{\sqrt{5}}{2} \end{pmatrix} \tag{13.34}$$

The matrix $\mathbf{V}$ is composed of the eigenvectors $\mathbf{v}_i = \left(v_{1i}\ v_{2i}\right)^T$ corresponding to the two eigenvalues and determined by

$$\begin{pmatrix} 1 - \lambda_i & -\frac{j}{2} \\ \frac{j}{2} & \frac{1}{4} - \lambda_i \end{pmatrix} \begin{pmatrix} v_{1i} \\ v_{2i} \end{pmatrix} = \mathbf{0} \tag{13.35}$$

For λ_1 we obtain $\mathbf{v}_1 = \left(\frac{j}{2} v_{21} \ v_{21} \right)^T$. In the same way we get $\mathbf{v}_2 = \left(-j2v_{22} \ v_{22} \right)^T$. The entries $v_{21} \in \mathbb{C}$ and $v_{22} \in \mathbb{C}$ are free parameters, however, as $\mathbf{R}_{HH}$ is a Hermiteian matrix, the two eigenvectors must be orthogonal, see Appendix B. Orthogonalization and normalization of these vectors will not be done at this point. Thus we obtain the transform matrix

$$\mathbf{V} = \left(\mathbf{v}_1 \ \mathbf{v}_2 \right) = \begin{pmatrix} \frac{j}{2} v_{21} & -j2v_{22} \\ v_{21} & v_{22} \end{pmatrix} \tag{13.36}$$

With (13.34) follows

$$\mathbf{G} = \mathbf{R}_{HH}^{\frac{1}{2}} = \Lambda^{\frac{1}{2}} \mathbf{V}^H = \sqrt{5} v_{22}^* \begin{pmatrix} 0 & 0 \\ j & \frac{1}{2} \end{pmatrix} \tag{13.37}$$

$\mathbf{R}_{HH}^{\frac{1}{2}}$ is not a Hermiteian matrix, as $\mathbf{R}_{HH}^{\frac{1}{2}} \neq \left(\mathbf{R}_{HH}^{\frac{1}{2}} \right)^H$ holds. With the condition $\mathbf{R}_{HH} = \mathbf{G}^H \mathbf{G}$ the parameter $v_{22} = |v_{22}| e^{-j\Phi}$ is determined as $v_{22} = \frac{1}{\sqrt{5}} e^{-j\Phi}$ with arbitrary Φ. Then the result of the matrix decomposition is

$$\mathbf{G} = \mathbf{R}_{HH}^{\frac{1}{2}} = e^{j\Phi} \begin{pmatrix} 0 & 0 \\ j & \frac{1}{2} \end{pmatrix} \tag{13.38}$$

Moreover, $\mathbf{UG}$ is the general solution, where $\mathbf{U}$ is a unitary matrix. The reader convinces oneself easily with $\mathbf{U}^H \mathbf{U} = \mathbf{I}$ that $\mathbf{G}^H \mathbf{G} = \mathbf{R}_{HH}$ is true.

13.4.4 MIMO Channel Model with Transmit and Receive Correlation (Kronecker Model)

The wireless transmitter and receiver may operate in environments with different scattering and reflections at the transmitter and the receiver side. Then a description of the correlation close to the antenna elements by dedicated transmit and receive covariance matrices, $\mathbf{R}_{tx}$ and $\mathbf{R}_{rx}$, rather than by only one channel correlation matrix $\mathbf{R}_{HH}$ is reasonable.

The resulting channel model is depicted in Fig. 13.6 as a block diagram, where the matrices $\mathbf{B} \in \mathbb{C}^{M \times M}$ and $\mathbf{A} \in \mathbb{C}^{N \times N}$ shall enhance the uncorrelated MIMO channel with matrix $\mathbf{H}_w(k)$ by the square root matrices $\mathbf{B} = \mathbf{R}_{tx}^{\frac{1}{2}}$ and $\mathbf{A} = \mathbf{R}_{rx}^{\frac{1}{2}}$, where

$$\mathbf{R}_{tx} = \mathbf{B}^H \mathbf{B} \tag{13.39}$$

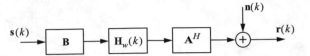

Fig. 13.6 Wireless link with an uncorrelated channel $\mathbf{H}_w(k)$ enhanced by matrices $\mathbf{B} = \mathbf{R}_{tx}^{1/2}$ and $\mathbf{A} = \mathbf{R}_{rx}^{1/2}$ introducing transmit and receive correlation, respectively

and

$$\mathbf{R}_{rx} = \mathbf{A}^H \mathbf{A} \qquad (13.40)$$

holds.

Obviously, the channel matrix for this model is then described by

$$\mathbf{H}(k) = \mathbf{A}^H \mathbf{H}_w(k) \mathbf{B} \qquad (13.41)$$

The correlation matrices $\mathbf{R}_{tx}$ and $\mathbf{R}_{rx}$ are determined from the channel matrix

$$\mathbf{H} = \begin{pmatrix} h_{11} & \cdots & h_{1j} & \cdots & h_{1M} \\ & & \vdots & & \\ h_{i1} & \cdots & h_{ij} & \cdots & h_{iM} \\ & & \vdots & & \\ h_{N1} & \cdots & h_{Nj} & \cdots & h_{NM} \end{pmatrix} = \left(\mathbf{h}_1 \ \cdots \ \mathbf{h}_j \ \cdots \ \mathbf{h}_M \right) = \begin{pmatrix} \mathbf{g}_1^T \\ \vdots \\ \mathbf{g}_i^T \\ \vdots \\ \mathbf{g}_N^T \end{pmatrix} \qquad (13.42)$$

using the column vector

$$\mathbf{h}_j = \left(h_{1j} \ \cdots \ h_{ij} \ \cdots \ h_{Nj} \right)^T \qquad (13.43)$$

and the row vector

$$\mathbf{g}_i^T = \left(h_{i1} \ \cdots \ h_{ij} \ \cdots \ h_{iM} \right) \qquad (13.44)$$

The channel vector $\mathbf{g}_i^T$ determines the multiple input single output (MISO) channel depicted in Fig. 13.7a and $\mathbf{h}_j$ describes the single input multiple output (SIMO) scheme in Fig. 13.7b.

Then the transpose of the transmit covariance matrix is defined as

$$\mathbf{R}_{tx}^T = \mathbf{E}\left[\mathbf{g}_i \mathbf{g}_i^H \right] \ ; \ i = 1, 2, ..., N \qquad (13.45)$$

which shall be unchanged irrespective of what antenna element i is receiving. Similarly, the receive covariance matrix is determined as

$$\mathbf{R}_{rx} = \mathbf{E}\left[\mathbf{h}_j \mathbf{h}_j^H \right] \ ; \ j = 1, 2, ..., M \qquad (13.46)$$

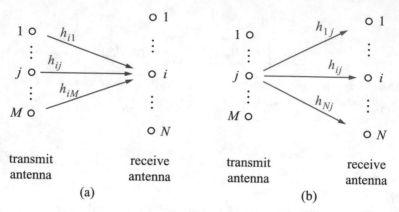

Fig. 13.7 a Multiple input single output (MISO) channel determined by row vector $\mathbf{g}_i^T = \left(h_{i1} \cdots h_{ij} \cdots h_{iM}\right)$. b Single input multiple output (SIMO) channel determined by column vector $\mathbf{h}_j = \left(h_{1j} \cdots h_{ij} \cdots h_{Nj}\right)^T$

and shall be the same irrespective of what transmit antenna element j is emitting. Furthermore, we assume that all main diagonal entries t_{ii} of $\mathbf{R}_{tx}$ and r_{jj} of $\mathbf{R}_{rx}$ are normalized equal to one.

Below we show that from the channel model (13.41) the stacked vector of the channel matrix can be derived as

$$\text{vec}\left(\mathbf{H}\right) = \left(\mathbf{B}^T \otimes \mathbf{A}^H\right) \text{vec}\left(\mathbf{H}_w\right) \tag{13.47}$$

With the covariance matrices (13.45) and (13.46) the covariance matrix $\mathbf{R}_{HH}$ of the channel follows as

$$\mathbf{R}_{HH} = \mathbf{R}_{tx}^* \otimes \mathbf{R}_{rx} \tag{13.48}$$

where $\otimes$ symbolizes the Kronecker product. The proof is also found below. Given the matrix $\mathbf{R}_{tx}^* \in \mathbb{C}^{M \times M}$ with elements t_{lm}^* ; $l, m = 1, 2, ..., M$ and $\mathbf{R}_{rx} \in \mathbb{C}^{N \times N}$ then the Kronecker product is defined as

$$\mathbf{R}_{HH} = \begin{pmatrix} t_{11}^* \mathbf{R}_{rx} & t_{12}^* \mathbf{R}_{rx} & \cdots & t_{1M}^* \mathbf{R}_{rx} \\ t_{21}^* \mathbf{R}_{rx} & t_{22}^* \mathbf{R}_{rx} & \cdots & t_{2M}^* \mathbf{R}_{rx} \\ & & \cdots & \\ t_{M1}^* \mathbf{R}_{rx} & t_{M2}^* \mathbf{R}_{rx} & \cdots & t_{MM}^* \mathbf{R}_{rx} \end{pmatrix} \in \mathbb{C}^{MN \times MN} \tag{13.49}$$

Furthermore it is straightforward to show with (13.45) that

$$\mathbf{R}_{tx}^* = \mathbf{R}_{tx}^T \tag{13.50}$$

holds. In the following proofs of (13.47) and (13.48) we use three Lemmas of linear algebra, [17]: Let $\mathbf{Q}$, $\mathbf{R}$, $\mathbf{S}$, and $\mathbf{T}$ be matrices of compatible dimensions. Then

$$\text{vec}\,(\mathbf{QRS}) = \left(\mathbf{S}^T \otimes \mathbf{Q}\right) \text{vec}\,(\mathbf{R}) \tag{13.51}$$

$$(\mathbf{Q} \otimes \mathbf{R})^H = \mathbf{Q}^H \otimes \mathbf{R}^H \tag{13.52}$$

$$(\mathbf{Q} \otimes \mathbf{R})\,(\mathbf{S} \otimes \mathbf{T}) = \mathbf{QS} \otimes \mathbf{RT} \tag{13.53}$$

Proof of (13.47)

With (13.51) follows directly from the channel model defined by (13.41) $\text{vec}\,(\mathbf{H}) = \text{vec}\,(\mathbf{A}^H \mathbf{H}_w \mathbf{B}) = \left(\mathbf{B}^T \otimes \mathbf{A}^H\right) \text{vec}\,(\mathbf{H}_w)$ and the proof is finished.

Proof of (13.48)

From (13.24) we obtain with (13.47) $\mathbf{R}_{HH} = \mathbf{E}\left[\left((\mathbf{B}^T \otimes \mathbf{A}^H)\,\text{vec}\,(\mathbf{H}_w)\right)\left((\mathbf{B}^T \otimes \mathbf{A}^H)\,\text{vec}\,(\mathbf{H}_w)\right)^H\right]$ and using (13.52) $\mathbf{R}_{HH} = \mathbf{E}\left[\left((\mathbf{B}^T \otimes \mathbf{A}^H)\,\text{vec}\,(\mathbf{H}_w)\right)(\text{vec}\,(\mathbf{H}_w))^H\,(\mathbf{B}^* \otimes \mathbf{A})\right] = \left(\mathbf{B}^T \otimes \mathbf{A}^H\right)\mathbf{E}\left[(\text{vec}\,(\mathbf{H}_w))\,(\text{vec}\,(\mathbf{H}_w))^H\right](\mathbf{B}^* \otimes \mathbf{A})$. With $\mathbf{E}\left[(\text{vec}\,(\mathbf{H}_w))\,(\text{vec}\,(\mathbf{H}_w))^H\right] = \mathbf{I}_{MN}$ and (13.53) follows $\mathbf{R}_{HH} = \mathbf{B}^T \mathbf{B}^* \otimes \mathbf{A}^H \mathbf{A}$. Inserting (13.39) and (13.40) eventually yields the result $\mathbf{R}_{HH} = \mathbf{R}_{tx}^* \otimes \mathbf{R}_{rx}$ and the proof is finished.

Proof of (13.46)

We proof in the following that the channel model determined by (13.41) results in a receive correlation matrix $\mathbf{R}_{rx}$ given by (13.46), which is identical for all transmit antenna indices j. First we write $\mathbf{R}_{HH}$ from (13.26) for the general case

$$\mathbf{R}_{HH} = \mathbf{E}\left[\begin{pmatrix} \mathbf{h}_1 \\ \vdots \\ \mathbf{h}_M \end{pmatrix}\begin{pmatrix} \mathbf{h}_1^H \cdots \mathbf{h}_M^H \end{pmatrix}\right] = \begin{pmatrix} \mathbf{E}\left[\mathbf{h}_1\mathbf{h}_1^H\right] & \cdots & \mathbf{E}\left[\mathbf{h}_1\mathbf{h}_M^H\right] \\ \vdots & \cdots & \vdots \\ \mathbf{E}\left[\mathbf{h}_M\mathbf{h}_1^H\right] & \cdots & \mathbf{E}\left[\mathbf{h}_M\mathbf{h}_M^H\right] \end{pmatrix} \tag{13.54}$$

When we compare the main diagonal entries of (13.49) and (13.54) we see that $\mathbf{E}\left[\mathbf{h}_1\mathbf{h}_1^H\right] = \mathbf{R}_{rx}, ..., \mathbf{E}\left[\mathbf{h}_M\mathbf{h}_M^H\right] = \mathbf{R}_{rx}$ are independent of j, because we have assumed that the main diagonal elements of $\mathbf{R}_{tx}$ are $t_{11} = \cdots = t_{MM} = 1$. This finalizes the proof.

Proof of (13.45)

$\mathbf{R}_{tx}^T$ is defined in (13.45) through the vectors $\mathbf{g}_i$ and we can determine the transpose of the channel matrix (13.42) as

$$\mathbf{H}^T = \begin{pmatrix} \mathbf{g}_1 & \cdots & \mathbf{g}_N \end{pmatrix} \tag{13.55}$$

Please note that $\mathbf{g}_i$ is the column vector associated to $\mathbf{g}_i^T$. Then follows

$$\mathbf{R}_{H^T H^T} = \mathbf{E}\left[\text{vec}\,(\mathbf{H}^T)\,(\text{vec}\,(\mathbf{H}^T))^H\right] = \begin{pmatrix} \mathbf{E}\left[\mathbf{g}_1\mathbf{g}_1^H\right] & \cdots & \mathbf{E}\left[\mathbf{g}_1\mathbf{g}_N^H\right] \\ & \cdots & \\ \mathbf{E}\left[\mathbf{g}_N\mathbf{g}_1^H\right] & \cdots & \mathbf{E}\left[\mathbf{g}_N\mathbf{g}_N^H\right] \end{pmatrix} \tag{13.56}$$

Next we show that from the channel model (13.41) follows

$$\mathbf{R}_{H^T H^T} = \mathbf{R}_{rx} \otimes \mathbf{R}_{tx}^* \tag{13.57}$$

(13.41) provides $\mathrm{vec}\,(\mathbf{H}^T) = \mathrm{vec}\,(\mathbf{B}^T \mathbf{H}_w^T \mathbf{A}^*)$ and with Lemma (13.51) follows $\mathrm{vec}\,(\mathbf{H}^T) = (\mathbf{A}^H \otimes \mathbf{B}^T)\,\mathrm{vec}\,(\mathbf{H}_w^T)$. Then we obtain with Lemma (13.52) $\left(\mathrm{vec}\,(\mathbf{H}^T)\right)^H = \left(\mathrm{vec}\,(\mathbf{H}_w^T)\right)^H (\mathbf{A} \otimes \mathbf{B}^*)$, and for the covariance matrix of $\mathbf{H}^T$ follows $\mathbf{R}_{H^T H^T} = \mathbf{E}\left[\left(\mathrm{vec}\,(\mathbf{H}^T)\right)\left(\mathrm{vec}\,(\mathbf{H}^T)\right)^H\right] = (\mathbf{A}^H \otimes \mathbf{B}^T)\,\mathbf{I}_{MN}\,(\mathbf{A} \otimes \mathbf{B}^*) = \mathbf{A}^H \mathbf{A} \otimes \mathbf{B}^T \mathbf{B}^*$, where we have used Lemma (13.53) in the last step. With (13.39) and (13.40) we obtain eventually $\mathbf{R}_{H^T H^T} = \mathbf{R}_{rx} \otimes \mathbf{R}_{tx}^*$ and the proof of (13.57) ends.

Let $\mathbf{R}_{rx} \in \mathbb{C}^{N \times N}$ be equipped with the entries r_{pq} ; $p, q = 1, 2, ..., N$, then the Kronecker product in (13.57) yields

$$\mathbf{R}_{H^T H^T} = \begin{pmatrix} r_{11}\mathbf{R}_{tx}^* & r_{12}\mathbf{R}_{tx}^* & \cdots & r_{1N}\mathbf{R}_{tx}^* \\ r_{21}\mathbf{R}_{tx}^* & \mathbf{R}_{tx}^* & \cdots & r_{2N}\mathbf{R}_{tx}^* \\ & & \cdots & \\ r_{N1}\mathbf{R}_{tx}^* & r_{N2}\mathbf{R}_{tx}^* & \cdots & r_{NN}\mathbf{R}_{tx}^* \end{pmatrix} \in \mathbb{C}^{MN \times MN} \tag{13.58}$$

When comparing the main diagonals of (13.56) and (13.58) taking the prerequisite $r_{11} = ... = r_{NN} = 1$ into account we conclude $\mathbf{R}_{tx}^* = \mathbf{E}\left[\mathbf{g}_1 \mathbf{g}_1^H\right] = \cdots = \mathbf{E}\left[\mathbf{g}_N \mathbf{g}_N^H\right]$, which is independent of the index i of the receive antenna element. With $\mathbf{R}_{tx}^* = \mathbf{R}_{tx}^T$ from (13.50) the proof is finished.

Example 3

We come back to the Example 2 in Sect. 13.4.3 and ask whether the channel meets the conditions of the Kronecker model. We find from (13.46) for the receive covariance matrix $\mathbf{R}_{rx} = \mathbf{E}\left[\mathbf{h}_1 \mathbf{h}_1^H\right] = \mathbf{R}_{HH}$ given in (13.31). The transmit covariance matrices defined in (13.45) are $\mathbf{R}_{tx,1}^T = \mathbf{E}\left[\mathbf{g}_1 \mathbf{g}_1^H\right] = \mathbf{E}\left[h_{11}h_{11}^*\right] = 1$ and $\mathbf{R}_{tx,2}^T = \mathbf{E}\left[\mathbf{g}_2 \mathbf{g}_2^H\right] = \mathbf{E}\left[h_{21}h_{21}^*\right] = \frac{1}{4}$. Obviously, they are different and we conclude that the channel matrix in Example 2 does not fulfill the conditions of the Kronecker channel model.

13.4.5 Exponential Covariance Matrix Model

This model also follows the idea of separate covariance matrices at the transmitter and the receiver. However, instead of determining these matrices by a MISO and a SIMO scheme, as applied by the Kronecker model, fixed covariance matrices with an exponential decay of the magnitudes of the correlation coefficients outside of the main diagonals is presumed.

Let

$$(\mathbf{R}_{tx})_{pq} = \rho_{tx}^{|p-q|^\beta} \; ; \; p, q = 1, 2, ..., M \; ; \; q \geq p \; ; \; \mathbf{R}_{tx} = \mathbf{R}_{tx}^H \tag{13.59}$$

be the covariance coefficients with $|\rho_{tx}| \leq 1$ and be the entries of the transmit covariance matrix $\mathbf{R}_{tx}$. Similarly,

$$(\mathbf{R}_{rx})_{pq} = \rho_{rx}^{|p-q|^{\beta}} \; ; \; p, q = 1, 2, ..., N \; ; \; q \geq p \; ; \; \mathbf{R}_{rx} = \mathbf{R}_{rx}^{H} \qquad (13.60)$$

are the entries of the receive covariance matrix $\mathbf{R}_{rx}$ with $|\rho_{rx}| \leq 1$ and β typically is in the range from 1 to 2. Only for the main diagonal elements the equal sign holds. Small magnitudes $|\rho_{tx}|$ and $|\rho_{rx}|$ characterize low correlation and values close to 1 indicate a strong correlation of the delay spread functions between the antenna elements p and q. Obviously, with this model the magnitudes of the covariance coefficients decay exponentially with the distance $|p - q|$ between the antenna elements with numbers p and q. This model is motivated from measurements of the local transmit and receive region. $\mathbf{R}_{HH}$ is then given by (13.48).

Finally, the exponential covariance model is also applicable directly to the entries of $\mathbf{R}_{HH}$ in (13.24) without the differentiation between the transmit and the receive correlation, [18].

Example 4

An example of a transmit correlation matrix for a transmit antenna with $M = 4$ elements follows from (13.59) assuming $\beta = 1$,

$$\mathbf{R}_{tx} = \begin{pmatrix} 1 & \rho_{tx} & \rho_{tx}^2 & \rho_{tx}^3 \\ \rho_{tx}^* & 1 & \rho_{tx} & \rho_{tx}^2 \\ \left(\rho_{tx}^2\right)^* & \rho_{tx}^* & 1 & \rho_{tx} \\ \left(\rho_{tx}^3\right)^* & \left(\rho_{tx}^2\right)^* & \rho_{tx}^* & 1 \end{pmatrix} \qquad (13.61)$$

References

1. Kaye, A., George, D.: Transmission of multiplexed PAM signals over multiple channel and diversity systems. In: IEEE Transactions on Communication Technology (1970)
2. Brandenburg, L., Wyner, A.: Capacity of the Gaussian channel with memory: the multivariate case. Bell Syst. Tech. J. **53**, 745–778 (1974)
3. Van Etten, W.: Maximum likelihood receiver for multiple channel transmission systems. IEEE Trans. Commun. **24**, 276–283 (1976)
4. Wikipedia. MIMO. https://en.wikipedia.org/wiki/MIMO
5. Salz, J.: Digital transmission over cross-coupled linear channels. AT&T Tech. J. **64**, 1147–1159 (1985)
6. Raleigh, G., Cioffi, J.: Spatio-temporal coding for wireless communications. In: IEEE Global Telecommunications Conference (1996)
7. Foschini, G.: Layered space-time architecture for wireless communication in a fading environment when using multi-element antennas. Bell Syst. Tech. J. **1**, 41–59 (1996)
8. Golden, G., Foschini, G., Valenzuela, R., Wolniansky, P.: Detection algorithm and initial laboratory results using V-BLAST space-time communication architecture. Elect. Lett. **35** (1999)

9. Schneider, D., Speidel, J., Stadelmeier, L., Schill, D., Schwager, A.: MIMO for inhome power line communications. In: International Conference on Source and Channel Coding (SCC), ITG Fachberichte (2008)
10. Berger, L.T., Schwager, A., Pagani, P., Schneider, D.: MIMO power line communications—Narrow and broadband standards EMC and advanced processing. CRC Press, Boca Raton (2014)
11. G.993.5: Self-FEXT cancellation (vectoring) for use with VDSL2 transceivers
12. Khalid, F., Speidel, J.: Advances in MIMO techniques for mobile communications—a survey. Int. J. Commun. Netw. Sys. Sci. **3** (2010)
13. Zadeh, L.A.: Frequency analysis of variable networks. Proceedings of the IRE (1950)
14. Bello, P.A.: Characterization of randomly time-variant linear channels. IEEE Trans. Commun. Syst. (1963)
15. Molisch, A.F.: Wireless Communications. Wiley and IEEE Press, Hoboken (2009)
16. Spatial channel model for MIMO simulations. Technical report, 3GPP TR 25996 V 9.0.0 (2010)
17. Horn, R.A., Johnson, C.R.: Matrix analysis. Cambridge University Press, Cambridge (2013)
18. Loyka, S.: Channel capacity of MIMO architecture using the exponential correlation matrix. IEEE Commun. Lett. **5** (2001)

Chapter 14
Principles of Linear MIMO Receivers

14.1 Introduction

As depicted in the block diagram of Fig. 14.1, we consider a MIMO system with frequency flat and in general time-varying channel with channel matrix $\mathbf{H}(k) \in \mathbb{C}^{N \times M}$, input signal vector $\mathbf{s}(k) \in \mathbb{C}^{M \times 1}$, noise vector $\mathbf{n}(k) \in \mathbb{C}^{N \times 1}$, and receive vector $\mathbf{r}(k) \in \mathbb{C}^{N \times 1}$. At the receiver a linear filter described by a matrix $\mathbf{W}(k) \in \mathbb{C}^{M \times N}$ is employed to obtain at its output a good replica $\mathbf{y}(k)$ of the transmit signal vector $s(k)$. We assume that a channel estimator not shown in Fig. 14.1 has provided perfect channel state information so that the instantaneous channel matrix $\mathbf{H}(k)$ is known for every discrete-time instant k at the receiver.

From Fig. 14.1 we obtain

$$\mathbf{r}(k) = \mathbf{H}(k)\mathbf{s}(k) + \mathbf{n}(k) \tag{14.1}$$

and

$$\mathbf{y}(k) = \mathbf{W}(k)\mathbf{r}(k) \tag{14.2}$$

By inserting (14.1) into (14.2) we get

$$\mathbf{y}(k) = \mathbf{W}(k)\mathbf{H}(k)\mathbf{s}(k) + \mathbf{W}(k)\mathbf{n}(k) = \mathbf{G}(k)\mathbf{s}(k) + \mathbf{n}'(k) \tag{14.3}$$

with the interference matrix

$$\mathbf{G}(k) = \mathbf{W}(k)\mathbf{H}(k) \tag{14.4}$$

and the noise

$$\mathbf{n}'(k) = \mathbf{W}(k)\mathbf{n}(k) \tag{14.5}$$

Let $\mathbf{G}(k) = \left(g_{ij}(k)\right)_{M \times M}$ and $\mathbf{W}(k) = \left(w_{ij}(k)\right)_{M \times N}$, then we obtain the ith line of the system of Eqs. (14.3)

© Springer Nature Switzerland AG 2021
J. Speidel, *Introduction to Digital Communications*,
Signals and Communication Technology,
https://doi.org/10.1007/978-3-030-67357-4_14

Fig. 14.1 Block diagram of
a MIMO system with
channel matrix $\mathbf{H}(k)$ and
linear receiver with matrix
$\mathbf{W}(k)$

$$y_i(k) = g_{ii}(k)s_i(k) + \sum_{j=1,\, j \neq i}^{M} g_{ij}(k)s_j(k) + \sum_{j=1}^{M} w_{ij}(k)n_j(k) \ ; \ \ i = 1, ..., M \quad (14.6)$$

The first term in (14.6) is the desired signal $s_i(k)$ of the transmit antenna i with some
weighting coefficient $g_{ii}(k)$. The second term is denoted as inter-channel interference
composed of the signals from all other transmit antennas and the last contribution is
the noise component at the receiver output. Please note that the $g_{ij}(k)$ are functions
of the receive matrix elements $w_{ij}(k)$. The task of the linear receiver is to adjust
its matrix $\mathbf{W}$ in such a way that the inter-channel interference is reduced or even
completely removed.

14.2   Operation Modes for MIMO Systems

Before we are going into more details of the MIMO receivers we first take the
opportunity with (14.1) to get some inside into the options provided by MIMO
systems. For that purpose we first take a simple example with $M = N = 2$ and
rewrite (14.1) in detail.

$$r_1(k) = h_{11}(k)s_1(k) + h_{12}(k)s_2(k) + n_1(k) \quad (14.7)$$

$$r_2(k) = h_{21}(k)s_1(k) + h_{22}(k)s_2(k) + n_2(k) \quad (14.8)$$

Spatial Multiplexing to Increase Symbol Rate
In this MIMO transmission mode two independent signals $s_1(k)$ and $s_2(k)$ from
antenna 1 and antenna 2 can be sent, respectively. Consequently, we can increase
the symbol rate provided that the receiver is able to recover $s_1(k)$ and $s_2(k)$ from
the receive signals $r_1(k)$ and $r_2(k)$, and the spatial correlation of the delay spread
functions h_{ij} is small.

Spatial Diversity to Improve Transmission Quality
As a starting point, let us consider that only one antenna at the transmitter and one
antenna at receiver are available. Then, the receive signal is

$$r_1(k) = h_{11}(k)s_1(k) + n_1(k) \quad (14.9)$$

If the delay spread function $h_{11}(k)$ is zero due to fading at time instant k, it is impossible to recover $s_1(k)$, because we just have received noise $n_1(k)$. This can be overcome by providing more than one antenna either at the transmitter or at the receiver or at both sides provided that the spatial correlation of the delay spread functions is small.

Simple Transmit Diversity

Let us consider the case of a simple transmit diversity scheme with $M = 2$ transmit and $N = 1$ receive antennas and that on each of the transmit antennas the same signal is sent, thus $s_2(k) = s_1(k)$. The receive signal then is

$$r_1(k) = (h_{11}(k) + h_{12}(k))\, s_1(k) + n_1(k) \tag{14.10}$$

We see even if one of the channel coefficients is zero due to fading the transmit signal $s_1(k)$ can still be recovered, owing to the redundant channel path established by the other transmit antenna. The probability that both paths $h_{11}(k)$ and $h_{12}(k)$ are zero at the same time instant k due to fading is rather small, if the two channels are almost independent. Therefore, transmit diversity helps to increase transmission quality by providing spatial redundancy.

Simple Receive Diversity

Consider now the case of a simple receive diversity with $N = 2$ receive and $M = 1$ transmit antennas, which allows us to transmit one signal $s_1(k)$. Then we obtain at the receiver

$$r_1(k) = h_{11}(k)s_1(k) + n_1(k) \tag{14.11}$$

$$r_2(k) = h_{21}(k)s_1(k) + r_2(k) \tag{14.12}$$

and we can argue in a similar way as before. Therefore, receive diversity can also help to improve transmission quality. However, we see that in both cases of simple transmit and receive diversity only one signal $s_1(k)$ can be transmitted and no increase of the symbol rate is achieved.

Beamforming

With more than one antenna at the transmitter the emitted electro-magnetic beam can be steered in a certain direction, e.g., to a "hot spot" of many mobile users. The principle is that a matrix $\mathbf{A}$ changes the direction of a vector $\mathbf{x}$ after multiplication $\mathbf{A}\mathbf{x}$. This technique is called beamforming and in most cases only one receive antenna is used.

Combining Spatial Multiplex and Diversity or Beamforming

The general MIMO approach can take advantage of all these features, namely, spatial multiplex to increase data rate, diversity to improve transmission quality, or beamforming in the sense of a compromise. In case of $M = N = 2$ the Eqs. (14.7) and (14.8) describe the input-output relation of the MIMO system.

In the next sections, we will discuss methods how to recover the transmit from the receive signals. For that purpose, we assume perfect channel knowledge at the receiver, i.e., we impose the prerequisite that the complete channel matrix $\mathbf{H}(k)$ is known at every time instant k. Furthermore, we drop the discrete-time variable k to simplify the notation.

14.3   Zero-Forcing Receiver for Equal Number of Transmit and Receive Antennas

In this section we derive the matrix $\mathbf{W}$ of a zero-forcing (ZF) receiver for the case that the transmitter and the receiver are equipped with the same number of antennas, i.e.,

$$M = N \tag{14.13}$$

Consequently, $\mathbf{H} \in \mathbb{C}^{M \times M}$ is a square matrix. Furthermore, assume a channel matrix with full rank

$$\text{rank}\,(\mathbf{H}) = M \tag{14.14}$$

Then, the inverse matrix $\mathbf{H}^{-1}$ exists. Moreover, if we consider very small noise at the receiver, $\mathbf{n} \approx \mathbf{0}$, we get from (14.3) approximately

$$\mathbf{y} = \mathbf{WHs} \tag{14.15}$$

With $\mathbf{WH} = \mathbf{I}_M$ the inverse matrix

$$\mathbf{W} = \mathbf{H}^{-1} \tag{14.16}$$

provides the solution. Then, we obtain the output signal of the receiver from (14.3)

$$\mathbf{y} = \mathbf{s} + \mathbf{H}^{-1}\mathbf{n} \tag{14.17}$$

We see the inter-channel interference (14.3) is completely removed, $\mathbf{G}(k) = \mathbf{I}_M$, and the receiver output signal $\mathbf{y}$ is just corrupted by the noise $\mathbf{Wn} = \mathbf{H}^{-1}\mathbf{n}$. To check how the noise has changed, we calculate the mean noise power

$$\mathbf{E}\left[\|\mathbf{Wn}\|^2\right] \tag{14.18}$$

and find that in some cases this value can be larger than the original mean noise power $\mathbf{E}\left[\|\mathbf{n}\|^2\right]$ making the detection of the bit sequence in the receive signal more inaccurate.

14.4 Zero-Forcing Receiver for Unequal Number of Transmit and Receive Antennas

14.4.1 Receiver with More Antennas Than Transmitter, $N > M$

Solving the System of Linear Equations

Example 1

Let us first consider a simple example with $M = 2$ transmit and $N = 3$ receive antennas and assume small noise $\mathbf{n} \approx \mathbf{0}$. Then we obtain from the basic input-output relation (14.1)

$$r_1 = h_{11}s_1 + h_{12}s_2 \qquad (14.19)$$

$$r_2 = h_{21}s_1 + h_{22}s_2 \qquad (14.20)$$

$$r_3 = h_{31}s_1 + h_{32}s_2 \qquad (14.21)$$

Obviously, this is an over-determined system with $N = 3$ equations for the $M = 2$ unknowns. The channel matrix $\mathbf{H}$ is non-square and consequently an inverse matrix does not exist. We know from linear algebra that such a system of equations can only be solved exactly, if and only if at least $N - M$ equations are linearly depending on each other. In other words, for the rank of the matrix $\mathbf{H}$, which is the number of linearly independent lines or columns, $\text{rank}(\mathbf{H}) \leq M$ must be true. Consequently, we can cancel $N - M$ superfluous equations. The remaining system with M equations for the M unknowns can be solved conventionally. Whether this remaining system has one unique solution or an infinite manifold of a parametric solution depends on the rank of the remaining $M \times M$ square matrix.

If the three equations are all independent, no solution exists. This fact is easily understood, because in our example we could take two equations out of the three and solve them. Then we insert the resulting s_1 and s_2 into the third equation and are faced with a contradiction indicating that the obtained "solution" is not feasible for the original 3×2 system of equations.

This is the mathematical view. But what happens in a real system? Are the three equations linearly depending or are they independent? For the answer we can think of an experiment, where we sent two known signals $s_1(k)$ and $s_2(k)$ from the two antennas over the channel with the known coefficients h_{11} and h_{12}. Then we measure the received signals $r_1(k), ..., r_3(k)$. Equations (14.19)–(14.21) must hold because every element in these equations is physically valid. Thus, we conclude that in reality the three equations must be linearly depending on each other. However, if additive noises, which are independent of each other and independent of all signals come into play, then the resulting equations may no longer be linearly depending and the over-determined system will have no solution.

Solution with the Pseudo Inverse

The fact that a system of equations can have no solution was somehow disappointing in mathematics. Therefore, in early 1920 the mathematician E. H. Moore came up with an approximate solution, [1]. However, it is reported that his notation was hard to understand, because of the telegraphic style and the idiosyncratic notation. Therefore his work was not recognized enough, until a formulation by Penrose [2, 3] appeared. Penrose formulated four axioms for a special kind of an inverse matrix, which is denoted as Moore–Penrose inverse in the literature. It is applicable in general for finding an inverse of rectangular matrices of any dimension approximately and is also called pseudo inverse matrix.

Our problem can be formulated as follows. We are looking for a general not necessarily exact solution, which gives us the transmit signal vector **s** or an estimate of it from the receive signal vector **r** at every time instant k. As the system of Eq. (14.1) is over-determined for $N > M$, we can obtain an approximate solution in the sense that the left-hand side of (14.1) is just approximately equal to the right-hand side. As a prerequisite, the channel matrix $\mathbf{H} \in \mathbb{C}^{N \times M}$ shall have full rank, i.e.

$$\text{rank}\,(\mathbf{H}) = M \tag{14.22}$$

In the following we also assume small noise, i.e. $\mathbf{n} \approx \mathbf{0}$, and minimize the difference between the left and the right-hand side of (14.1),

$$\Delta = \mathbf{r} - \mathbf{H}\mathbf{s} \tag{14.23}$$

in the mean squared error sense,

$$\|\Delta\|^2 = \Delta^H \Delta = \min_{\mathbf{s}} \tag{14.24}$$

The result is

$$\mathbf{s} = \left(\mathbf{H}^H \mathbf{H}\right)^{-1} \mathbf{H}^H \mathbf{r} \tag{14.25}$$

and we find the matrix of the zero-forcing receiver as

$$\mathbf{H}^+ = \left(\mathbf{H}^H \mathbf{H}\right)^{-1} \mathbf{H}^H = \mathbf{W} \in \mathbb{C}^{M \times N} \tag{14.26}$$

The proof is given at the end of this section. Plugging (14.26) into (14.4) yields

$$\mathbf{G} = \mathbf{H}^+ \mathbf{H} = \left(\mathbf{H}^H \mathbf{H}\right)^{-1} \mathbf{H}^H \mathbf{H} = \mathbf{I}_M \tag{14.27}$$

Hence, $\mathbf{H}^+$ acts as a kind of inverse matrix with respect to $\mathbf{H}$. Therefore $\mathbf{H}^+$ is called (left-hand sided) *pseudo inverse* or *Moore–Penrose inverse matrix*. From the perspective of linear algebra we are solving the modified system of equations $\mathbf{H}^+\mathbf{H}\mathbf{s} = \mathbf{H}^+\mathbf{r}$ rather than the original one $\mathbf{H}\mathbf{s} = \mathbf{r}$ given in (14.19)–(14.21). This provides the exact solution **s**. As a consequence for our MIMO system, we have to equip the receiver with the matrix $\mathbf{W}$ in (14.26) yielding the output signal

$$\mathbf{y} = \mathbf{W}\mathbf{r} = \mathbf{s} + \mathbf{H}^+\mathbf{n} \qquad (14.28)$$

Again we see that the output signal vector $\mathbf{y}$ of the zero-forcing receiver contains the original transmit signal vector, which is corrupted by additive noise. The inter-channel interference is completely canceled. As in the previous case $M = N$, the mean power of the resulting noise

$$\mathbf{n}' = \mathbf{H}^+\mathbf{n}. \qquad (14.29)$$

can be increased compared to $\mathbf{n}$ by the receiver matrix $\mathbf{H}^+$.

From our assumption (14.22) follows that $\mathbf{H}^H\mathbf{H}$ has also full rank M and consequently $\left(\mathbf{H}^H\mathbf{H}\right)^{-1}$ exists.

Example 2
Show that the minimum squared error is

$$\|\Delta\|_{\min}^2 = \mathbf{r}^H\left(\mathbf{I}_N - \mathbf{H}\mathbf{H}^+\right)\mathbf{r} \qquad (14.30)$$

Solution:
From (14.24) follows with (14.23) $\|\Delta\|^2 = (\mathbf{r}^H - \mathbf{s}^H\mathbf{H}^H)(\mathbf{r} - \mathbf{H}\mathbf{s})$. Owing to the solution $\mathbf{s} = \mathbf{H}^+\mathbf{r}$ from (14.25) with (14.26) and knowing that $\left(\mathbf{H}\mathbf{H}^+\right)^H = \mathbf{H}\mathbf{H}^+$ follows $\|\Delta\|^2 = \|\Delta\|_{\min}^2 = \mathbf{r}^H\mathbf{r} - \mathbf{r}^H\mathbf{H}\mathbf{H}^+\mathbf{r}$. This is equal to $\|\Delta\|_{\min}^2 = \mathbf{r}^H\left(\mathbf{I}_N - \mathbf{H}\mathbf{H}^+\right)\mathbf{r}$, which finalizes the proof. As expected, the minimal squared error depends on the matrix $\mathbf{H}$, the left-hand side $\mathbf{r}$ of the system of equations, and is in general not zero.

Example 3
To illustrate the problem of an over-determined system of equations let us consider the following example of three equations with two unknowns s_1 and s_2.

$$\mathbf{H}\mathbf{s} = \mathbf{r} \; ; \quad \begin{pmatrix} 1 & 1 \\ 1 & -1 \\ 2 & 1 \end{pmatrix}\begin{pmatrix} s_1 \\ s_2 \end{pmatrix} = \begin{pmatrix} 0 \\ 1 \\ 1 \end{pmatrix}$$

Obviously, the three equations are linearly independent. We can follow the path outlined above and derive a solution from the first and the second equation yielding $s_1 = \frac{1}{2}$ and $s_2 = -\frac{1}{2}$. By plugging those into the third equation results in the contradiction $\frac{1}{2} = 1$. Alternatively, we can start with another pair of equations and proceed as before ending up with a contradiction again. Consequently, there is no solution. If we calculate the squared error between the left and the right-hand side of the system of equations, we obtain $\|\Delta\|^2 = \frac{1}{4}$.

As shown before, the pseudo inverse (14.26) can provide an approximation by minimizing $\|\Delta\|^2$. We calculate $\mathbf{H}^H\mathbf{H} = \begin{pmatrix} 6 & 2 \\ 2 & 3 \end{pmatrix}$ and the inverse[1] $(\mathbf{H}^H\mathbf{H})^{-1} = \frac{1}{14}\begin{pmatrix} 3 & -2 \\ -2 & 6 \end{pmatrix}$. Then follows $\mathbf{H}^+ = \frac{1}{14}\begin{pmatrix} 1 & 5 & 4 \\ 4 & -8 & 2 \end{pmatrix}$. The minimal squared error is obtained from (14.30) as $\|\Delta\|_{\min}^2 = \frac{1}{14}$ which is well below $\|\Delta\|^2 = \frac{1}{4}$.

For MIMO transmission the reduction of inter-channel interference is decisive. The reader assures oneself easily that the cascade of the channel and receive matrix results in $\mathbf{H}^+\mathbf{H} = \frac{1}{14}\begin{pmatrix} 1 & 5 & 4 \\ 4 & -8 & 2 \end{pmatrix}\begin{pmatrix} 1 & 1 \\ 1 & -1 \\ 2 & 1 \end{pmatrix} = \begin{pmatrix} 1 & 0 \\ 0 & 1 \end{pmatrix}$ and the inter-channel interference is completely removed by the pseudo inverse $\mathbf{H}^+$.

This example verifies that the exact solution $\mathbf{s}$ is obtained by multiplying the receive vector $\mathbf{r}$ by the pseudo inverse matrix of the channel. However, the linear system of equations $\mathbf{r} = \mathbf{Hs}$ is solved only approximately with a squared error between left and right-hand side of $\|\Delta\|^2 = \frac{1}{4}$.

Example 4
Is $\mathbf{W} = \mathbf{H}^+$ also valid for $N = M$?

The answer is yes, because for $N = M$ the channel matrix is a square matrix and the prerequisite (14.22) guarantees that $\mathbf{H}^{-1}$ exists. Consequently, we obtain for the pseudo inverse from (14.26)

$$\mathbf{H}^+ = \mathbf{H}^{-1}(\mathbf{H}^H)^{-1}\mathbf{H}^H = \mathbf{H}^{-1}\mathbf{I}_M = \mathbf{H}^{-1} \tag{14.31}$$

and the pseudo inverse boils down to the inverse matrix (14.16). Furthermore, we conclude from (14.31) $\mathbf{HH}^+ = \mathbf{I}_N$ and from (14.30) we find that the squared error then is zero, $\|\Delta\|_{\min}^2 = 0$, as expected.

Proof of the Pseudo Inverse Matrix $\mathbf{H}^+$ in (14.26)
For the proof we set the first partial derivative of our target function with respect to the free parameters equal to zero. To follow this line we use the trace of a matrix, which we differentiate with respect to a matrix or a vector, as outlined in the Appendix B, and rewrite the squared error of (14.24) as

$$\|\Delta\|^2 = \text{tr}\left(\Delta\Delta^H\right) \tag{14.32}$$

Please note that the product of a column vector Δ and a row vector Δ^H results in a matrix. With (14.23) we obtain

$$\Delta\Delta^H = \mathbf{rr}^H - \mathbf{rs}^H\mathbf{H}^H - \mathbf{Hsr}^H + \mathbf{Hss}^H\mathbf{H}^H \tag{14.33}$$

---

[1] Given $\mathbf{A} = \begin{pmatrix} a & b \\ c & d \end{pmatrix}$, then $\mathbf{A}^{-1} = \begin{pmatrix} d & -b \\ -c & a \end{pmatrix} 1/\det\mathbf{A}$ with $\det\mathbf{A} = ad - bc$

We recognize that all terms are elements of $\mathbb{C}^{N \times N}$. According to the Appendix B the trace of the sum of matrices is identical to the sum of the traces. To find the minimum we set the partial derivative equal to zero, $\frac{\partial \mathrm{tr}(\Delta\Delta^H)}{\partial s^*} = 0$, which is a necessary and sufficient condition here, because our quadratic error $\mathrm{tr}\left(\Delta\Delta^H\right)$ is a convex function of $\mathbf{s}$ or $\mathbf{s}^*$. The partial derivation yields

$$\frac{\partial \mathrm{tr}\left(\Delta\Delta^H\right)}{\partial \mathbf{s}^*} = \frac{\partial \mathrm{tr}\left(\mathbf{r}\mathbf{r}^H\right)}{\partial \mathbf{s}^*} - \frac{\partial \mathrm{tr}\left(\mathbf{r}\mathbf{s}^H\mathbf{H}^H\right)}{\partial \mathbf{s}^*} - \frac{\partial \mathrm{tr}\left(\mathbf{H}\mathbf{s}\mathbf{r}^H\right)}{\partial \mathbf{s}^*} + \frac{\partial \mathrm{tr}\left(\mathbf{H}\mathbf{s}\mathbf{s}^H\mathbf{H}^H\right)}{\partial \mathbf{s}^*} = 0 \tag{14.34}$$

and with the differentiation Lemmas for traces shown in the Appendix B we obtain for the individual terms

$$\frac{\partial \mathrm{tr}\left(\mathbf{r}\mathbf{r}^H\right)}{\partial \mathbf{s}^*} = 0 \tag{14.35}$$

$$\frac{\partial \mathrm{tr}\left(\mathbf{r}\mathbf{s}^H\mathbf{H}^H\right)}{\partial \mathbf{s}^*} = \mathbf{H}^H\mathbf{r} \tag{14.36}$$

$$\frac{\partial \mathrm{tr}\left(\mathbf{H}\mathbf{s}\mathbf{r}^H\right)}{\partial \mathbf{s}^*} = 0 \tag{14.37}$$

$$\frac{\partial \mathrm{tr}\left(\mathbf{H}\mathbf{s}\mathbf{s}^H\mathbf{H}^H\right)}{\partial \mathbf{s}^*} = \mathbf{H}^H\mathbf{H}\mathbf{s} \tag{14.38}$$

Inserting (14.35)–(14.38) into (14.34) yields

$$\mathbf{H}^H\mathbf{r} = \mathbf{H}^H\mathbf{H}\mathbf{s} \tag{14.39}$$

from which we conclude by multiplication from the left-hand side with $\left(\mathbf{H}^H\mathbf{H}\right)^{-1}$

$$\mathbf{s} = \left(\mathbf{H}^H\mathbf{H}\right)^{-1}\mathbf{H}^H\mathbf{r} \tag{14.40}$$

and consequently $\mathbf{H}^+ = \left(\mathbf{H}^H\mathbf{H}\right)^{-1}\mathbf{H}^H = \mathbf{W}$ in (14.26) follows, which finalizes the proof.

An Alternative Derivation of the Pseudo Inverse Matrix
Motivated by the fact that $\mathbf{H}^H\mathbf{H}$ is a square matrix with full rank, M, we can come to a straightforward derivation of the pseudo inverse as follows. We consider (14.1) with $\mathbf{n} = \mathbf{0}$

$$\mathbf{r} = \mathbf{H}\mathbf{s} \tag{14.41}$$

and multiply from the left with $\mathbf{H}^H$ resulting in

$$\mathbf{H}^H\mathbf{r} = \mathbf{H}^H\mathbf{H}\mathbf{s} \tag{14.42}$$

In mathematics (14.42) is called the Gaussian normal equation. $\mathbf{H}^H \mathbf{H}$ is a square matrix with the dimension $M \times M$. Then the task is to solve a system of linear equations defined by a non-singular square matrix. The solution is straightforward. We multiply from the left with the inverse matrix $\left(\mathbf{H}^H \mathbf{H}\right)^{-1}$, which exists due to the fact that $\mathbf{H}^H \mathbf{H}$ has full rank, resulting in

$$\left(\mathbf{H}^H \mathbf{H}\right)^{-1} \mathbf{H}^H \mathbf{r} = \left(\mathbf{H}^H \mathbf{H}\right)^{-1} \mathbf{H}^H \mathbf{H} \mathbf{s} \tag{14.43}$$

from which we conclude

$$\mathbf{s} = \left(\mathbf{H}^H \mathbf{H}\right)^{-1} \mathbf{H}^H \mathbf{r} \tag{14.44}$$

and finally the pseudo inverse

$$\mathbf{H}^+ = \left(\mathbf{H}^H \mathbf{H}\right)^{-1} \mathbf{H}^H \tag{14.45}$$

As a final remark, if $\mathbf{H}$ does not have full rank M, a special pseudo inverse can be determined from a singular value decomposition, [4].

14.4.2 Receiver with Less Antennas Than Transmitter, $N < M$

Solving the System of Linear Equations

Example 5
Let us consider a simple example with $M = 3$ transmit and $N = 2$ receive antennas. Assuming small noise, $\mathbf{n} \approx \mathbf{0}$, we obtain from the basic input-output Eq. (14.1)

$$r_1 = h_{11}s_1 + h_{12}s_2 + h_{13}s_3 \tag{14.46}$$

$$r_2 = h_{21}s_1 + h_{22}s_2 + h_{23}s_3 \tag{14.47}$$

and recognize an under-determined system of equations. Mathematically this can be solved by assuming any value for one variable, e.g., s_3. Then we can move $h_{13}s_3$ and $h_{23}s_3$ to the left-hand side and obtain a 2×2 system of equations with a square matrix, which can be solved conventionally. If this matrix has full rank two, the solutions for s_1 and s_2 contain one free parameter s_3, which can be set to any value. However, from a communications point of view the selection of an arbitrary s_3 at the receiver is useless and the third transmit antenna could be dropped. Another way would be to keep $M = 3$ and to introduce redundancy by sending one signal s_1 or s_2 also from the third antenna as a kind of transmit diversity. However, the total symbol rate is reduced.

Solution with the Pseudo Inverse

In general, the $N \times M$ under-determined system of linear equations can have an exact solution with at least $M - N$ free parameters, which give rise to an $(M - N)$ infinite linear manifold. From a technical point of view, $M - N$ transmit antennas are redundant. Alternatively, we look for an approximation, as for the over-determined system of equations discussed in the previous section and follow the line of a pseudo inverse here as well. For that purpose we take the prerequisite that $\mathbf{H}$ has full rank, which is

$$\text{rank}\,(\mathbf{H}) = N \le \min\{M, N\} \tag{14.48}$$

Then the $M \times M$ matrix $\mathbf{H}^H \mathbf{H}$ in (14.26) can only have rank N. Because $N < M$, the inverse $(\mathbf{H}^H \mathbf{H})^{-1}$ does not exist and consequently $\mathbf{H}^+$ given by (14.26) is not applicable. However, a Moore–Penrose pseudo inverse also exists and is given by

$$\mathbf{H}^{++} = \mathbf{H}^H \left(\mathbf{H}\mathbf{H}^H\right)^{-1} \in \mathbb{C}^{M \times N} \tag{14.49}$$

$\mathbf{H}\mathbf{H}^H$ is an $N \times N$ matrix with full rank N, if $\mathbf{H}$ exhibits the full rank N. Hence, $\left(\mathbf{H}\mathbf{H}^H\right)^{-1}$ exists and the pseudo inverse $\mathbf{H}^{++}$ can be calculated. Equation (14.49) is proven by checking the four Moore–Penrose axioms for which we refer to [3, 5]. We will give an alternative proof in the next section using some results from the minimum mean squared error receiver. An approach similar to (14.24) is not successful. For $M = N$ follows $\mathbf{H}^{++} = \mathbf{H}^{-1}$.

As a final remark, if $\mathbf{H}$ does not have full rank N, a special pseudo inverse can be determined from a singular value decomposition, [4].

Receiver Output with $\mathbf{H}^{++}$ as Receive Filter

The output signal $\mathbf{y}(k) = \mathbf{G}(k)\mathbf{s}(k) + \mathbf{n}'(k)$ of the receiver is given by (14.3). With $\mathbf{W} = \mathbf{H}^{++}$, we calculate the inter-channel interference term from (14.4) resulting in $\mathbf{G} = \mathbf{H}^H \left(\mathbf{H}\mathbf{H}^H\right)^{-1} \mathbf{H}$ and see that in general $\mathbf{G} \ne \mathbf{I}_M$ holds, which means that the receive matrix $\mathbf{H}^{++}$ cannot remove inter-channel interference in general.

Receive Signal with $\mathbf{H}^{++}$ as Transmit Prefilter

We recognize that $\mathbf{H}\mathbf{H}^{++} = \mathbf{I}_N$ holds. Hence, if we allocate a filter with matrix $\mathbf{H}^{++}$ at the transmitter as a prefilter, we obtain the transmit signal $\mathbf{s} = \mathbf{H}^{++}\mathbf{c}$, where $\mathbf{c} = \mathbf{c}(k)$ defines the vector of QAM transmit symbols. The receive signal follows from (14.1)

$$\mathbf{r} = \mathbf{c} + \mathbf{n} \tag{14.50}$$

Obviously, the inter-channel interference is completely removed. Prefilters in general are discussed in Chap. 18 in more detail. Consequently, we do not consider $\mathbf{H}^{++}$ as a receive filter for practical applications any more.

Example 6

Given a system with $M = 3$ transmit and $N = 2$ receive antennas. The channel with matrix $\mathbf{H} = \begin{pmatrix} 1 & 0 & \frac{j}{2} \\ -\frac{j}{2} & 1 & 0 \end{pmatrix}$ shall be time-invariant in the time interval under consideration. (a) Find the matrix $\mathbf{H}^{++}$. Calculate the inter-channel interference matrix $\mathbf{G}$, if $\mathbf{H}^{++}$ is used (b) as receive filter and (c) as transmit prefilter.

Solution:

(a) We calculate $\mathbf{H}^{++}$ in several steps: $\mathbf{H}\mathbf{H}^{H} = \frac{1}{4}\begin{pmatrix} 5 & 2j \\ -2j & 5 \end{pmatrix}$, $(\mathbf{H}\mathbf{H}^{H})^{-1} = \frac{4}{21}\begin{pmatrix} 5 & -2j \\ 2j & 5 \end{pmatrix}$

and finally obtain $\mathbf{H}^{++} = \frac{2}{21}\begin{pmatrix} 8 & j \\ j4 & 10 \\ -j5 & -2 \end{pmatrix}$.

(b) If we apply $\mathbf{H}^{++}$ as a receive filter, the inter-channel interference term follows from (14.4) as $\mathbf{G} = \mathbf{H}^{++}\mathbf{H} = \frac{17}{21}\begin{pmatrix} 1 & j\frac{2}{17} & j\frac{8}{17} \\ -j\frac{2}{17} & \frac{20}{17} & -\frac{4}{17} \\ -j\frac{8}{17} & -\frac{4}{17} & \frac{5}{17} \end{pmatrix}$. Disregarding the factor, $\mathbf{G}$ devi-

ates from the identity matrix $\mathbf{I}_3$. Thus we confirm that in general the zero-forcing receiver with pseudo inverse $\mathbf{H}^{++}$ is not able to remove inter-channel interference. (c) Now we consider a transmit prefilter with the matrix $\mathbf{H}^{++}$ and obtain the corresponding inter-channel interference term $\mathbf{H}\mathbf{H}^{++} = \mathbf{I}_2$. The prefilter $\mathbf{H}^{++}$ completely removes inter-channel interference.

14.5 Signal-to-Noise Ratio of Linear Receivers

14.5.1 General Relations

We consider Fig. 14.1 and are going to calculate the signal-to-noise ratio γ_y of the signal $\mathbf{y}(k)$ at the output of the receive filter $\mathbf{W}$, which is the input of the decision device. As is well known from the theory of digital communications, the higher γ_y the lower the bit error ratio can be at the output of the decision device. The signal part of $\mathbf{y}(k)$ in (14.3) is

$$\mathbf{y}_s(k) = \mathbf{W}(k)\mathbf{H}(k)\mathbf{s}(k) \tag{14.51}$$

which is superimposed by the noise part

$$\mathbf{y}_n(k) = \mathbf{W}(k)\mathbf{n}(k)$$

In the following we define the covariance matrix of the signal $\mathbf{s}(k)$ and the noise $\mathbf{n}(k)$, respectively,

$$\mathbf{R}_{ss} = \mathbf{E}\left[\mathbf{s}\mathbf{s}^{H}\right] \in \mathbb{C}^{M \times M} \; ; \; \mathbf{R}_{nn} = \mathbf{E}\left[\mathbf{n}\mathbf{n}^{H}\right] \in \mathbb{C}^{N \times N} \tag{14.52}$$

Furthermore, we drop k to simplify the notation. As shown in the Appendix B, the mean power of a signal is given by the trace of its covariance matrix. Consequently, for the mean power of $\mathbf{y}_s$ follows

$$E\left[\|\mathbf{y}_s\|^2\right] = \mathrm{tr}\left(E\left[\mathbf{y}_s\mathbf{y}_s^H\right]\right) \tag{14.53}$$

With (14.51) we get $\mathrm{tr}\left(E\left[\mathbf{y}_s\mathbf{y}_s^H\right]\right) = \mathrm{tr}\left(E\left[\mathbf{WHss}^H\mathbf{H}^H\mathbf{W}^H\right]\right)$ and using (14.52) provides the result

$$E\left[\|\mathbf{y}_s\|^2\right] = \mathrm{tr}\left(\mathbf{R}_{ss}\left(\mathbf{WH}\right)^H\mathbf{WH}\right) \tag{14.54}$$

where we have used the cyclic permutation rule for the trace outlined in the Appendix B. Similarly, we obtain for the noise part $\mathbf{y}_n$ the mean power

$$E\left[\|\mathbf{y}_n\|^2\right] = \mathrm{tr}\left(\mathbf{R}_{nn}\mathbf{W}^H\mathbf{W}\right) \tag{14.55}$$

and finally the signal-to-noise ratio

$$\gamma_y = \frac{E\left[\|\mathbf{y}_s\|^2\right]}{E\left[\|\mathbf{y}_n\|^2\right]} = \frac{\mathrm{tr}\left(\mathbf{R}_{ss}\mathbf{H}^H\mathbf{W}^H\mathbf{WH}\right)}{\mathrm{tr}\left(\mathbf{R}_{nn}\mathbf{W}^H\mathbf{W}\right)} \tag{14.56}$$

For comparison the signal-to-noise ratio at the receiver input is

$$\gamma_r = \frac{E\left[\|\mathbf{Hs}\|^2\right]}{E\left[\|\mathbf{n}\|^2\right]} = \frac{\mathrm{tr}\left(\mathbf{R}_{ss}\mathbf{H}^H\mathbf{H}\right)}{\mathrm{tr}\left(\mathbf{R}_{nn}\right)} \tag{14.57}$$

14.5.2 Normalization of the Channel Matrix $\mathbf{H}$

To facilitate a fair comparison of the signal-to-noise ratios of different MIMO schemes a normalization of the channel matrix $\mathbf{H}$ is adequate. For that purpose we define the power gain g_P of the channel as the relation between its mean output power $E\left[\|\mathbf{Hs}\|^2\right]$ and mean input power $E\left[\|\mathbf{s}\|^2\right]$. The input signal shall be uncorrelated, thus $E\left[\|\mathbf{s}\|^2\right] = \mathrm{tr}\left(\mathbf{R}_{ss}\right) = ME_S$ and the mean output power then is $E\left[\|\mathbf{Hs}\|^2\right] = E_S\mathrm{tr}\left(\mathbf{H}^H\mathbf{H}\right)$. We know from Appendix B that $\mathrm{tr}\left(\mathbf{H}^H\mathbf{H}\right) = \|\mathbf{H}\|_F^2 = \sum_{j=1}^{M}\sum_{i=1}^{N}\left|h_{ij}\right|^2$ is the Frobenius norm of a matrix. Thus, we can define the power gain as

$$g_P = \frac{E\left[\|\mathbf{Hs}\|^2\right]}{E\left[\|\mathbf{s}\|^2\right]} = \frac{\mathrm{tr}\left(\mathbf{H}^H\mathbf{H}\right)}{M} = \frac{\|\mathbf{H}\|_F^2}{M} \tag{14.58}$$

Please remember that the entries of $\mathbf{H}$ are delay spread functions representing the link from the transmitter to the receiver including all building blocks, such as radio channel, filters, modulator, amplifier etc. Therefore the overall gain can be adjusted to achieve the normalized channel matrix

$$\mathbf{H}_N = \frac{1}{\sqrt{g_P}}\mathbf{H} = \sqrt{\frac{M}{\|\mathbf{H}\|_F^2}}\mathbf{H} \tag{14.59}$$

The entries of $\mathbf{H}$ represent amplitudes rather than powers, which gives rise to apply the square root of g_P. If a given channel owns the power gain $g_p = 1$, then its channel matrix is normalized and fullfills tr $\left(\mathbf{H}^H \mathbf{H}\right) = M$ in (14.58).

With (14.59) we find $\mathbf{H}^H \mathbf{H} = g_P \mathbf{H}_N^H \mathbf{H}_N$, $\left(\mathbf{H}^H \mathbf{H}\right)^{-1} = \frac{1}{g_P} \left(\mathbf{H}_N^H \mathbf{H}_N\right)^{-1}$, and similar relations for the traces. Then follows from (14.56) the signal-to-noise ratio of $\mathbf{H}_N$

$$\gamma_{y,N} = \frac{\gamma_y}{g_P} \tag{14.60}$$

14.5.3 Signal-to-Noise Ratio with Zero-Forcing Receiver for $M \le N$

We have found that in general the zero-forcing receiver with $\mathbf{H}^{++}$ for $M > N$ in (14.49) cannot reduce inter-channel interference. Hence, we focus on systems with receivers $\mathbf{H}^+$ for $M \le N$ in (14.26) in the following and obtain

$$\mathbf{W}\mathbf{H} = \mathbf{H}^+ \mathbf{H} = \mathbf{I}_M \; ; \; M \le N \tag{14.61}$$

Then the mean power (14.54) of the signal part at the receiver output is

$$\mathbf{E}\left[\|\mathbf{y}_s\|^2\right] = \text{tr}\,(\mathbf{R}_{ss}) \; ; \; M \le N \tag{14.62}$$

For the mean power (14.55) of the noise part we have to determine

$$\mathbf{W}^H \mathbf{W} = \left(\mathbf{H}^+\right)^H \mathbf{H}^+ = \mathbf{H}\left(\mathbf{H}^H\mathbf{H}\right)^{-1}\left(\mathbf{H}^H\mathbf{H}\right)^{-1}\mathbf{H}^H \; ; \; M \le N \tag{14.63}$$

and mean power of the noise part follows as

$$\mathbf{E}\left[\|\mathbf{y}_n\|^2\right] = \text{tr}\left(\mathbf{R}_{nn}\mathbf{H}\left(\mathbf{H}^H\mathbf{H}\right)^{-1}\left(\mathbf{H}^H\mathbf{H}\right)^{-1}\mathbf{H}^H\right) \; ; \; M \le N \tag{14.64}$$

Plugging (14.62) and (14.64) into (14.56) yields the signal-to-noise ratio at the zero-forcing receiver output

$$\gamma_y = \frac{\mathbf{E}\left[\|\mathbf{y}_s\|^2\right]}{\mathbf{E}\left[\|\mathbf{y}_n\|^2\right]} = \frac{\text{tr}(\mathbf{R}_{ss})}{\text{tr}\left(\mathbf{R}_{nn}\mathbf{H}(\mathbf{H}^H\mathbf{H})^{-1}(\mathbf{H}^H\mathbf{H})^{-1}\mathbf{H}^H\right)} \; ; \; M \le N \tag{14.65}$$

The mean noise power $\mathbf{E}\left[\|\mathbf{y}_n\|^2\right]$ in (14.64) is worth to be commentated. If the determinant of $\mathbf{H}^H \mathbf{H}$ owns rather small absolute values, then the entries of the corresponding inverse matrix can take on large numbers. As a consequence, the noise power at the output of the receive filter can get much larger than that of the input, see Example 8. A matrix with such an unfavorable property is called ill conditioned. It gets even worse, if the channel matrix $\mathbf{H}$ does not have full rank, because $\det(\mathbf{H}^H \mathbf{H})$

approaches zero. Note that $\text{rank}(\mathbf{H}) = \text{rank}(\mathbf{H}^H\mathbf{H}) \leq \min\{M, N\}$. Similarities are discussed in Chap. 18 for the output signal of a precoder.

Example 7: Uncorrelated Signal and Uncorrelated Noise, $M \leq N$
Consider a MIMO scheme with $M \leq N$ and a zero-forcing (ZF) receiver for the usual case that transmit signal s and also noise n are uncorrelated. Thus, the covariance matrices are $\mathbf{R}_{ss} = E_S\mathbf{I}_M$ and $\mathbf{R}_{nn} = \sigma_n^2\mathbf{I}_N$. Find the mean signal and noise power as well as the singal-to-noise ratio at input and output of the zero-forcing receiver.

Solution:
From (14.62) we get the *mean signal power at the ZF receiver output*

$$\mathbf{E}\left[\|\mathbf{y}_s\|^2\right] = \text{tr}\,(\mathbf{R}_{ss}) = \mathbf{E}\left[\|\mathbf{s}\|^2\right] = ME_S \; ; \; M \leq N \qquad (14.66)$$

With the cyclic permutation rule we otain from (14.64)
$\sigma_n^2\text{tr}\left((\mathbf{H}^H\mathbf{H})^{-1}(\mathbf{H}^H\mathbf{H})^{-1}\mathbf{H}^H\mathbf{H}\right)$ and the *mean noise power at the ZF receiver output*

$$\mathbf{E}\left[\|\mathbf{y}_n\|^2\right] = \sigma_n^2\text{tr}\left((\mathbf{H}^H\mathbf{H})^{-1}\right) \; ; \; M \leq N \qquad (14.67)$$

Then follows the *SNR at the ZF receiver output*

$$\gamma_y = \frac{\mathbf{E}\left[\|\mathbf{y}_s\|^2\right]}{\mathbf{E}\left[\|\mathbf{y}_n\|^2\right]} = \frac{ME_S}{\sigma_n^2\text{tr}\left((\mathbf{H}^H\mathbf{H})^{-1}\right)} \; ; \; M \leq N \qquad (14.68)$$

and (14.57) yields the *SNR at the ZF receiver input*

$$\gamma_r = \frac{\mathbf{E}\left[\|\mathbf{H}\mathbf{s}\|^2\right]}{\mathbf{E}\left[\|\mathbf{n}\|^2\right]} = \frac{E_S\text{tr}\left(\mathbf{H}^H\mathbf{H}\right)}{\sigma_n^2 N} \qquad (14.69)$$

Example 8
(a) Given a transmission scheme with $M = 2$ transmit, $N = 3$ receive antennas, a zero-forcing (ZF) receiver, and a normalized channel matrix $\mathbf{H} = \sqrt{\frac{32}{37}}\begin{pmatrix} 1 & 0 \\ 1 & j/2 \\ 0 & -j/4 \end{pmatrix}$, which is constant in a considered time interval. Signal s and noise n are uncorrelated, and $E_S/\sigma_n^2 = 4$ holds. Calculate the mean power of signal and noise at the input and output of the receiver. Find the signal-to-noise ratios (SNR) γ_r and γ_y at the receiver input and output, respectively.

Solution:
We calculate $\mathbf{H}^H\mathbf{H} = \frac{32}{37}\begin{pmatrix} 2 & j/2 \\ -j/2 & 5/16 \end{pmatrix}$ from which follows $(\mathbf{H}^H\mathbf{H})^{-1} = \frac{37}{12}\begin{pmatrix} 5/16 & -j/2 \\ j/2 & 2 \end{pmatrix}$. At the receiver input we obtain the mean signal power $\mathbf{E}\left[\|\mathbf{H}\mathbf{s}\|^2\right] =$

$E_S \mathrm{tr}\left(\mathbf{H}^H \mathbf{H}\right) = 2E_S$ and the mean noise power $\mathbf{E}\left[\|\mathbf{n}\|^2\right] = N\sigma_n^2 = 3\sigma_n^2$. At the receiver output one gets the mean signal power $\mathbf{E}\left[\|\mathbf{y}_s\|^2\right] = 2E_S$ and the mean noise power $\mathbf{E}\left[\|\mathbf{y}_n\|^2\right] = 7.1\sigma_n^2$. Finally, the SNR are $\gamma_r = \frac{2}{3}\frac{E_S}{\sigma_n^2} = 2.7$ and $\gamma_y = 0.28\frac{E_S}{\sigma_n^2} = 1.1$. Apparently, with this channel the ZF receiver enhances the mean noise power at the output significantly, whereas the mean signal power at input and output of a ZF receiver are the same. Thus, the output SNR γ_y gets smaller than γ_r at the input. This is the burden payed for the perfect rejection of inter-channel interference.

The design goal of the mean squared error receiver, which will be discussed in the next section, takes the noise into account and hence can provide a compromise between inter-channel interference and output noise.

14.6   Minimum Mean Squared Error Receiver

14.6.1   Prerequisites

In contrast to the zero-forcing receiver we now follow a strategy to minimize the quadratic error between the transmit signal $\mathbf{s}$ and the receiver output signal $\mathbf{y}$ and thus include the noise in our design criterion from the beginning. Hence, the target function of the Minimum Mean Squared Error (MMSE) receiver is

$$J = \mathbf{E}\left[\|\Delta\|^2\right] = \min_{\mathbf{W}} \tag{14.70}$$

with the error vector

$$\Delta = \mathbf{s} - \mathbf{y} \tag{14.71}$$

The squared error then is

$$\|\Delta\|^2 = \Delta^H \Delta = \mathrm{tr}\left(\Delta\Delta^H\right) \tag{14.72}$$

From (14.70) follows

$$J = \mathbf{E}\left[\mathrm{tr}\left(\Delta\Delta^H\right)\right] = \mathrm{tr}\left(\mathbf{E}\left[\Delta\Delta^H\right]\right) = \min_{\mathbf{W}} \tag{14.73}$$

where we have used the fact that the trace and the expectation operator are linear and thus can be applied in reverse order. In the course of the computation of the optimal receiver matrix $\mathbf{W}$ we will need the covariance matrix $\mathbf{R}_{ss}$ of the transmit signal $\mathbf{s}$ and the covariance matrix $\mathbf{R}_{nn}$ of the noise $\mathbf{n}$ defined in (14.52). We state the following prerequisites. In the first place the noise has zero mean, i.e.

$$\mathbf{E}\left[\mathbf{n}\right] = \mathbf{0} \tag{14.74}$$

and secondly, the transmit signal **s** is statistically independent of the noise **n**, i.e. the cross-correlation matrix is

$$\mathbf{R}_{sn} = \mathbf{E}\left[\mathbf{s}\mathbf{n}^{H}\right] = \mathbf{E}\left[\mathbf{s}\right]\mathbf{E}\left[\mathbf{n}^{H}\right] \tag{14.75}$$

and with (14.74) follows

$$\mathbf{R}_{sn} = \mathbf{0} \tag{14.76}$$

Similarly

$$\mathbf{R}_{ns} = \mathbf{E}\left[\mathbf{n}\mathbf{s}^{H}\right] = \mathbf{R}_{sn}^{H} = \mathbf{0} \tag{14.77}$$

We further assume that the channel matrix **H** and the receiver matrix **W** are deterministic but still can be functions of discrete time k.

14.6.2 Receiver Matrix

The partial derivative of J with respect to $\mathbf{W}^{*}$ set to zero provides the necessary and sufficient condition for the minimum, because the quadratic error is a convex function of **W** or $\mathbf{W}^{*}$. The optimal receiver matrix, which minimizes the squared error, then is

$$\mathbf{W} - \mathbf{R}_{ss}\mathbf{H}^{H}\left(\mathbf{H}\mathbf{R}_{ss}\mathbf{H}^{H} + \mathbf{R}_{nn}\right)^{-1} \in \mathbb{C}^{M\times N} \tag{14.78}$$

provided that the matrix $\mathbf{H}\mathbf{R}_{ss}\mathbf{H}^{H} + \mathbf{R}_{nn}$ is non-singular. Before we outline the proof a special case is considered.

In many applications we can assume that the signals emitted by the transmit antenna elements are spatially uncorrelated and each with mean power E_S, thus $\mathbf{R}_{ss} = E_S\mathbf{I}_M$ holds. Also the noise at each receive antenna element can be considered to be spatially uncorrelated each with mean power σ_n^2 resulting in the covariance matrix $\mathbf{R}_{nn} = \sigma_n^2\mathbf{I}_N$. Then we get the optimal MMSE receiver matrix $\mathbf{W} = E_S\mathbf{H}^{H}\left(\mathbf{H}E_S\mathbf{H}^{H} + \sigma_n^2\mathbf{I}_N\right)^{-1}$ and finally[2]

$$\mathbf{W} = \mathbf{H}^{H}\left(\mathbf{H}\mathbf{H}^{H} + \alpha\mathbf{I}_N\right)^{-1} \in \mathbb{C}^{M\times N} \tag{14.79}$$

with the ratio of the mean noise power per receive antenna and the mean signal power per transmit antenna

$$\alpha = \frac{\sigma_n^2}{E_S} \tag{14.80}$$

According to communication theory the receive matrix describes a MIMO matched filter. **W** is applicable, if the inverse matrix in (14.79) exists.

---

[2]Note $(\alpha\mathbf{A})^{-1} = \frac{1}{\alpha}\mathbf{A}^{-1}$; $\alpha \neq 0$.

Now we apply the following matrix inversion lemma (Woodbury identiy), [4, 6, 7]. Let $\mathbf{A}$, $\mathbf{B}$, $\mathbf{C}$, and $\mathbf{D}$ be non-singular matrices with compatible dimensions. Then

$$(\mathbf{A} + \mathbf{BCD})^{-1} = \mathbf{A}^{-1} - \mathbf{A}^{-1}\mathbf{B}\left(\mathbf{C}^{-1} + \mathbf{DA}^{-1}\mathbf{B}\right)^{-1}\mathbf{DA}^{-1} \qquad (14.81)$$

holds and we obtain the alternative MMSE receive matrix

$$\mathbf{W} = \left(\mathbf{H}^{H}\mathbf{H} + \alpha\mathbf{I}_{M}\right)^{-1}\mathbf{H}^{H} \in \mathbb{C}^{M \times N} ; \quad \alpha \neq 0 \qquad (14.82)$$

The proof is given at the end of this section. We see that both solutions (14.79) and (14.82) respect the noise. For low noise, $\alpha \to 0$, we obtain from (14.79)

$$\mathbf{H}^{H}\left(\mathbf{HH}^{H}\right)^{-1} = \mathbf{H}^{++} \qquad (14.83)$$

and from (14.82)

$$\left(\mathbf{H}^{H}\mathbf{H}\right)^{-1}\mathbf{H}^{H} = \mathbf{H}^{+} \qquad (14.84)$$

Obviously, these are the identical matrices, which we achieved also for the zero-forcing receiver. Hence, we conclude that the MMSE receiver degrades to the zero-forcing receiver for high signal-to-noise ratio. With the MMSE receiver matrix (14.82) the receiver output signal (14.3) is

$$\mathbf{y} = \mathbf{Gs} + \mathbf{n}' \qquad (14.85)$$

with

$$\mathbf{G} = \mathbf{WH} = \left(\mathbf{H}^{H}\mathbf{H} + \alpha\mathbf{I}_{M}\right)^{-1}\mathbf{H}^{H}\mathbf{H} \qquad (14.86)$$

as the remaining inter-channel interference term and the output noise is

$$\mathbf{n}' = \mathbf{Wn} = \left(\mathbf{H}^{H}\mathbf{H} + \alpha\mathbf{I}_{M}\right)^{-1}\mathbf{H}^{H}\mathbf{n} \qquad (14.87)$$

Please note that $\mathbf{G}$ approaches $\mathbf{I}_{M}$ for low noise $\alpha \to 0$. Thus, the lower the noise the better the reduction of the inter-channel interference will be. We are not surprised, because the MMSE solution approaches the zero-forcing matrix. Similar considerations can be done, if the transmitter is furnished with a precoder owing to the matrix $\mathbf{W}$ in (14.79).

The signal-to-noise ratio at the MMSE receiver output can be found by inserting the receive matrix (14.82) or its variants into (14.56). In the same way we obtain the mean noise power at the receiver output with (14.55). However, the formulas remain not very transparent and therefore a discussion for specific α is more adequate.

For $\alpha \neq 0$ the MMSE receiver in (14.82) owns the additional term $\alpha\mathbf{I}_{M}$ compared to the zero-forcing receiver, which in many cases can enable a proper inverse $\left(\mathbf{H}^{H}\mathbf{H} + \alpha\mathbf{I}_{M}\right)^{-1}$ even in the case, where $\mathbf{H}$ and thus $\mathbf{H}^{H}\mathbf{H}$ do not have full rank or are ill conditioned. Hence, the MMSE receiver often provides an adequate solution in case of a rank deficient channel matrix, where the zero-forcing receiver fails.

Example 9
Given a MIMO system with $M = 2$, $N = 3$, a normalized channel matrix $\mathbf{H} = \sqrt{\frac{32}{37}} \begin{pmatrix} 1 & 0 \\ 1 & j/2 \\ 0 & -j/4 \end{pmatrix}$ as in Example 8, uncorrelated transmit signal $\mathbf{s}$ with mean power $2E_S$, uncorrelated receiver noise $\mathbf{n}$ with mean power $3\sigma_n^2$, and $E_S/\sigma_n^2 = 1/\alpha = 4$. (a) Determine the MMSE receiver in (14.82). (b) Calculate the inter-channel interference $\mathbf{G} = \mathbf{WH}$. (c) Find the mean power of the receiver input and output signal. Calculate the mean power of the receiver input and output noise. (d) Find the signal-to-noise ratio at input and output of the receiver. Compare the figures with the zero-forcing receiver of Example 8.

Solution:
Lengthy matrix calculation are done with computer algebra. $\mathbf{H}^H \mathbf{H}$ is the same as in Example 8. Furthermore, we obtain $\mathbf{H}^H \mathbf{H} + 0.25 \mathbf{I}_2 = \begin{pmatrix} 1.98 & 0.43j \\ -0.43j & 0.52 \end{pmatrix}$ and $\left(\mathbf{H}^H \mathbf{H} + 0.25 \mathbf{I}_2 \right)^{-1} = \begin{pmatrix} 0.62 & -0.51j \\ 0.51j & 2.34 \end{pmatrix}$.

(a) For the MMSE receiver matrix in (14.82) we get $\mathbf{W} = \begin{pmatrix} 0.6 & 0.3 & 0.1 \\ 0.5 & -0.6j & 0.5j \end{pmatrix}$.

(b) $\mathbf{G} = \mathbf{WH} = 0.9 \begin{pmatrix} 1 & 0.15j \\ -0.15j & 0.5 \end{pmatrix}$ approximates the unity matrix, but deviates noticeable.

(c) Mean signal power at receiver input is the same as in Example 8, $\mathbf{E}\left[\|\mathbf{Hs}\|^2 \right] = E_S \text{tr} \left(\mathbf{H}^H \mathbf{H} \right) = 2E_S$ and from (14.54) follows at the output $\mathbf{E}\left[\|\mathbf{y}_s\|^2 \right] = E_S \text{tr} \left((\mathbf{WH})^H \mathbf{WH} \right) = 0.95 E_S$. Mean power of the receiver input noise is $\mathbf{E}\left[\|\mathbf{n}\|^2 \right] = 3\sigma_n^2$ and (14.55) yields at the output $\mathbf{E}\left[\|\mathbf{y}_n\|^2 \right] = \sigma_n^2 \text{tr} \left(\mathbf{W}^H \mathbf{W} \right) = 1.3\sigma_n^2$.

(d) From (14.57) we obtain $\gamma_r = \frac{2}{3} \frac{E_S}{\sigma_n^2} = 2.7$ at the input and $\gamma_y = 0.7 \frac{E_S}{\sigma_n^2} = 2.8$ from (14.56) at the output. Compared to the ZF receiver in Example 8 the SNR at the output of the MMSE receiver has improved from 1.1 to 2.8, but inter-channel interference is not fully reduced, as expected.

Proof of MMSE Receiver Matrix (14.78)
We insert (14.71) into (14.73) and obtain with (14.2)

$$J = \text{tr} \left(\mathbf{E}\left[\mathbf{\Delta \Delta}^H \right] \right) = \text{tr} \left(\mathbf{E}\left[(\mathbf{s} - \mathbf{Wr})(\mathbf{s}^H - \mathbf{r}^H \mathbf{W}^H) \right] \right) \tag{14.88}$$

and with (14.1)

$$J = \text{tr} \left(\mathbf{E}\left[(\mathbf{s} - \mathbf{WHs} - \mathbf{Wn})(\mathbf{s}^H - \mathbf{s}^H \mathbf{H}^H \mathbf{W}^H - \mathbf{n}^H \mathbf{W}^H) \right] \right) \tag{14.89}$$

After expanding we get

$$J = \mathrm{tr}\left(\mathbf{R}_{ss} - \mathbf{R}_{ss}\mathbf{H}^H\mathbf{W}^H - \mathbf{R}_{sn}\mathbf{W}^H - \mathbf{W}\mathbf{H}\mathbf{R}_{ss} + \mathbf{W}\mathbf{H}\mathbf{R}_{ss}\mathbf{H}^H\mathbf{W}^H\right) +$$

$$+ \mathrm{tr}\left(\mathbf{W}\mathbf{H}\mathbf{R}_{sn}\mathbf{W}^H - \mathbf{W}\mathbf{R}_{ns} + \mathbf{W}\mathbf{R}_{ns}\mathbf{H}^H\mathbf{W}^H + \mathbf{W}\mathbf{R}_{nn}\mathbf{W}^H\right) \qquad (14.90)$$

Inserting (14.76) and (14.77) and employing the linear property of the trace operator yields

$$J = \mathrm{tr}\left(\mathbf{R}_{ss}\right) - \mathrm{tr}\left(\mathbf{R}_{ss}\mathbf{H}^H\mathbf{W}^H\right) - \mathrm{tr}\left(\mathbf{W}\mathbf{H}\mathbf{R}_{ss}\right) + \mathrm{tr}\left(\mathbf{W}\mathbf{H}\mathbf{R}_{ss}\mathbf{H}^H\mathbf{W}^H\right) + \mathrm{tr}\left(\mathbf{W}\mathbf{R}_{nn}\mathbf{W}^H\right)$$
$$(14.91)$$

We see that J is a quadratic and convex function of $\mathbf{W}$. Therefore setting the partial derivative with respect to $\mathbf{W}^*$ equal to zero is a necessary and sufficient condition for the minimum. With the cyclic permutation Lemma given in the Appendix B we move $\mathbf{W}$ to the end of the string of matrices in the traces and can directly apply the differentiation Lemmas outlined in the Appendix B. We obtain

$$\frac{\partial\,\mathrm{tr}\left(\mathbf{R}_{ss}\right)}{\partial\,\mathbf{W}^*} = 0 \; ; \quad \frac{\partial\,\mathrm{tr}\left(\mathbf{R}_{ss}\mathbf{H}^H\mathbf{W}^H\right)}{\partial\,\mathbf{W}^*} = \mathbf{R}_{ss}\mathbf{H}^H \; ; \quad \frac{\partial\,\mathrm{tr}\left(\mathbf{W}\mathbf{H}\mathbf{R}_{ss}\right)}{\partial\,\mathbf{W}^*} = 0 \qquad (14.92)$$

$$\frac{\partial\,\mathrm{tr}\left(\mathbf{W}\mathbf{H}\mathbf{R}_{ss}\mathbf{H}^H\mathbf{W}^H\right)}{\partial\,\mathbf{W}^*} = \mathbf{W}\mathbf{H}\mathbf{R}_{ss}\mathbf{H}^H \; ; \quad \frac{\partial\,\mathrm{tr}\left(\mathbf{W}\mathbf{R}_{nn}\mathbf{W}^H\right)}{\partial\,\mathbf{W}^*} = \mathbf{W}\mathbf{R}_{nn} \qquad (14.93)$$

Setting the partial derivative of J with respect to $\mathbf{W}^*$ equal to zero yields with (14.92) and (14.93)

$$\frac{\partial\,J}{\partial\,\mathbf{W}^*} = -\mathbf{R}_{ss}\mathbf{H}^H + \mathbf{W}\left(\mathbf{H}\mathbf{R}_{ss}\mathbf{H}^H + \mathbf{R}_{nn}\right) = 0 \qquad (14.94)$$

from which we conclude the proposition (14.78)

$$\mathbf{W} = \mathbf{R}_{ss}\mathbf{H}^H\left(\mathbf{H}\mathbf{R}_{ss}\mathbf{H}^H + \mathbf{R}_{nn}\right)^{-1} \qquad (14.95)$$

and finalize the proof.

Proof of (14.82)

Let $\mathbf{A} = \alpha\mathbf{I}_N$, $\mathbf{B} = \mathbf{H}$, $\mathbf{C} = \mathbf{I}_M$, and $\mathbf{D} = \mathbf{H}^H$, then follows from (14.79) with (14.81) $\mathbf{H}^H\left(\mathbf{H}\mathbf{H}^H + \alpha\mathbf{I}_N\right)^{-1} = \mathbf{H}^H\left(\mathbf{A}^{-1} - \mathbf{A}^{-1}\mathbf{H}\left(\mathbf{I}_M + \mathbf{H}^H\mathbf{A}^{-1}\mathbf{H}\right)^{-1}\mathbf{H}^H\mathbf{A}^{-1}\right) = \mathbf{H}^H\mathbf{A}^{-1} - \mathbf{H}^H\mathbf{A}^{-1}\mathbf{H}\left(\mathbf{I}_M + \mathbf{H}^H\mathbf{A}^{-1}\mathbf{H}\right)^{-1}\mathbf{H}^H\mathbf{A}^{-1}$. Now we multiply the first term from the left with the identity matrix $\left(\mathbf{I}_M + \mathbf{H}^H\mathbf{A}^{-1}\mathbf{H}\right)\left(\mathbf{I}_M + \mathbf{H}^H\mathbf{A}^{-1}\mathbf{H}\right)^{-1} = \mathbf{I}_M$ and obtain $\left(\mathbf{I}_M + \mathbf{H}^H\mathbf{A}^{-1}\mathbf{H}\right)\left(\mathbf{I}_M + \mathbf{H}^H\mathbf{A}^{-1}\mathbf{H}\right)^{-1}\mathbf{H}^H\mathbf{A}^{-1} - \mathbf{H}^H\mathbf{A}^{-1}\mathbf{H}\left(\mathbf{I}_M + \mathbf{H}^H\mathbf{A}^{-1}\mathbf{H}\right)^{-1}\mathbf{H}^H\mathbf{A}^{-1}$. After factoring out $\left(\mathbf{I}_M + \mathbf{H}^H\mathbf{A}^{-1}\mathbf{H}\right)^{-1}\mathbf{H}^H\mathbf{A}^{-1}$ the intermediate result is $\mathbf{I}_M\left(\mathbf{I}_M + \mathbf{H}^H\mathbf{A}^{-1}\mathbf{H}\right)^{-1}\mathbf{H}^H\mathbf{A}^{-1}$. With $\mathbf{A}^{-1} = \frac{1}{\alpha}\mathbf{I}_N$ follows $\left(\mathbf{I}_M + \mathbf{H}^H\frac{1}{\alpha}\mathbf{H}\right)^{-1}\mathbf{H}^H\frac{1}{\alpha} = \left(\mathbf{H}^H\mathbf{H} + \alpha\mathbf{I}_M\right)^{-1}\mathbf{H}^H$ and the proof is finished. Please note that $\alpha \neq 0$ must hold otherwise $\mathbf{A}^{-1}$ does not exist.

14.7 Linear Combiner for Single Input Multiple Output System

14.7.1 Principle of Linear Combining and the Signal-to-Noise Ratio

As depicted in Fig. 14.2, we are now considering a single input multiple output (SIMO) system with $M = 1$ transmit and $N > 1$ receive antennas. The transmit signal is denoted as s with the mean power $\mathbf{E}\left[|s|^2\right] = E_S$. The noise shall be spatially uncorrelated with mean power σ_n^2 per antenna and the covariance matrix $\mathbf{R}_{nn} = \sigma_n^2 \mathbf{I}_N$. The channel matrix of this SIMO system is just a column vector

$$\mathbf{h} = \begin{pmatrix} h_1 \\ h_2 \\ \vdots \\ h_N \end{pmatrix} \tag{14.96}$$

The linear combiner is described by its receive matrix, which is reduced to a row vector

$$\mathbf{W} = \begin{pmatrix} w_1 & w_2 & \cdots & w_N \end{pmatrix} \tag{14.97}$$

with complex components w_i ; $i = 1, 2, ..., N$. For the output signal of the linear combiner we obtain

$$y = \mathbf{W}\mathbf{h}s + \mathbf{W}\mathbf{n} \tag{14.98}$$

with the signal part

$$y_s = \mathbf{W}\mathbf{h}s \tag{14.99}$$

and the noise part

$$y_n = \mathbf{W}\mathbf{n} \tag{14.100}$$

Please note that the output of the receiver is a scalar rather than a vector. The signal-to-noise ratio is defined as

Fig. 14.2 Single input multiple output (SIMO) transmission with linear combiner receiver

$$\gamma = \frac{\mathbf{E}\left[|y_s|^2\right]}{\mathbf{E}\left[|y_n|^2\right]} \tag{14.101}$$

For the mean power of y_s we obtain

$$\mathbf{E}\left[|y_s|^2\right] = \mathbf{E}\left[(\mathbf{W}h s)^H \mathbf{W}h s\right] = E_S(\mathbf{W}h)^H \mathbf{W}h \tag{14.102}$$

and the mean power of the noise part y_n is

$$\mathbf{E}\left[|y_n|^2\right] = \mathbf{E}\left[y_n^H y_n\right] = \mathbf{W}\,\mathbf{E}\left[\mathbf{n}\mathbf{n}^H\right]\mathbf{W}^H = \sigma_n^2 \mathbf{W}\mathbf{W}^H \tag{14.103}$$

As y_n is just a complex scalar, $y_n^H = y_n^*$ holds. Then follows the signal-to-noise ratio from (14.101)

$$\gamma = \frac{\mathbf{E}\left[|y_s|^2\right]}{\mathbf{E}\left[|y_n|^2\right]} = \frac{E_S}{\sigma_n^2}\frac{(\mathbf{W}h)^H \mathbf{W}h}{\mathbf{W}\mathbf{W}^H} \tag{14.104}$$

14.7.2   MMSE Receiver for SIMO System (Maximum Ratio Combiner)

We would like to find the MMSE receiver matrix, which is a row vector. First we have to clarify, whether (14.79) or (14.82) is applicable. Our solution should hold also for low noise, $\alpha \to 0$, i.e. for the zero-forcing receiver. Because of $M = 1$, we have to select the zero-forcing receiver (14.84), which holds for $M \leq N$. Consequently we apply (14.82). With (14.96) we obtain

$$\mathbf{h}^H \mathbf{h} = |h_1|^2 + |h_2|^2 + \cdots + |h_N|^2 = \|\mathbf{h}\|^2 \tag{14.105}$$

which is a scalar and thus invertible, if $\|\mathbf{h}\|^2 \neq 0$. Then (14.82) yields the receive matrix

$$\mathbf{W} = \left(\|\mathbf{h}\|^2 + \alpha\right)^{-1} \mathbf{h}^H = \frac{\mathbf{h}^H}{\|\mathbf{h}\|^2 + \alpha} = \frac{1}{\|\mathbf{h}\|^2 + \alpha}\left(h_1^* \; h_2^* \; \cdots \; h_N^*\right) \tag{14.106}$$

with the entries

$$w_i = \frac{h_i^*}{\|\mathbf{h}\|^2 + \alpha}\,, \quad i = 1, 2, ..., N \tag{14.107}$$

Obviously, they show up as conjugate complex channel coefficients with some real-valued factor. We see that the optimal receiver is a matched filter known from communications theory. As is well known, a matched filter maximizes the signal-to-noise ratio at its output. This property has given rise to the designation "maximum ratio combiner" for this receiver type. The calculation of the signal-to-noise ratio is straightforward from (14.104). We find for the numerator

$$\mathbf{E}\left[|y_s|^2\right] = \left(\frac{\|\mathbf{h}\|^2}{\|\mathbf{h}\|^2 + \alpha}\right)^2 E_S \qquad (14.108)$$

and for the denominator

$$\mathbf{E}\left[|y_n|^2\right] = \frac{\|\mathbf{h}\|^2}{\left(\|\mathbf{h}\|^2 + \alpha\right)^2} \sigma_n^2 \qquad (14.109)$$

Then the signal-to-noise ratio of the maximum ratio combining scheme follows as

$$\gamma_{MRC} = \frac{\mathbf{E}\left[|y_s|^2\right]}{\mathbf{E}\left[|y_n|^2\right]} = \|\mathbf{h}\|^2 \frac{E_S}{\sigma_n^2} = \frac{1}{\alpha}\left(|h_1|^2 + |h_2|^2 + \dots + |h_N|^2\right) \qquad (14.110)$$

Finally we obtain the combiner output signal from (14.98)

$$y = \frac{\|\mathbf{h}\|^2}{\|\mathbf{h}\|^2 + \alpha} s + \frac{\mathbf{h}^H}{\|\mathbf{h}\|^2 + \alpha} \mathbf{n} \qquad (14.111)$$

We substantiate that the maximum ratio combiner can completely cancel the inter-channel interference, because the signal part of the combiner output is just composed of the transmit signal s associated with a real-valued factor. For low noise, $\alpha \ll \|\mathbf{h}\|^2$, follows

$$y = s + \frac{\mathbf{h}^H}{\|\mathbf{h}\|^2} \mathbf{n} \qquad (14.112)$$

Figure 14.3 shows the symbol error rate of two arrangements. Firstly, we consider a SIMO system with $N = 2$ receive antannas and a MRC receiver. Secondly, a SISO system with one antenna on each side serves for comparison. We consider the symbol error rate of two scenarios, in the presence of white Gaussian noise (AWGN) only and for AWGN together with Rayleigh fading. The modulation scheme is 2-PSK and the receiver employs a threshold detector. A computer simulation was done with the open online platform "webdemo" [8]. To verify the results, the error rates have also been calculated, indicated by "theory" in Fig. 14.3, and the curves coincide perfectly. As expected, the additional impairment of Rayleigh fading requires significantly more signal-to-noise ratio at the receiver to achieve an adequate symbol error rate in comparison to a channel, which is just corrupted by AWGN. It is widely appreciated that the introduction of the SIMO scheme with maximum ratio combining provides a large improvement compared to the SISO system.

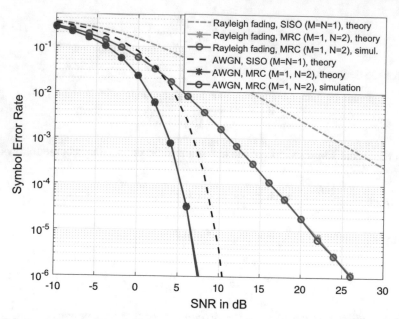

Fig. 14.3 Symbol error rate as a function of signal-to-noise ratio (SNR) for SISO system ($M = N = 1$) compared to SIMO system ($M = 1$, $N = 2$) with maximum ratio combining (MRC), channel with AWGN only and with additional Rayleigh fading, 2-PSK, theoretical and simulated results fit well. Source: Online platform "webdemo" [8]

14.7.3 Equal Gain Combiner

The channel vector of the SIMO system is given by (14.96) and by introducing magnitudes and phases follows

$$\mathbf{h} = \left(|h_1| \, \mathrm{e}^{\mathrm{j}\phi_1} \ |h_2| \, \mathrm{e}^{\mathrm{j}\phi_2} \ \cdots \ |h_N| \, \mathrm{e}^{\mathrm{j}\phi_N} \right)^T \tag{14.113}$$

The vector of the equal gain combiner is defined by

$$\mathbf{W} = \frac{1}{\|\mathbf{h}\|^2 + \alpha} \left(\mathrm{e}^{-\mathrm{j}\phi_1} \ \mathrm{e}^{-\mathrm{j}\phi_2} \ \cdots \ \mathrm{e}^{-\mathrm{j}\phi_N} \right) \tag{14.114}$$

which means that all receiver coefficients w_i have the same gain $\left(\|\mathbf{h}\|^2 + \alpha \right)^{-1}$. This has motivated the name "equal gain combiner". However, the receiver coefficients own different phases, which correspond to the negative phases of the respective channel coefficients. In other words, the receiver does not care about the absolute values of the individual channel coefficients. Consequently, $\mathbf{W}$ is technically rather simple to implement. To determine the signal-to-noise ratio at the receiver output we use (14.104). For that purpose we find

$$\mathbf{Wh} = \frac{1}{\|\mathbf{h}\|^2 + \alpha} \left(|h_1| + |h_2| + \ldots + |h_N|\right) \tag{14.115}$$

which is a real scalar. For the denominator of (14.104) we obtain

$$\sigma_n^2 \mathbf{WW}^H = \frac{\sigma_n^2 N}{\left(\|\mathbf{h}\|^2 + \alpha\right)^2} \tag{14.116}$$

Consequently, the signal-to-noise ratio is

$$\gamma_{EGC} = \frac{\mathbf{E}\left[|y_s|^2\right]}{\mathbf{E}\left[|y_n|^2\right]} = \frac{E_S}{N\sigma_n^2} \left(|h_1| + |h_2| + \ldots + |h_N|\right)^2 \tag{14.117}$$

Finally, we obtain for the output signal of the equal gain combiner from (14.98) making use of (14.115) and (14.114)

$$y = \frac{1}{\|\mathbf{h}\|^2 + \alpha} \left(|h_1| + |h_2| + \ldots + |h_N|\right) s +$$

$$+ \frac{1}{\|\mathbf{h}\|^2 + \alpha} \left(e^{-j\phi_1} \; e^{-j\phi_2} \; \cdots \; e^{-j\phi_N}\right) \mathbf{n} \tag{14.118}$$

As we can see, the equal gain combiner output contains the transmit signal s associated with a real-valued factor and thus no inter-channel interference is present.

Example 10
Compare the signal-to-noise ratios of the maximum ratio combiner, γ_{MRC}, and the equal gain combiner, γ_{EGC}.

Solution:
We find with (14.110) and (14.117)

$$\frac{\gamma_{MRC}}{\gamma_{EGC}} = N \frac{|h_1|^2 + |h_2|^2 + \ldots + |h_N|^2}{\left(|h_1| + |h_2| + \ldots + |h_N|\right)^2} \tag{14.119}$$

Obviously, the right-hand side is $\leq N$. Consequently, the signal-to-noise ratio of the maximum ratio combiner is at most by factor N larger than that of the equal gain combiner. The equal sign holds for a SISO channel (one coefficient, $N = 1$).

Example 11
Determine the signal-to-noise ratio for a normalized channel vector.

Solution:
We introduce a normalization of the channel matrix using the channel power gain defined in (14.58). With $M = 1$ and $\operatorname{tr}\left(\mathbf{h}^H \mathbf{h}\right) = \|\mathbf{h}\|^2$ from (14.105) we get the power gain for the SIMO channel as $g_P = \|\mathbf{h}\|^2$. To find the normalized results we have to replace $\mathbf{h}$ by $\frac{\mathbf{h}}{\sqrt{g_P}} = \frac{\mathbf{h}}{\|\mathbf{h}\|}$ in the relevant formulas, which means that all entries of the

vector are divided by $\|\mathbf{h}\|$. Then follows from (14.110) the normalized signal-to-noise ratio for a maximum ratio combining receiver

$$\gamma_{MRC,N} = \frac{E_S}{\sigma_n^2} = \frac{1}{\alpha} \tag{14.120}$$

Replacing $|h_i|$ by $\frac{|h_i|}{\|\mathbf{h}\|}$; $i = 1, 2, ..., N$ in (14.117) yields the normalized signal-to-noise ratio for the equal gain combining scheme

$$\gamma_{EGC,N} = \frac{1}{N\alpha} \frac{(|h_1| + |h_2| + ... + |h_N|)^2}{\|\mathbf{h}\|^2} \tag{14.121}$$

and finally from (14.119) the ratio

$$\frac{\gamma_{MRC,N}}{\gamma_{EGC,N}} = \frac{\gamma_{MRC}}{\gamma_{EGC}} \tag{14.122}$$

which remains unchanged as in (14.119), because the normalization factor g_P drops out.

14.8   Decision of Receiver Output Signal

As we have seen, a linear receiver with matrix $\mathbf{W}$ tries to minimize the inter-channel interference and to some extend also the noise in the signal components $y_i(k)$ of the output vector (14.2)

$$\mathbf{y}(k) = \left(y_1(k) \ y_2(k) \ \cdots \ y_M(k) \right)^T \tag{14.123}$$

However, the signal components are still corrupted by some noise. As depicted in Fig. 14.1, a decision device following the linear receiver is required to recover the QAM symbols in each component $y_i(k)$. This process is also called signal detection and there are various detection strategies known from communications theory and outlined in Part I. The simplest one is threshold decision, where the complex signal plane can be structured by a rectangular grid with the grid lines as the decision thresholds. The decision device allocates to $y_i(k)$ an estimate $\hat{s}_i(k)$ of the transmit signal alphabet. Depending on the noise and the fading of the channel coefficients, quite some false decisions may occur, which can be quantified by a symbol error probability. Another strategy is maximum likelihood detection of the $y_i(k)$ or of the whole vector $\mathbf{y}(k)$. This method can minimize the error probability either on a "symbol by symbol" basis at every time instant k or by considering finite sequences of symbols, after which a "sequence detection" using the Viterbi algorithm is executed.

We know from communications theory that a cascade of a linear receiver and a decision device is sub-optimal for achieving minimal symbol error probability. An optimal solution is the a-posterior, in special cases also the maximum likelihood detector, which are applied directly on the receive signal $\mathbf{r}(k)$ in (14.1). This will lead to a nonlinear receiver and will be presented in the next Chapter.

References

1. Moore, E.H.: On the reciprocal of the general algebraic matrix. Bull. Am. Math. Soc. **26** (1920)
2. Penrose, R.: A generalized inverse for matrices. Proc. Camb. Philos. Soc. **51** (1955)
3. Albert, A.: Regression and the Moore–Penrose pseudoinverse, vol. 94. Elesvier, Academic Press, Cambridge (1972)
4. Petersen, K.B., Pedersen, M.S.: The Matrix Cookbook. Technical University of Denmark (2012)
5. Ben-Israel, A., Greville, T.N.E.: Generalized Inverses. Springer, Berlin (2003)
6. Hjorungnes, A.: Complex-valued Matrix Derivatives with Applications in Signal Processing and Communications. Cambridge University Press, Cambridge (2011)
7. Horn, R.A., Johnson, C.R.: Matrix Analysis. Cambridge University Press, Cambridge (2013)
8. Chaoudhry, B.B.: Diversity combining, webdemo. Technical report, Institute of Telecommunications, University of Stuttgart, Germany (2018). http://webdemo.inue.uni-stuttgart.de

Chapter 15
Principles of Nonlinear MIMO Receivers

15.1 Maximum Likelihood MIMO Receiver

Principle

As we have seen in the previous chapter, a linear receiver tries to reduce the impact of inter-channel interference and partially of the noise in the receive signal $\mathbf{y}(k)$ of Fig. 14.1. Next, the signal is subject to a decision also called detection to recover the QAM symbols in each component $y_i(k)$. Various decision strategies are known from communications theory and outlined in Part I. In this section we will consider a Maximum Likelihood (ML) detector as a receiver. In contrast to the linear receiver the signal $\hat{\mathbf{s}}(k)$ will be estimated directly from the receive vector

$$\mathbf{r}(k) = \left(r_1(k)\ r_2(k)\ \cdots\ r_N(k) \right)^T \tag{15.1}$$

Hence, a receive matrix $\mathbf{W}$ is not present. In the following we drop the discrete-time k to simplify the notation. The observed receive vector

$$\mathbf{r} = \mathbf{H}\mathbf{s} + \mathbf{n} \tag{15.2}$$

is corrupted by additive noise $\mathbf{n}$, where $\mathbf{H}\mathbf{s}$ is the receive signal in case of a noise-free channel. As for the linear receivers, we assume that the channel matrix $\mathbf{H}$ is precisely known to the receiver. In a practical system the entries of $\mathbf{H}$ have to be estimated by a separate channel estimator, which is not considered here. The transmit signal vector is given by

$$\mathbf{s} = \left(s_1\ s_2\ \cdots\ s_M \right)^T \tag{15.3}$$

in which each component s_j is taken from a finite QAM symbol alphabet $\mathcal{B}$, e.g., $\mathcal{B} = \{1, j, -1, -j\}$ for 4-ary phase shift keying (4-PSK) or $\mathcal{B} = \{1, -1\}$ for 2-PSK. We assume an additive white Gaussian noise (AWGN) vector $\mathbf{n} = (n_1\ n_2 \ldots n_N)^T$ with the following properties,

© Springer Nature Switzerland AG 2021
J. Speidel, *Introduction to Digital Communications*,
Signals and Communication Technology,
https://doi.org/10.1007/978-3-030-67357-4_15

- statistically independent with covariance matrix

$$\mathbf{R}_{nn} = \sigma_n^2 \mathbf{I}_N \tag{15.4}$$

- all noise components n_i possess the same mean power σ_n^2 and zero mean $\mathbf{E}[n_i] = 0$; $i = 1, 2, \ldots, N$.
- the real part $n_{R,i}$ and the imaginary part $n_{I,i}$ of the noise $n_i = n_{R,i} + jn_{I,i}$ are statistically independent, have the same mean power $\frac{\sigma_n^2}{2}$, and the same Gaussian probability density function

$$p_x(x) = \frac{1}{\sqrt{2\pi}\,\sigma_x}\,e^{-\frac{x^2}{2\sigma_x^2}} \; ; \; \sigma_x^2 = \frac{\sigma_n^2}{2} \tag{15.5}$$

where x stands for $n_{R,i}$ and $n_{I,i}$, $i = 1, 2, \ldots, N$. Consequently, the density function of the noise n_i is given by the product

$$p_{n_i}(n_i) = p_{n_{R,i}}(n_{R,i})p_{n_{I,i}}(n_{I,i}) = \frac{1}{\pi\sigma_n^2}\,e^{-\frac{|n_i|^2}{\sigma_n^2}} \; ; \; i = 1, 2, \ldots, N \tag{15.6}$$

- the multivariate probability density function of the noise vector $\mathbf{n}$ then follows as the product

$$p_{\mathbf{n}}(n_1, n_2, \ldots, n_N) = \left(\frac{1}{\pi\sigma_n^2}\right)^N e^{-\frac{|n_1|^2+|n_2|^2+\cdots+|n_N|^2}{\sigma_n^2}} \tag{15.7}$$

or with short hand notation

$$p_{\mathbf{n}}(\mathbf{n}) = \left(\frac{1}{\pi\sigma_n^2}\right)^N e^{-\frac{\|\mathbf{n}\|^2}{\sigma_n^2}} \tag{15.8}$$

For the decision process we first define the following conditional probability density function

$$p_L(\mathbf{r}\,|\mathbf{Hs}) \tag{15.9}$$

which is also called likelihood probability density function. It can be interpreted as the density function of $\mathbf{r}$ under the condition that $\mathbf{s}$ was sent, knowing $\mathbf{H}$. Please note that $p_L(\mathbf{r}\,|\mathbf{Hs})$ describes a finite set of probability density functions generated by all possible transmit vectors $\mathbf{s} \in \mathcal{A}$, where $\mathcal{A}$ is the set of all possible transmit vectors. Be it that each of the M components of $\mathbf{s}$ can take on L_Q different QAM symbol values, then $\mathcal{A}$ contains L_Q^M different vectors $\mathbf{s}_m$, $m = 1, 2, \ldots, L_Q^M$. The maximum likelihood detector selects out of all possible $\mathbf{Hs}$ that estimate $\mathbf{s} = \hat{\mathbf{s}}$, which is maximal likely to the receive vector $\mathbf{r}$, i.e. which has the largest $p_L(\mathbf{r}\,|\mathbf{Hs})$. Hence, the detection criterion is

$$p_L(\mathbf{r}\,|\mathbf{Hs}) = \max_{\mathbf{s} \in \mathcal{A}} \tag{15.10}$$

from which the optimal estimate

$$\hat{\mathbf{s}} = \arg\max_{\mathbf{s} \in \mathcal{A}} \; p_L \left(\mathbf{r} \, | \mathbf{Hs} \right) \tag{15.11}$$

results. As is well known from communications theory, if the transmit vectors $\mathbf{s} \in \mathcal{A}$ are equally distributed, then $\hat{\mathbf{s}}$ also maximizes the a-posterior probability and thus minimizes the symbol error probability. With (15.2) and (15.8) we obtain from (15.9)

$$p_L \left(\mathbf{r} \, | \mathbf{Hs} \right) = p_{\mathbf{n}} \left(\mathbf{r} - \mathbf{Hs} \right) = \left(\frac{1}{\pi \sigma_n^2} \right)^N e^{-\frac{\|\mathbf{r} - \mathbf{Hs}\|^2}{\sigma_n^2}} \tag{15.12}$$

The argument of the exponential function is always negative. Consequently, the maximal $p_L \left(\mathbf{r} \, | \mathbf{Hs} \right)$ must fulfill the condition

$$\|\mathbf{r} - \mathbf{Hs}\|^2 = \min_{\mathbf{s} \in \mathcal{A}} \tag{15.13}$$

and the solution formally is

$$\hat{\mathbf{s}} = \arg\min_{\mathbf{s} \in \mathcal{A}} \; \|\mathbf{r} - \mathbf{Hs}\|^2 \tag{15.14}$$

Obviously, the statistical detection problem (15.10) translates into the minimization of the Euclidean distance between two vectors, namely the receive vector $\mathbf{r}$ and the vector $\mathbf{Hs}$, which is the transmit signal $\mathbf{s}$ having passed through the known channel $\mathbf{H}$. Hence, a maximum likelihood detector can be implemented as an algorithm, which calculates a squared error according to (15.13) for all possible transmit signal vectors $\mathbf{s} \in \mathcal{A}$ and selects that $\mathbf{s} = \hat{\mathbf{s}}$, which yields the minimal quadratic error. Of course, the receiver has to know the transmit vector alphabet $\mathcal{A}$, which is quite normal for the design of a digital communications system.

Just a few words about the computational complexity. As already mentioned, if the transmitter is equipped with M antennas and each antenna output signal can take on L_Q different values, then there are L_Q^M different vectors $\mathbf{s}$, for which the detector has to execute (15.13). We conclude that the number of operations in the maximum likelihood detector grows exponentially with the number M of transmit antennas.

Example 1

As a simple example we take a MIMO transmitter with $M = 2$ antennas. The modulation scheme shall be 2-PSK with the symbol alphabet $\mathcal{B} = \{1, -1\}$. Consequently $L_Q = 2$ and each component of $\mathbf{s}$ can take on one value out of $\mathcal{B}$ at time instant k. The channel matrix shall be given as

$$\mathbf{H} = \begin{pmatrix} 1 & 0.5 \\ 0 & 1 \\ 1 & 1 \end{pmatrix}$$

Table 15.1 Example, calculation steps for maximum likelihood detection

s	$\begin{pmatrix}1\\1\end{pmatrix}$	$\begin{pmatrix}1\\-1\end{pmatrix}$	$\begin{pmatrix}-1\\1\end{pmatrix}$	$\begin{pmatrix}-1\\-1\end{pmatrix}$
Hs	$\begin{pmatrix}1.5\\1.0\\2.0\end{pmatrix}$	$\begin{pmatrix}0.5\\-1.0\\0\end{pmatrix}$	$\begin{pmatrix}-0.5\\1.0\\0\end{pmatrix}$	$\begin{pmatrix}-1.5\\-1.0\\-2.0\end{pmatrix}$
r − Hs	$\begin{pmatrix}-0.4\\-2.1\\-1.1\end{pmatrix}$	$\begin{pmatrix}0.6\\-0.1\\0.9\end{pmatrix}$	$\begin{pmatrix}1.6\\-2.1\\0.9\end{pmatrix}$	$\begin{pmatrix}2.6\\-0.1\\2.9\end{pmatrix}$
$\lVert\mathbf{r}-\mathbf{Hs}\rVert^2$	5.78	1.18	7.78	15.81

and the noisy receive vector is observed as $\mathbf{r} = (1.1\ -1.1\ 0.9)^T$. The receiver knows the set $\mathcal{A}$ of all $L_Q^M = 4$ different transmit vectors.

$$\mathcal{A} = \left\{\begin{pmatrix}1\\1\end{pmatrix},\begin{pmatrix}1\\-1\end{pmatrix},\begin{pmatrix}-1\\1\end{pmatrix},\begin{pmatrix}-1\\-1\end{pmatrix}\right\} \tag{15.15}$$

Then the maximum likelihood receiver calculates all vectors $\mathbf{Hs}$ and $\mathbf{r} - \mathbf{Hs}$ as well as the squared error $\lVert\mathbf{r}-\mathbf{Hs}\rVert^2$ in Table 15.1. Finally, the minimal $\lVert\mathbf{r}-\mathbf{Hs}\rVert^2$, which is 1.18 in our example, is selected and the detector concludes that most likely

$$\hat{\mathbf{s}} = \begin{pmatrix}1\\-1\end{pmatrix}$$

was sent.

15.2   Receiver with Ordered Successive Interference Cancellation

Prerequisites

We are now coming back to the transmission system depicted in Fig. 14.1 and are going to combine the linear receiver with the decision device. Our target is to successively detect the transmit signal components $s_j(k)$, $j = 1, 2, \ldots, M$ of the transmit vector $\mathbf{s}(k)$. Again, to simplify notation we drop k. The starting point of our considerations is the linear receiver. The receive signal is given by

$$\mathbf{r} = \mathbf{H}\mathbf{s} + \mathbf{n} \tag{15.16}$$

with the channel matrix

$$\mathbf{H} = \begin{pmatrix} \mathbf{h}_1 & \mathbf{h}_2 & \cdots & \mathbf{h}_M \end{pmatrix} \tag{15.17}$$

in which $\mathbf{h}_j \in \mathbb{C}^{N \times 1}$, $j = 1, \ldots, M$ are the column vectors. The receiver matrix

$$\mathbf{W} = \begin{pmatrix} \mathbf{w}_1 \\ \mathbf{w}_2 \\ \vdots \\ \mathbf{w}_M \end{pmatrix} \tag{15.18}$$

is structured by its row vectors $\mathbf{w}_i \in \mathbb{C}^{1 \times N}$, $i = 1, \ldots, M$ and can be calculated as the pseudo inverse or the MMSE receive matrix of the channel matrix. Finally, we get the output of the receiver filter

$$\mathbf{y} = \begin{pmatrix} y_1 \\ y_2 \\ \vdots \\ y_M \end{pmatrix} \tag{15.19}$$

by multiplication

$$\mathbf{y} = \mathbf{W}\mathbf{r} \tag{15.20}$$

Hence, the output signal component y_i is obtained as

$$y_i = \mathbf{w}_i \mathbf{r} \; ; \; i = 1, 2, \ldots, M \tag{15.21}$$

According to Fig. 14.1, a decision device follows and we characterize the input-output relation by the decision function $q(\ldots)$ yielding

$$\hat{s}_i = q(y_i) \; ; \; i = 1, 2, \ldots, M \tag{15.22}$$

The decision device can be a simple threshold detector but also a more sophisticated maximum likelihood detector. If the receiver applies the receive matrix $\mathbf{W}$, the system of equations is solved for all $y_1, y_2, \ldots, y_M$ in one step and the decided signal components $\hat{s}_1, \hat{s}_2, \ldots, \hat{s}_M$ are obtained in parallel. Now we are going to discuss a method, in which the system of linear equations (15.20) is solved successively in several steps, where in each step the decision operation (15.22) is applied.

Ordered Successive Interference Cancellation (OSIC)

As outlined before, we are looking for an algorithm, with which the $\hat{s}_i = q(y_i)$, $i = 1, 2, \ldots, M$ are calculated one after the other rather than in one step. For example, the natural "order" $r_1, r_2, \ldots, r_N$ of equations is taken in the following. In principle, operations (15.20) and (15.22) are merged. With (15.17) we obtain from (15.16)

$$\left(\mathbf{h}_1 \cdots \mathbf{h}_\nu \cdots \mathbf{h}_M \right) \begin{pmatrix} s_1 \\ \vdots \\ s_\nu \\ \vdots \\ s_M \end{pmatrix} + \mathbf{n} = \mathbf{r} \tag{15.23}$$

which is equivalent to

$$\mathbf{h}_1 s_1 + \cdots + \mathbf{h}_{\nu-1} s_{\nu-1} + \mathbf{h}_{\nu+1} s_{\nu+1} + \cdots + \mathbf{h}_M s_M + \mathbf{n} = \mathbf{r} - \mathbf{h}_\nu s_\nu \tag{15.24}$$

and is the key equation, in which we have moved $\mathbf{h}_\nu s_\nu$ to the right-hand side. The idea is first to find a solution for s_ν, e.g., s_1, and then reduce the dimension of the system of linear equations by one. It matters to introduce ν as an index for the iteration step ν.

The algorithm is best explained with $M = 3$ as an example. Let $\mathbf{H} = \left(\mathbf{h}_1 \ \mathbf{h}_2 \ \mathbf{h}_3 \right)$, $\mathbf{r}$, and the decision rule $q(\ldots)$ be given. In the course of the iterations the matrix $\mathbf{H}$ will change and therefore it will be indicated as $\mathbf{H}^{(\nu)}$. In each step only the first row vector $\mathbf{w}_1^{(\nu)}$ of the receive matrix

$$\mathbf{W}^{(\nu)} = \begin{pmatrix} \mathbf{w}_1^{(\nu)} \\ \mathbf{w}_2^{(\nu)} \\ \mathbf{w}_3^{(\nu)} \end{pmatrix} \tag{15.25}$$

has to be calculated either from the pseudo inverse or the MMSE matrix with respect to $\mathbf{H}^{(\nu)}$. The iteration steps are as follows,

step $\nu = 1$:

Let $\quad \mathbf{H}^{(1)} = \left(\mathbf{h}_1 \ \mathbf{h}_2 \ \mathbf{h}_3 \right), \quad \mathbf{r}^{(1)} = \mathbf{r}$

then $\quad \mathbf{H}^{(1)} \begin{pmatrix} s_1 \\ s_2 \\ s_3 \end{pmatrix} + \mathbf{n} = \mathbf{r}^{(1)}$

calculate $\quad \mathbf{w}_1^{(1)}$ from $\mathbf{H}^{(1)}$, using (15.25)

calculate $\quad y_1 = \mathbf{w}_1^{(1)} \mathbf{r}^{(1)}$ using (15.21)

decide $\quad \hat{s}_1 = q(y_1)$ using (15.22)

set $\quad s_1 = \hat{s}_1$ and the new equation system is

$$\left(\mathbf{h}_2 \ \mathbf{h}_3 \right) \begin{pmatrix} s_2 \\ s_3 \end{pmatrix} + \mathbf{n} = \mathbf{r}^{(1)} - \mathbf{h}_1 \hat{s}_1$$

step $\nu = 2$:

$$\text{Let} \quad \mathbf{H}^{(2)} = \left(\mathbf{h}_2 \ \mathbf{h}_3 \right), \quad \mathbf{r}^{(2)} = \mathbf{r}^{(1)} - \mathbf{h}_1 \hat{s}_1$$

$$\text{then} \quad \mathbf{H}^{(2)} \begin{pmatrix} s_2 \\ s_3 \end{pmatrix} + \mathbf{n} = \mathbf{r}^{(2)}$$

calculate $\mathbf{w}_1^{(2)}$ from $\mathbf{H}^{(2)}$ using (15.25)

calculate $y_2 = \mathbf{w}_1^{(2)} \mathbf{r}^{(2)}$ using (15.21)

decide $\hat{s}_2 = q(y_2)$ using (15.22)

set $s_2 = \hat{s}_2$ and the new equation system is

$$\mathbf{h}_3 s_3 + \mathbf{n} = \mathbf{r}^{(2)} - \mathbf{h}_2 \hat{s}_2$$

step $\nu = 3$:

$$\text{Let} \quad \mathbf{H}^{(3)} = \mathbf{h}_3, \quad \mathbf{r}^{(3)} = \mathbf{r}^{(2)} - \mathbf{h}_2 \hat{s}_2$$

$$\text{then} \quad \mathbf{H}^{(3)} s_3 + \mathbf{n} = \mathbf{r}^{(3)}$$

calculate $\mathbf{w}_1^{(3)}$ from $\mathbf{H}^{(3)}$ using (15.25)

calculate $y_3 = \mathbf{w}_1^{(3)} \mathbf{r}^{(3)}$ using (15.21)

decide $\hat{s}_3 = q(y_3)$ using (15.22)

set $s_3 = \hat{s}_3$

end In our example with $N = 3$ the algorithm terminates.

The advantage of this algorithm is its low computational complexity and the feature that it reduces inter-channel interference in every decision step. However, decision errors, which may occur at low signal-to-noise ratios are critical, because they can impact the next decision and thus may cause an error propagation for the following steps. This effect can be mitigated by utilizing an appropriate order, justifying the name OSIC, e.g., by taking the r_i with the largest SNR first. We notice that the algorithm is in principle based on the triangulation of a matrix into a lower or an upper triangular form also called L-U decomposition [1], which is continuously applied from one step to the next. This is in principle a linear operation. However, the described OSIC algorithm gets nonlinear owing to the decision made in each step. The algorithm has been practically used in several systems, such as the layered space-time architecture.

15.3 Comparison of Different Receivers

The design criteria for linear and nonlinear receivers have quite some similarities which are now going to be discussed for the zero forcing (ZF), the minimum mean squared error (MMSE), the OSIC, and the maximum likelihood (ML) receiver. Table 15.2 shows a survey of the different design criteria.

The design of the zero forcing receiver with and without OSIC does not include the noise at the receiver. Computation of the receiver matrix $\mathbf{W}$ for the MMSE receiver

Table 15.2 Comparison of design criteria for various receivers

Receiver	Target function	Noise	Result	Output	Method
Zero-forcing (ZF)	$\|\mathbf{r} - \mathbf{Hs}\|^2$ $= \min_{\mathbf{s} \in \mathbb{C}^{M \times 1}}$	Not included	Matrix $\mathbf{W}$	$\mathbf{y} = \mathbf{Wr}$	Linear
MMSE	$\mathbf{E}\left[\|\mathbf{Wr} - \mathbf{s}\|^2\right]$ $= \min_{\mathbf{s} \in \mathbb{C}^{M \times 1}}$	Included	Matrix $\mathbf{W}$	$\mathbf{y} = \mathbf{Wr}$	Linear
Maximum likelihood	$\|\mathbf{r} - \mathbf{Hs}\|^2$ $= \min_{\mathbf{s} \in \mathcal{A}}$	Included	Symbol $\hat{\mathbf{s}}$	$\hat{\mathbf{s}} \in \mathcal{A}$	Non-linear
OSIC ZF	As ZF	Not included	Symbol $\hat{\mathbf{s}}$	$\hat{\mathbf{s}} \in \mathcal{A}$	Non-linear
OSIC MMSE	As MMSE	Included	Symbol $\hat{\mathbf{s}}$	$\hat{\mathbf{s}} \in \mathcal{A}$	Non-linear

requires the knowledge of the signal-to-noise ratio $\frac{1}{\alpha}$, which is not needed for the maximum likelihood detection. This method operates without any receiver matrix. Moreover, on the first glance the target functions of the zero forcing algorithm using the pseudo inverse matrix and the maximum likelihood receiver look the same. Both receivers minimize the squared error $\|\mathbf{r} - \mathbf{Hs}\|^2$. However, the zero forcing receiver provides a "soft" output signal $\mathbf{y} \in \mathbb{C}^{M \times 1}$ with continuous amplitude and phase compared to the output of the maximum likelihood receiver, which is a discrete vector $\hat{\mathbf{s}} \in \mathcal{A}$. Hence, the maximum likelihood scheme minimizes the same target function as the zero forcing receiver, however, as the result of a discrete minimization problem with the constraint $\hat{\mathbf{s}} \in \mathcal{A}$. This can be formulated as an integer least squares problem for which several mathematical algorithms from the area of integer programming are known, [2, 3]. Such methods have been used for lattice decoding and are summarized as sphere decoding algorithm, because they search in a limited hyper sphere of the complex vector space rather than performing an overall brute search [4–6] and thus do not always provide the global optimum. In principle, the hyper sphere is centered around the receive vector $\mathbf{r}$ and for an efficient search the sphere should cover the lattice points given by the vectors ($\mathbf{Hs}$; $\mathbf{s} \in \mathcal{A}$) located in the vicinity of $\mathbf{r}$. As a result, the complexity of the maximum likelihood algorithm can be significantly reduced and sphere decoding became an important alternative to the much simpler but sub-optimal linear receivers.

On the other hand the complexity of the zero forcing and the minimum mean squared error (MMSE) receiver can also be considerably reduced by introducing successive interference cancellation (OSIC), because only parts of an inverse matrix

Fig. 15.1 Typical symbol error rate of various receivers for a 2x2 MIMO channel corrupted by white Gaussian noise and Rayleigh fading using 2-PSK, *Source* Online platform "webdemo" [7]

have to be calculated rather than a full inverse or pseudo inverse matrix. However, it should be kept in mind that in case of ill conditioned matrices the calculation of the inverse matrices may turn out to be numerically not stable.

Figure 15.1 shows a rough comparison of the symbol error rate for various receivers as the result of a computer simulation using the platform "webdemo" [7]. According to our expectations, the maximum likelihood detector (ML) demonstrates the best performance followed by the nonlinear receiver with OSIC. Compared to the zero forcing receiver (ZF) the minimum mean squared error approach (MMSE) takes the noise into account and thus outperforms the ZF receiver in general.

References

1. Horn, R.A., Johnson, C.R.: Matrix Analysis. Cambridge University Press, Cambridge (2013)
2. Kannan, R.: Improved algorithms on integer programming and related lattice problems. In: Proceedings of ACM Symposium on Theory of Computation (1983)
3. Finke, U., Pohst, M.: Improved methods for calculating vectors of short length in a lattice, including a complexity analysis. Math. Comput. **44** (1985)
4. Viterbo, E., Boutros, J.: A universal lattice code decoder for fading channels. IEEE Trans. Inf. Theory **45** (1999)
5. Damen, O., Chkeif, A., Belfiore, J.: Lattice code decoder for space-time codes. IEEE Commun. Lett. **4** (2000)
6. Kailath, T., Vikalo, H., Hassibi, B.: MIMO receive algorithms. In: Space-Time Wireless Systems: From Array Processing to MIMO Communications. Cambridge University Press, Cambridge (2008)
7. Zhao, N.: MIMO detection algorithms, webdemo. Technical report. Institute of Telecommunications, University of Stuttgart, Germany (2018). http://webdemo.inue.uni-stuttgart.de

Chapter 16
MIMO System Decomposition into Eigenmodes

In this chapter we allude to a topic, which gives quite some inside into the functionality of a MIMO system. As we have seen, the MIMO channel matrix $\mathbf{H}(k)$ introduces inter-channel interference to the receive signal

$$\mathbf{r}(k) = \mathbf{H}(k)\mathbf{s}(k) + \mathbf{n}(k) \tag{16.1}$$

We are now interested in the decoupling of the receive signal. To achieve this goal $\mathbf{H}(k)$ has to be transformed into a matrix, in which only one diagonal is covered by entries unequal to zero and all remaining elements have to be zero. In the following we drop the discrete time k to simplify the notation.

16.1 MIMO System Transformation Using Singular Value Decomposition

A first idea to transform the channel matrix $\mathbf{H}$ into diagonal form is the application of the eigenvalue decomposition, as outlined in the Appendix B.

$$\mathbf{H} = \mathbf{V}\Lambda\mathbf{V}^{-1} \tag{16.2}$$

where Λ is a diagonal matrix containing the eigenvalues of $\mathbf{H}$ in its main diagonal. $\mathbf{V}$ is composed of eigenvectors associated with the respective eigenvalues. However, this approach has various drawbacks both from a mathematical and a technical point of view. First of all $\mathbf{H}$ must be a square matrix and this fact would restrict the approach to a system with the same number of antennas at transmitter and receiver, $M = N$. Secondly, not all square matrices can be transformed to diagonal form mathematically. An alternative is the Singular Value Decomposition (SVD), which

© Springer Nature Switzerland AG 2021
J. Speidel, *Introduction to Digital Communications*,
Signals and Communication Technology,
https://doi.org/10.1007/978-3-030-67357-4_16

can be applied to any matrix and which is outlined in detail in the Appendix B. In this chapter we recall the principle steps.

The singular value decomposition of the channel matrix $\mathbf{H} \in \mathbb{C}^{N \times M}$ is given by

$$\mathbf{H} = \mathbf{U}\mathbf{D}\mathbf{V}^H \tag{16.3}$$

with

$$\mathbf{D} = \begin{pmatrix} \sqrt{\lambda_1} & 0 & 0 \cdots & 0 & 0 \cdots 0 \\ 0 & \sqrt{\lambda_2} & 0 \cdots & 0 & 0 \cdots 0 \\ & & \ddots & & \\ 0 & 0 & 0 \cdots \sqrt{\lambda_P} & 0 \cdots 0 \\ 0 & 0 & 0 \cdots & 0 & 0 \cdots 0 \\ & \vdots & \cdots & & \ddots \\ 0 & 0 & 0 \cdots & 0 & 0 \cdots 0 \\ 0 & 0 & 0 \cdots & 0 & 0 \cdots 0 \end{pmatrix} = \begin{pmatrix} \Lambda_P^{\frac{1}{2}} & 0 \cdots 0 \\ 0 & 0 \cdots 0 \\ & \ddots \\ 0 & 0 \cdots 0 \\ 0 & 0 \cdots 0 \end{pmatrix} \in \mathbb{R}^{N \times M} \tag{16.4}$$

where $\lambda_1 \geq \lambda_2 \geq \cdots \lambda_P > 0$ and $\lambda_{P+1} = \lambda_{P+2} = \cdots = \lambda_N = 0$ are the N eigenvalues of the matrix

$$\mathbf{Q}_N = \mathbf{H}\mathbf{H}^H \in \mathbb{C}^{N \times N} \tag{16.5}$$

Please note that $\mathbf{Q}_N$ is a Hermiteian matrix, because $\mathbf{Q}_N^H = \mathbf{Q}_N$ and consequently all eigenvalues are positive or zero.

$$P = \text{rank}\,(\mathbf{Q}_N) \tag{16.6}$$

is the rank of the matrix $\mathbf{Q}_N$. In general the rank of a matrix $\mathbf{H} \in \mathbb{C}^{N \times M}$ is defined as the number of linearly independent rows or columns and thus

$$\text{rank}\,(\mathbf{H}) \leq \min\{M, N\} \tag{16.7}$$

holds. With (16.5) follows

$$P \leq N \tag{16.8}$$

$\sqrt{\lambda_i}$; $i = 1, ..., P$ are called the singular values of the matrix $\mathbf{H}$.
$\mathbf{U} \in \mathbb{C}^{N \times N}$ and $\mathbf{V} \in \mathbb{C}^{M \times M}$ are unitary matrices, thus

$$\mathbf{U}^{-1} = \mathbf{U}^H \; ; \; \mathbf{V}^{-1} = \mathbf{V}^H \tag{16.9}$$

hold. Furthermore, $\mathbf{U}$ is the matrix of the normalized eigenvectors with respect to the eigenvalues $\lambda_1, \lambda_2, \cdots, \lambda_N$. Let

$$\Lambda_N = \text{diag}\,(\lambda_1, \lambda_2, ..., \lambda_P, 0, ..., 0) \in \mathbb{R}^{N \times N} \tag{16.10}$$

be a diagonal matrix composed of the eigenvalues of $\mathbf{Q}_N$. Then the eigenvalue decomposition of $\mathbf{Q}_N$ is

$$\mathbf{U}^H \mathbf{Q}_N \mathbf{U} = \Lambda_N \qquad (16.11)$$

One method to find the matrix $\mathbf{V}$ in (16.3) is the eigenvalue decomposition of the matrix

$$\mathbf{Q}_M = \mathbf{H}^H \mathbf{H} \in \mathbb{C}^{M \times M} \qquad (16.12)$$

which is

$$\mathbf{V}^H \mathbf{Q}_M \mathbf{V} = \Lambda_M \qquad (16.13)$$

with the diagonal matrix

$$\Lambda_M = \mathrm{diag}\ (\lambda_1, \lambda_2, ..., \lambda_P, 0, ..., 0) \in \mathbb{R}^{M \times M} \qquad (16.14)$$

$\mathbf{V}$ is the matrix of eigenvectors with respect to the eigenvalues $\lambda_1, \lambda_2, \cdots, \lambda_M$. Note that the eigenvalues $\lambda_1, \lambda_2, ..., \lambda_P$ are the same as for the matrix $\mathbf{Q}_N$. Furthermore

$$\mathrm{rank}\ (\mathbf{Q}_M) = \mathrm{rank}\ (\mathbf{Q}_N) = \mathrm{rank}\ (\mathbf{H}) = P \qquad (16.15)$$

holds and the matrices Λ_M and Λ_N contain the same diagonal matrix

$$\Lambda_P - \mathrm{diag}\ (\lambda_1, \lambda_2, ..., \lambda_P) \in \mathbb{R}^{P \times P} \qquad (16.16)$$

of the P eigenvalues, which are unequal to zero. Please note in (16.4)

$$\Lambda_P^{\frac{1}{2}} = \mathrm{diag}\ \left(\sqrt{\lambda_1}, \sqrt{\lambda_2}, ..., \sqrt{\lambda_P} \right) \qquad (16.17)$$

holds. We obtain from the input-output relation (16.1) with (16.3)

$$\mathbf{r} = \mathbf{U} \mathbf{D} \mathbf{V}^H \mathbf{s} + \mathbf{n} \qquad (16.18)$$

and by multiplication of this equation from the left with $\mathbf{U}^H$ follows

$$\mathbf{U}^H \mathbf{r} = \mathbf{U}^H \mathbf{U} \mathbf{D} \mathbf{V}^H \mathbf{s} + \mathbf{U}^H \mathbf{n} \qquad (16.19)$$

With $\mathbf{U}^H \mathbf{U} = \mathbf{I}_N$ and the transformed signals

$$\tilde{\mathbf{r}} = \mathbf{U}^H \mathbf{r} \qquad (16.20)$$

$$\tilde{\mathbf{s}} = \mathbf{V}^H \mathbf{s} \qquad (16.21)$$

and the transformed noise

$$\tilde{\mathbf{n}} = \mathbf{U}^H \mathbf{n} \qquad (16.22)$$

a new description of the system referred to as *eigenmode system* based on the eigenmodes can be given as

$$\tilde{\mathbf{r}} = \mathbf{D}\tilde{\mathbf{s}} + \tilde{\mathbf{n}} \qquad (16.23)$$

The new *eigenmode channel* is described by the matrix $\mathbf{D}$, which has entries unequal to zero only in a diagonal. All other entries are zero. Equation (16.23) clearly reveals that the eigenmode branches and thus the signal and noise components are decoupled, which is the basic idea of the approach. This can be even better seen when we write down the ith equation of (16.23)

$$\tilde{r}_i = \sqrt{\lambda_i}\tilde{s}_i + \tilde{n}_i \; ; \; i = 1, 2, ..., P \qquad (16.24)$$

and

$$\tilde{r}_i = \tilde{n}_i \; ; \; i = P + 1, P + 2, ..., N \qquad (16.25)$$

16.2 Implementation of the MIMO Eigenmode Decomposition

Both Eqs. (16.24) and (16.25) can be implemented with the block diagram depicted in Fig. 16.1a.

(a)   (b)

Fig. 16.1 a Block diagram of MIMO system decomposed into parallel eigenmodes. **b** Block diagram of the MIMO eigenmodes using vector and matrix notation

Fig. 16.2 Block diagram of the MIMO system with eigenmodes $\tilde{\mathbf{s}}$, $\tilde{\mathbf{n}}$, and $\tilde{\mathbf{r}}$ as input and output, respectively

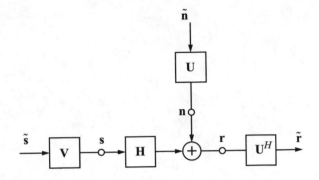

In essence the branches with indices $i = 1, 2, ..., P$ are called the eigenmodes of the MIMO system, because they carry the information from the transmitter to the receiver. The remaining branches $i = P + 1, P + 2, ..., N$ contain only noise rather than information and are therefore often not included in the definition of the MIMO eigenmodes in the narrow sense. Hence, these branches do not contribute to the MIMO channel capacity. Figure 16.1b represents the implementation of (16.18) as a block diagram.

An alternative implementation of (16.23) is shown in Fig. 16.2, where the kernel is given by the original MIMO system. From (16.20)–(16.22) follows

$$\mathbf{r} = \mathbf{U}\tilde{\mathbf{r}} \tag{16.26}$$

$$\mathbf{s} = \mathbf{V}\tilde{\mathbf{s}} \tag{16.27}$$

$$\mathbf{n} = \mathbf{U}\tilde{\mathbf{n}} \tag{16.28}$$

Chapter 17
Channel Capacity of Single-User Transmission Systems

In this chapter, we allude to a topic which is important for the design of a communications system. We will answer the question how many bit/s can be transmitted per symbol or equivalently per channel use. For a certain bandwidth of the channel the interesting point is how many bit/s per Hz bandwidth can be achieved as a maximum. The channel capacity was introduced by Shannon in his pioneering work [1] in the year 1948 for single input single output (SISO) channels. The extension to MIMO channels was given by Telatar [2]. The capacity of various models for stochastic MIMO channels have been intensively studied, e.g., in [3]. In the following we start with the SISO channel capacity for a real-valued signal, extend it to complex signals, and finally derive the channel capacity of a MIMO scheme.

17.1 Channel Capacity of SISO System

17.1.1 AWGN Channel with Real Signals and Noise

We shortly review the basics of the channel capacity C given by Shannon [1]. In general C is formulated as the maximal mutual information $I(X, Y)$ between an input X and an output Y in Fig. 17.1a.

$$C = \max I(X, Y) \tag{17.1}$$

where X and Y are random variables or stochastic processes. Maximization is done over all degrees of freedom of the system. The mutual information is a measure, which quantifies the information we can get about X by the observation of Y. This is actually the situation in information transmission, as we can just observe the signal Y at the receiver and try to retrieve as much as possible information about X. The

© Springer Nature Switzerland AG 2021
J. Speidel, *Introduction to Digital Communications*,
Signals and Communication Technology,
https://doi.org/10.1007/978-3-030-67357-4_17

Fig. 17.1 **a** Noisy channel
with input X and output Y,
b Additive White Gaussian
Noise (AWGN) channel

capacity C is measured in bit/channel use or bit/symbol. In Fig. 17.1b we consider the
simple arrangement, which is the additive white Gaussian noise (AWGN) channel.
First we assume that the transmit signal $s(t) = s_R(t)$, the noise $n(t) = n_R(t)$, and
the receive signal $r(t) = r_R(t)$ are real-valued. Obviously $r(t) = s(t) + n(t)$ holds.
For the AWGN channel the maximization operation in (17.1) can be done only over
the probability density function of the input signal $s(t)$, because it is the only degree
of freedom here. Often the maximization is dropped, and just $I(X, Y)$ is called
channel capacity, which is not precise enough, and should be better denoted just as
a system capacity. However, we will still use the term "channel capacity" and the
actual meaning will be understood from the context. The signal and the noise shall
have the following mean power

$$\mathbf{E}[s_R^2] = \frac{E_S}{2} \tag{17.2}$$

$$\mathbf{E}[n_R^2] = \frac{\sigma_n^2}{2} \tag{17.3}$$

The transmit signal contains discrete symbols with a spacing T on the time axis.
Hence, the symbol rate is

$$v_S = \frac{1}{T} \tag{17.4}$$

Consequently, we can define alternatively the channel capacity C' in bit/s as

$$C' = v_S C = \frac{1}{T} C \tag{17.5}$$

We persuade ourselves that C' is measured in $\frac{\text{bit}}{\text{s}}$ simply by checking the dimension
on the right-hand side of (17.5), $\frac{\text{symbol}}{\text{s}} \frac{\text{bit}}{\text{symbol}} = \frac{\text{bit}}{\text{s}}$.

Now we introduce a strict bandwidth limitation with cut-off frequency f_c to the
channel and assume that the receive signal $s_R(t)$ and also the noise $n_R(t)$ are strictly
band-limited with f_c. We know from the first Nyquist condition that we can transmit
without intersymbol interference over an ideal lowpass channel with the frequency
response

$$G(f) = \begin{cases} 1 \; ; \; f \leq f_c \\ 0 \; ; \; f > f_c \end{cases} \tag{17.6}$$

with cut-off frequency f_c at the maximal symbol rate

$$\frac{1}{T} = 2f_c \tag{17.7}$$

Obviously the same condition results, if we limit the spectrum of the signal and the noise to f_c with the ideal lowpass filter (17.6) and use a sampling frequency $\frac{1}{T}$, which just satisfies the sampling theorem. Then we get from (17.5)

$$C' = 2f_cC \tag{17.8}$$

If we would sample above the lower limit given by the sampling theorem, i.e. $\frac{1}{T} > 2f_c$, the discrete-time signal would contain additional samples. However, they are redundant, as they contain no additional information. Shannon has shown [1] that C is given by the logarithm of the signal-to-noise ratio as

$$C = \frac{1}{2} \log_2 \left(1 + \frac{\mathbf{E}[s_R^2]}{\mathbf{E}[n_R^2]} \right) \tag{17.9}$$

from which we conclude with (17.2) and (17.3)

$$C = \frac{1}{2} \log_2 \left(1 + \frac{E_S}{\sigma_n^2} \right) \tag{17.10}$$

and (17.8) yields

$$C' = f_c \log_2 \left(1 + \frac{E_S}{\sigma_n^2} \right) \tag{17.11}$$

Without proof the maximum of C occurs if the signal $s_R(t)$ has a Gaussian probability density function. We clearly see from (17.11) that C' increases linearly with the cut-off frequency f_C of the signal, however only logarithmic with respect to the signal-to-noise ratio $\frac{E_S}{\sigma_n^2}$.

17.1.2 AWGN Channel with Complex Signals and Noise

Now we are going to extend the previous considerations to the case of a complex modulation scheme known as quadrature amplitude modulation QAM with a complex carrier $e^{j2\pi f_0 t}$ and complex signals. We know from the modulation theory that we can describe this scheme by means of complex baseband signals and noise after demodulation. We define the signal and the noise by its real parts (index R), its imaginary parts (index I), and with the following properties

Signal

$$s = s_R + js_I \tag{17.12}$$

- s_R and s_I are statistically independent and both with zero mean

$$\mathbf{E}[s_R] = \mathbf{E}[s_I] = 0 \tag{17.13}$$

- Mean power

$$\mathbf{E}[s_R^2] = \mathbf{E}[s_I^2] = \frac{E_S}{2} \tag{17.14}$$

Then follows

$$\mathbf{E}\left[|s|^2\right] = \mathbf{E}\left[ss^*\right] = \mathbf{E}\left[s_R^2 + s_I^2\right] = \mathbf{E}\left[s_R^2\right] + \mathbf{E}\left[s_I^2\right] = E_S \tag{17.15}$$

A complex stochastic process with this property is called cyclic symmetrical.

Noise

$$n = n_R + jn_I \tag{17.16}$$

- n_R and n_I are statistically independent AWGN and both with zero mean

$$\mathbf{E}[n_R] = \mathbf{E}[n_I] = 0 \tag{17.17}$$

- Mean power

$$\mathbf{E}\left[n_R^2\right] = \mathbf{E}\left[n_I^2\right] = \frac{\sigma_n^2}{2} \tag{17.18}$$

Similar to (17.15) follows

$$\mathbf{E}\left[|n|^2\right] = \mathbf{E}\left[nn^*\right] = \mathbf{E}\left[n_R^2 + n_I^2\right] = \mathbf{E}\left[n_R^2\right] + \mathbf{E}\left[n_I^2\right] = \sigma_n^2 \tag{17.19}$$

We also see that n is a cyclic symmetrical stochastic process.

Channel Capacity

The real and imaginary part carry independent information. Therefore we get twice the capacity as in (17.9)

$$C = \log_2\left(1 + \frac{\mathbf{E}\left[|s|^2\right]}{\mathbf{E}\left[|n|^2\right]}\right) = \log_2\left(1 + \frac{E_S}{\sigma_n^2}\right) \tag{17.20}$$

and with (17.7) follows

$$C' = \frac{1}{T}C = 2f_c \log_2\left(1 + \frac{E_S}{\sigma_n^2}\right) \tag{17.21}$$

Please note that $2f_c$ is the bandwidth of the transmit signal spectrum after modulation measured in the frequency range $|f| \geq 0$ and f_c is the cut-off frequency of the equivalent baseband system.

17.2 Channel Capacity of MIMO Systems with Statistically Independent Transmit Signals and Noise

17.2.1 Prerequisites

As depicted in Fig. 17.2 we are now considering a MIMO system with the input signal vector $\mathbf{s}(k) \in \mathbb{C}^{M \times 1}$, a frequency flat fading channel with matrix $\mathbf{H}(k) \in \mathbb{C}^{N \times M}$, an additive noise vector $\mathbf{n}(k) \in \mathbb{C}^{N \times 1}$, and the receive vector $\mathbf{r}(k) \in \mathbb{C}^{N \times 1}$. In the course of our investigation we will introduce a prefilter with matrix $\mathbf{A} \in \mathbb{C}^{M \times M}$. Then the transmit signal changes from $\mathbf{s}(k)$ to $\underline{\mathbf{s}}(k)$. However, if not otherwise stated, we first drop the prefilter, thus in Fig. 17.2

$$\mathbf{A} = \mathbf{I}_M \tag{17.22}$$

holds and we will come back to the prefilter in Sects. 17.3 and 17.5. To simplify the notation we drop k in the following.

The signal and the noise shall comply with the following prerequisites,

Original Transmit Signal

$$\mathbf{s} = \left(s_1 \ s_2 \ \cdots \ s_M \right)^T \ ; \ s_j = s_{R,j} + js_{I,j} \ ; \ j = 1, 2, \ldots, M \tag{17.23}$$

$$s_{R,j} \ ; \ s_{I,j} \ ; \ j = 1, 2, \ldots, M \ \text{ statistically independent, zero mean} \tag{17.24}$$

$$E\left[s_{R,j}^2\right] = E\left[s_{I,j}^2\right] = \frac{E_S}{2} \ \Rightarrow \ E\left[|s_j|^2\right] = E_S \ ; \ j = 1, 2, \ldots, M \tag{17.25}$$

$$s_j \ ; \ s_m \ ; \ j, m = 1, 2, \ldots, M \ ; \ j \neq m \ ; \ \text{statistically independent} \tag{17.26}$$

From (17.26) follows the spatial covariance matrix of $\mathbf{s}$

$$\mathbf{R}_{ss} = E\left[\mathbf{s}\mathbf{s}^H\right] = E_S \mathbf{I}_M \tag{17.27}$$

Fig. 17.2 MIMO system with statistically independent input signal $\mathbf{s}$, noise $\mathbf{n}$, prefilter $\mathbf{A}$, and channel $\mathbf{H}$

Consequently, the total mean transmit power is

$$\mathbf{E}\left[\|\mathbf{s}\|^2\right] = \text{tr } (\mathbf{R}_{ss}) = ME_S \tag{17.28}$$

Noise

$$\mathbf{n} = \left(n_1 \; n_2 \; \cdots \; n_N\right)^T \; ; \; n_i = n_{R,i} + jn_{I,i} \; ; \; i = 1, 2, \ldots, N \tag{17.29}$$

$$n_{R,i} \; ; \; n_{I,i} \; ; \; i = 1, 2, \ldots, N \; \text{ statistically independent, zero mean} \tag{17.30}$$

$$\mathbf{E}\left[n_{R,i}^2\right] = \mathbf{E}\left[n_{I,i}^2\right] = \frac{\sigma_n^2}{2} \; \Rightarrow \; \mathbf{E}\left[|n_i|^2\right] = \sigma_n^2 \; ; \; i = 1, 2, \ldots, N \tag{17.31}$$

$$n_i \; ; \; n_m \; ; \; i, m = 1, 2, \ldots, N \; ; \; i \neq m \; ; \text{statistically independent} \tag{17.32}$$

From (17.32) follows the spatial covariance matrix of $\mathbf{n}$

$$\mathbf{R}_{nn} = \mathbf{E}\left[\mathbf{nn}^H\right] = \sigma_n^2 \mathbf{I}_N \tag{17.33}$$

and the total mean power of the noise at the receiver is

$$\mathbf{E}\left[\|\mathbf{n}\|^2\right] = \text{tr } (\mathbf{R}_{nn}) = N\sigma_n^2 \tag{17.34}$$

Let us give some remarks on these prerequisites. Statistical independence of real and imaginary part of the transmit signal is mostly in compliance with reality, because the user allocates to the QAM symbols independent data streams to fully exploit channel capacity. The statistical independence of the output signals of the different antenna elements strongly depends on the spatial correlation condition at the transmitter of a wireless link, such as a rich scattering environment. To allocate the same mean signal power E_S to every transmit antenna output signal, is a good starting point. However, we will see in Sect. 17.5 that a power allocation prefilter with matrix $\mathbf{A}$ at the transmitter can maximize the channel capacity. The assumption of the noise with zero mean and equal mean noise power at each receive antenna element fits well with reality. This also holds for the statistical independence of the real and the imaginary part of the noise and the noise at different antennas.

17.2.2 Instantaneous MIMO Channel Capacity

We have seen in the previous section that the output signals of the transmit antennas can be assumed as statistically independent. The same holds for the noise at the receiver. If the receive signals of the N branches would also be statistically independent, then the total capacity could be calculated easily as the sum of the individual capacities of the branches. However, the transmit signals are passing through the channel, undergo inter-channel interference, and thus the receive signals are spatially correlated. Nevertheless, the idea of independent receive signals is attractive and we follow this approach by applying the decomposition of the MIMO scheme into independent eigenmode branches.

Capacity of the MIMO Eigenmodes

We briefly review the main results of the eigenmode decomposition outlined in Chap. 16. The singular value decomposition (SVD) of the channel matrix $\mathbf{H} \in \mathbb{C}^{N \times M}$ is given by

$$\mathbf{H} = \mathbf{UDV}^H \tag{17.35}$$

with

$$\mathbf{D} = \begin{pmatrix} \sqrt{\lambda_1} & 0 & 0 \cdots & 0 & 0 \cdots 0 \\ 0 & \sqrt{\lambda_2} & 0 \cdots & 0 & 0 \cdots 0 \\ & & \ddots & & \\ 0 & 0 & 0 \cdots & \sqrt{\lambda_P} & 0 \cdots 0 \\ 0 & 0 & 0 \cdots & 0 & 0 \cdots 0 \\ & & \vdots \cdots & & \ddots \\ 0 & 0 & 0 \cdots & 0 & 0 \cdots 0 \\ 0 & 0 & 0 \cdots & 0 & 0 \cdots 0 \end{pmatrix} \in \mathbb{R}^{N \times M} \tag{17.36}$$

$\lambda_1 \geq \lambda_2 \geq \cdots \lambda_P > 0$, and $\lambda_{P+1} = \lambda_{P+2} = \cdots = \lambda_N = 0$ are the N eigenvalues of the Hermitian matrix $\mathbf{Q}_N = \mathbf{HH}^H \in \mathbb{C}^{N \times N}$ and $P - \text{rank}(\mathbf{Q}_N) \leq N$ is the rank of $\mathbf{Q}_N$.

$\sqrt{\lambda_i}$, $i = 1, \ldots, P$ are called the singular values of $\mathbf{H}$. The unitary matrix $\mathbf{U} \in \mathbb{C}^{N \times N}$ is the matrix of the normalized eigenvectors with respect to the eigenvalues $\lambda_1, \lambda_2, \cdots, \lambda_N$ of $\mathbf{Q}_N$. The matrix $\mathbf{Q}_N$ can be decomposed into the diagonal matrix

$$\mathbf{\Lambda}_N = \text{diag}(\lambda_1, \lambda_2, \ldots, \lambda_P, 0, \ldots, 0) \in \mathbb{R}^{N \times N} \tag{17.37}$$

The unitary matrix $\mathbf{V} \in \mathbb{C}^{M \times M}$ in (17.35) can be found by the eigenvalue decomposition of the matrix $\mathbf{Q}_M = \mathbf{H}^H \mathbf{H} \in \mathbb{C}^{M \times M}$ using the diagonal matrix

$$\mathbf{\Lambda}_M = \text{diag}(\lambda_1, \lambda_2, \ldots, \lambda_P, 0, \ldots, 0) \in \mathbb{R}^{M \times M} \tag{17.38}$$

with the eigenvalues $\lambda_1, \lambda_2, \cdots, \lambda_M$ of $\mathbf{Q}_M$. Then $\mathbf{V}$ represents the matrix of eigenvectors with respect to the eigenvalues of $\mathbf{Q}_M$. Note that rank $(\mathbf{Q}_M) = \text{rank}(\mathbf{Q}_N) = \text{rank}(\mathbf{H}) = P$ holds and that the matrices $\mathbf{\Lambda}_M$ and $\mathbf{\Lambda}_N$ contain the same diagonal matrix

$$\mathbf{\Lambda}_P = \text{diag}(\lambda_1, \lambda_2, \ldots, \lambda_P) \in \mathbb{R}^{P \times P} \tag{17.39}$$

of the P eigenvalues, which are unequal to zero. Then we obtain from the input-output relation of the MIMO system (no prefilter)

$$\mathbf{r} = \mathbf{Hs} + \mathbf{n} \tag{17.40}$$

Fig. 17.3 MIMO system decomposition into eigenmodes with statistically independent input signal s, noise n, prefilter **A**, and channel **H** (for **n** = **0**)

the input-output relation of the eigenmode system

$$\tilde{\mathbf{r}} = \mathbf{D}\tilde{\mathbf{s}} + \tilde{\mathbf{n}} \tag{17.41}$$

with the eigenmode signals

$$\tilde{\mathbf{r}} = \mathbf{U}^H \mathbf{r} \tag{17.42}$$

$$\tilde{\mathbf{s}} = \mathbf{V}^H \mathbf{s} \tag{17.43}$$

and the eigenmode noise

$$\tilde{\mathbf{n}} = \mathbf{U}^H \mathbf{n} \tag{17.44}$$

Equation (17.41) represents the linear system of equations

$$\tilde{r}_i = \sqrt{\lambda_i}\tilde{s}_i + \tilde{n}_i \; ; \; i = 1, 2, \ldots, P \tag{17.45}$$

$$\tilde{r}_i = \tilde{n}_i \; ; \; i = P + 1, P + 2, \ldots, N \tag{17.46}$$

where (17.45) determines the eigenmodes in the narrow sense. The remaining eigenmodes in (17.46) do not carry information and thus provide no contribution to the channel capacity. Figure 17.3 illustrates the eigenmode decomposition of the MIMO system as a block diagram.

Before we calculate the total capacity, let us first check the statistical properties of $\tilde{\mathbf{s}}$ and $\tilde{\mathbf{n}}$. We obtain with (17.43) and by assuming that **V** is deterministic

$$\mathbf{R}_{\tilde{s}\tilde{s}} = \mathbf{E}\left[\tilde{\mathbf{s}}\tilde{\mathbf{s}}^H\right] = \mathbf{E}\left[\mathbf{V}^H \mathbf{s}\mathbf{s}^H \mathbf{V}\right] = \mathbf{V}^H \mathbf{R}_{ss} \mathbf{V} \tag{17.47}$$

As we would like to have statistically independent eigenmode branches to follow our goal to calculate the total channel capacity from the sum of the P individual eigenmodes, we have to make sure that the components $\tilde{s}_j$ of $\tilde{\mathbf{s}}$ are independent and have the same mean power.

$$E\left[|\tilde{s}_j|^2\right] = E_S \ ; \quad j = 1, 2, \ldots, M \tag{17.48}$$

Hence the covariance matrix must be

$$\mathbf{R}_{\tilde{s}\tilde{s}} = E_S \mathbf{I}_M \tag{17.49}$$

This is fulfilled if $\mathbf{R}_{ss} = E_S \mathbf{I}_M$, which is the prerequisite given by (17.27). Furthermore, we derive the mean value from (17.43) with (17.24)

$$E\left[\tilde{\mathbf{s}}\right] = \mathbf{V}^H E\left[\mathbf{s}\right] = \mathbf{0} \tag{17.50}$$

which indicates that all components $\tilde{s}_j$ have zero mean in compliance with the requirement (17.24). In a similar way we get for the eigenmode of the noise assuming that $\mathbf{U}$ is deterministic

$$\mathbf{R}_{\tilde{n}\tilde{n}} = \mathbf{U}^H \mathbf{R}_{nn} \mathbf{U} \tag{17.51}$$

and with (17.33)

$$\mathbf{R}_{\tilde{n}\tilde{n}} = \sigma_n^2 \mathbf{I}_N \tag{17.52}$$

from which we see that the components $\tilde{n}_i$ of $\tilde{\mathbf{n}}$ are independent, and have the same mean power σ_n^2. Also the mean value is zero, because we get from (17.44) with (17.30)

$$E\left[\tilde{\mathbf{n}}\right] = \mathbf{U}^H E\left[\mathbf{n}\right] = \mathbf{0} \tag{17.53}$$

Furthermore, if $\mathbf{n}$ is AWGN with zero mean, we can substantiate that the same holds for $\tilde{\mathbf{n}}$, because a linear operation with a matrix $\mathbf{U}$ in (17.44) keeps the noise Gaussian.

As a conclusion the individual eigenmodes are statistically independent and fulfill the same requirements as $\mathbf{s}$ and $\mathbf{n}$ given in Sect. 17.1.2. Consequently, they are adequate for the calculation of the total channel capacity as the sum of the individual and independent eigenmode branches. For branch i we can apply (17.20), which requires the mean power of the signal and noise part of $\tilde{r}_i$ in (17.45), which are

$$E\left[\left|\sqrt{\lambda_i}\tilde{s}_i\right|^2\right] = \lambda_i E\left[|\tilde{s}_i|^2\right] = \lambda_i E_S \tag{17.54}$$

and

$$E\left[|\tilde{n}_i|^2\right] = \sigma_n^2 \tag{17.55}$$

respectively. Then we obtain the channel capacity of the eigenmode branch i from (17.20)

$$C_i = \log_2 \left(1 + \frac{\lambda_i E_S}{\sigma_n^2}\right) \tag{17.56}$$

and the total MIMO system capacity becomes

$$C = \sum_{i=1}^{P} C_i = \sum_{i=1}^{P} \log_2 \left(1 + \frac{\lambda_i E_S}{\sigma_n^2} \right) \tag{17.57}$$

measured in bit/symbol or in bit/channel use. Applying (17.21) yields

$$C' = 2 f_c \sum_{i=1}^{P} \log_2 \left(1 + \frac{\lambda_i E_S}{\sigma_n^2} \right) \tag{17.58}$$

and is measured in bit/s.

Discussion of MIMO Channel Capacity (17.57)

- We see that only P eigenmode paths contribute to the channel capacity, although there are M transmit and N receive antennas. As $P = \text{rank}\left(\mathbf{HH}^H\right) = \text{rank}\left(\mathbf{H}^H\mathbf{H}\right) = \text{rank}\left(\mathbf{H}\right)$, the number of linearly independent lines or columns in the channel matrix determine the number of contributing eigenmodes rather than the number of transmit and receive antennas.
- Let $\lambda_1 = \lambda_2 = \cdots = \lambda_P = \lambda$ and $\mathbf{H}$ with full rank. Then $C = P \log_2 \left(1 + \frac{\lambda E_S}{\sigma_n^2} \right)$ is proportional to $P = \min \{M, N\}$. In case of $M = N$ and a full rank channel matrix with $P = M$ the capacity is $C = M \log_2 \left(1 + \frac{\lambda E_S}{\sigma_n^2} \right)$ and hence proportional to the number of transmit (and receive) antennas.
- Let λ_{max} be the largest eigenvalue of $\mathbf{HH}^H$ and assume that the remaining eigenvalues are very small compared to λ_{max}. Then the eigenmode corresponding to λ_{max} dominates and the channel capacity approximately is $C = \log_2 \left(1 + \frac{\lambda_{max} E_S}{\sigma_n^2} \right)$.
- As $H(k)$ is time varying, also all eigenvalues and thus the capacity $C(k)$ depend on time. This is the reason why $C(k)$ is called instantaneous capacity.

17.2.3 Alternative Formulas for the MIMO Channel Capacity

Now we are going to derive some useful alternative formulas for the MIMO channel capacity.

Channel Capacity as a Function of Eigenvalues

We start with (17.57) and use the basic relation of the logarithm, $\log(x_1 x_2) = \log(x_1) + \log(x_2)$ yielding

$$C = \log_2 \left[\prod_{i=1}^{P} \left(1 + \frac{\lambda_i E_S}{\sigma_n^2} \right) \right] \tag{17.59}$$

Moreover, the argument of the logarithm can be interpreted as the determinant of the diagonal matrix $\mathbf{I}_P + \frac{E_S}{\sigma_n^2}\Lambda_P$, where Λ_P is given in (17.39). Accordingly we find

$$C = \log_2\left[\det\left(\mathbf{I}_P + \frac{E_S}{\sigma_n^2}\Lambda_P\right)\right] \qquad (17.60)$$

Now we introduce Λ_M or Λ_N, which have additional zeros in their main diagonal compared to Λ_P. However, the argument of the logarithm in (17.59) will just get some additional factors 1, which do not change the result. Hence (17.60) is identical to

$$C = \log_2\left[\det\left(\mathbf{I}_M + \frac{E_S}{\sigma_n^2}\Lambda_M\right)\right] \qquad (17.61)$$

and

$$C = \log_2\left[\det\left(\mathbf{I}_N + \frac{E_S}{\sigma_n^2}\Lambda_N\right)\right] \qquad (17.62)$$

Channel Capacity as a Function of the Channel Matrix

We are now looking for a relation between the channel capacity and the channel matrix $\mathbf{H}$. Such a formula would be useful, because no singular values of the channel matrix would have to be calculated. We first give the result and then the proof:

$$C = \log_2\left[\det\left(\mathbf{I}_N + \frac{E_S}{\sigma_n^2}\mathbf{H}\mathbf{H}^H\right)\right] \qquad (17.63)$$

and

$$C = \log_2\left[\det\left(\mathbf{I}_M + \frac{E_S}{\sigma_n^2}\mathbf{H}^H\mathbf{H}\right)\right] \qquad (17.64)$$

Proof of (17.63)

To this end we consider (17.61). We know that the eigenvalue decomposition of $\mathbf{Q}_M = \mathbf{H}^H\mathbf{H}$ is given by $\mathbf{V}^H\mathbf{H}^H\mathbf{H}\mathbf{V} = \Lambda_M$. Then we obtain from (17.61) $C = \log_2\left[\det\left(\mathbf{I}_M + \frac{E_S}{\sigma_n^2}\mathbf{V}^H\mathbf{H}^H\mathbf{H}\mathbf{V}\right)\right]$. Next we use the cyclic permutation rule for determinants from Appendix B.

Given $\mathbf{A} \in \mathbb{C}^{M\times N}$ and $\mathbf{B} \in \mathbb{C}^{N\times M}$, then $\det(\mathbf{I}_M + \mathbf{A}\mathbf{B}) = \det(\mathbf{I}_N + \mathbf{B}\mathbf{A})$ is true.

Thus, we can rewrite the capacity as $C = \log_2\left[\det\left(\mathbf{I}_N + \frac{E_S}{\sigma_n^2}\mathbf{H}\mathbf{V}\mathbf{V}^H\mathbf{H}^H\right)\right]$ and use the fact that $\mathbf{V}$ is a unitary matrix to get the final result $C = \log_2\left[\det\left(\mathbf{I}_N + \frac{E_S}{\sigma_n^2}\mathbf{H}\mathbf{H}^H\right)\right]$. In a similar way (17.64) can be proven.

17.3 MIMO Channel Capacity for Correlated Transmit Signals

We are now going to abandon the requirement that the input signal of the channel is spatially uncorrelated. According to Fig. 17.2 we assume an original signal $\mathbf{s}$ as in Sect. 17.2 with $\mathbf{R}_{ss} = E_S \mathbf{I}_M$, however we introduce the prefilter matrix $\mathbf{A} \in \mathbb{C}^{M \times M}$ that generates a channel input signal $\underline{\mathbf{s}} \in \mathbb{C}^{M \times 1}$ with covariance matrix

$$\mathbf{R}_{\underline{s}\underline{s}} = \mathbf{E}\left[\underline{\mathbf{s}}\,\underline{\mathbf{s}}^H\right] \tag{17.65}$$

Please note that $\mathbf{R}_{\underline{s}\underline{s}}$ is a Hermitian matrix, else arbitrary. The input signal $\mathbf{s}$ complies with the prerequisites for the calculation of the total channel capacity using singular value decomposition, as outlined in the Sect. 17.2. The output signal is given by

$$\mathbf{r} = \mathbf{H}\mathbf{A}\mathbf{s} + \mathbf{n} \tag{17.66}$$

With $\underline{\mathbf{s}} = \mathbf{A}\mathbf{s}$ the covariance matrix $\mathbf{R}_{\underline{s}\underline{s}}$ is given by

$$\mathbf{R}_{\underline{s}\underline{s}} = \mathbf{A}\mathbf{R}_{ss}\mathbf{A}^H = E_S \mathbf{A}\mathbf{A}^H \tag{17.67}$$

where the last step holds for $\mathbf{R}_{ss} = E_S \mathbf{I}_M$. If we partition

$$\mathbf{R}_{\underline{s}\underline{s}} = \mathbf{R}_{\underline{s}\underline{s}}^{\frac{1}{2}} \left(\mathbf{R}_{\underline{s}\underline{s}}^{\frac{1}{2}}\right)^H \tag{17.68}$$

a comparison with (17.67) yields the precoding matrix

$$\mathbf{A} = \frac{1}{\sqrt{E_S}} \mathbf{R}_{\underline{s}\underline{s}}^{\frac{1}{2}} \tag{17.69}$$

$\mathbf{R}_{\underline{s}\underline{s}}^{\frac{1}{2}}$ is the square root matrix of $\mathbf{R}_{\underline{s}\underline{s}}$.

As we recognize from (17.66), the system capacity with prefilter can be easily determined from (17.63) and (17.64), if $\mathbf{H}$ is replaced there by $\mathbf{H}\mathbf{A}$ yielding

$$C = \log_2\left[\det\left(\mathbf{I}_N + \frac{E_S}{\sigma_n^2}\mathbf{H}\mathbf{A}\mathbf{A}^H\mathbf{H}^H\right)\right] \tag{17.70}$$

and

$$C = \log_2 \left[\det \left(\mathbf{I}_M + \frac{E_S}{\sigma_n^2} \mathbf{A} \mathbf{A}^H \mathbf{H}^H \mathbf{H} \right) \right] \qquad (17.71)$$

where we have used the cyclic permutation rule for determinants in (17.71). Hence, with the prefilter matrix $\mathbf{A}$ the covariance matrix $\mathbf{R}_{\underaccent{\sim}{s}\underaccent{\sim}{s}}$ of the channel input signal $\underaccent{\sim}{s}$

and the system capacity can be adjusted. Then we obtain with (17.67) from (17.70) and (17.71) two equivalent forms for the channel capacity

$$C = \log_2 \left[\det \left(\mathbf{I}_N + \frac{1}{\sigma_n^2} \mathbf{H} \mathbf{R}_{\underaccent{\sim}{s}\underaccent{\sim}{s}} \mathbf{H}^H \right) \right] \qquad (17.72)$$

$$C = \log_2 \left[\det \left(\mathbf{I}_M + \frac{1}{\sigma_n^2} \mathbf{R}_{\underaccent{\sim}{s}\underaccent{\sim}{s}} \mathbf{H}^H \mathbf{H} \right) \right] \qquad (17.73)$$

17.4 Channel Capacity for Correlated MIMO Channel

We are going to determine the channel capacity of a MIMO channel with transmit and receive correlation. To this end we can use the Kronecker model from Sect. 13.4.4 with the channel matrix given by (13.41) resulting in $\mathbf{H} = \left(\mathbf{R}_{rr}^{\frac{1}{2}} \right)^H \mathbf{H}_w \mathbf{R}_{tx}^{\frac{1}{2}}$. Then we obtain from (17.63)

$$C = \log_2 \left[\det \left(\mathbf{I}_N + \frac{E_S}{\sigma_n^2} \mathbf{H}_w \mathbf{R}_{tx}^{\frac{1}{2}} \left(\mathbf{R}_{tx}^{\frac{1}{2}} \right)^H \mathbf{H}_w^H \mathbf{R}_{rx}^{\frac{1}{2}} \left(\mathbf{R}_{rx}^{\frac{1}{2}} \right)^H \right) \right] \qquad (17.74)$$

where we have used the cyclic permutation rule. In general, correlation of the delay spread functions of the channel results in a loss of the channel capacity, as detailed investigations with short term and long term statistical parameters have shown, [3]. The reader can convince oneself by experimenting some scenarios using the platform "web demo" provided by [4]. An example is depicted in Fig. 17.4 for a MIMO system with two transmit and two receive antennas.

The MIMO channel is corrupted by Gaussian noise. Transmit and receive correlation is present with the correlation matrices $\mathbf{R}_{tx} = \mathbf{R}_{rx} = \begin{pmatrix} 1 & \rho \\ \rho & 1 \end{pmatrix}$ and $\rho = 0.8$. The degradation of the channel capacity under channel correlation (lower curve) is clearly visible.

Fig. 17.4 Capacity of a MIMO system with $M = 2$ transmit and $N = 2$ receive antennas under Gaussian noise, transmit and receive correlation $\rho = 0.8$. Upper curve without (w/o) and lower curve with (w/) transmit and receive correlation, equal transmit power (without water filling WF). *Source* Online platform "webdemo" [4]

17.5   Maximizing MIMO System Capacity Using the Water Filling Algorithm

17.5.1   Prefilter for Transmit Power Allocation

In the previous sections we have not maximized the MIMO system capacity using the degrees of freedom. Among others we have assumed that the mean output power E_S is the same for all M transmit antenna elements. Now we are going to distribute the total mean transmit power ME_S individually over the antenna elements with the goal to maximize the capacity of the MIMO system. In Fig. 17.2 we have already taken precaution by introducing a filter with the matrix $\mathbf{A}$, which we will call power allocation filter. The block diagram in Fig. 17.3 shows also the decomposition of the MIMO channel matrix $\mathbf{H}$ into eigenmodes according to (17.35). We assume that the input signal $\mathbf{s}$ is still spatially uncorrelated with mean power ME_S and covariance matrix $\mathbf{R}_{ss} = E_S \mathbf{I}_M$. The matrix $\mathbf{A}$ together with $\mathbf{V}^H$ shall generate an input signal $\tilde{\mathbf{s}}$ of the eigenmode channel $\mathbf{D}$ with a covariance matrix

$$\mathbf{R}_{\tilde{s}\tilde{s}} = E_S \operatorname{diag}(a_1, a_2, \ldots, a_M) \tag{17.75}$$

where

Fig. 17.5 ith branch of the
eigenmode decomposition of
the MIMO system with
individual transmit power
allocation coefficient a_i

$$\sum_{j=1}^{M} a_j = M \ , \ \ \text{and } a_j \geq 0 \ ; \ \ j = 1, 2, \dots, M \qquad (17.76)$$

So, we are going to weight the input power of each eigenmode with a dedicated positive coefficient $a_j, \ j = 1, 2, \dots, M$. We substantiate from (17.75) that the vector $\tilde{s}$ remains spatially uncorrelated, as required for the calculation of the total capacity as the sum of the individual and independent capacities of the eigenmodes. It will be shown that a precoding matrix $\mathbf{A}$, which is generating the covariance matrix (17.75) is given by

$$\mathbf{A} = \frac{1}{\sqrt{E_S}} \mathbf{V} \mathbf{R}_{\tilde{s}\tilde{s}}^{\frac{1}{2}} = \mathbf{V} \text{diag} \left(\sqrt{a_1}, \sqrt{a_2}, \dots, \sqrt{a_M} \right) \qquad (17.77)$$

From Fig. 17.3 we see that

$$\tilde{s} = \mathbf{V}^H \mathbf{A} s \qquad (17.78)$$

and with (17.77) follows

$$\tilde{s} = \text{diag} \left(\sqrt{a_1}, \sqrt{a_2}, \dots, \sqrt{a_M} \right) s \qquad (17.79)$$

Please note that the power allocation matrix $\mathbf{A}$ shows up as a full rather than a diagonal matrix due to $\mathbf{V}$. We can write (17.79) in component form as

$$\tilde{s}_i = \sqrt{a_i} s_i \ ; \ \ i = 1, 2, \dots, M \qquad (17.80)$$

and from the input-output relation (17.45) of the eigenmode channel $\mathbf{D}$ we obtain for the i th branch of the eigenmode decomposition

$$\tilde{r}_i = \sqrt{\lambda_i} \sqrt{a_i} s_i + \tilde{n}_i \ ; \ \ i = 1, 2, \dots, P \qquad (17.81)$$

This is portrait in Fig. 17.5. The resulting total capacity is given by

$$C = \log_2 \left[\det \left(\mathbf{1}_M + \frac{E_S}{\sigma_n^2} \text{diag} \left(a_1, a_2, \dots, a_M \right) \Lambda_M \right) \right] \qquad (17.82)$$

The proof is given at the end of this section.
    In the next section we are going to calculate the optimal coefficients a_i.

Proof of (17.77)

With (17.78) and (17.27) we obtain $\mathbf{R}_{\tilde{s}\tilde{s}} = \mathbf{E}\left[\tilde{s}\tilde{s}^H\right] = \mathbf{V}^H \mathbf{A}\mathbf{A}^H \mathbf{V} E_S$. From the decomposition $\mathbf{R}_{\tilde{s}\tilde{s}} = \mathbf{R}_{\tilde{s}\tilde{s}}^{\frac{1}{2}} \left(\mathbf{R}_{\tilde{s}\tilde{s}}^{\frac{1}{2}}\right)^H = \sqrt{E_S}\mathbf{V}^H \mathbf{A}\left(\mathbf{V}^H \mathbf{A}\right)^H \sqrt{E_S}$ we conclude

$\mathbf{R}_{\tilde{s}\tilde{s}}^{\frac{1}{2}} = \sqrt{E_S}\mathbf{V}^H \mathbf{A}$ yielding with (17.75) the result

$\mathbf{A} = \frac{1}{\sqrt{E_S}}\mathbf{V}\mathbf{R}_{\tilde{s}\tilde{s}}^{\frac{1}{2}} = \mathbf{V}\text{diag}\left(\sqrt{a_1}, \sqrt{a_2}, \ldots, \sqrt{a_M}\right)$ and the proof is finished.

Proof of (17.82)

We apply (17.71) and insert $\mathbf{A}$ from (17.77) yielding with

$\mathbf{A}\mathbf{A}^H = \mathbf{V}\text{diag}(a_1, a_2 \ldots, a_M)\mathbf{V}^H$ and with the cyclic permutation rule for determinants $C = \log_2\left[\det\left(\mathbf{I}_M + \frac{E_S}{\sigma_n^2}\text{diag}(a_1, a_2 \ldots, a_M)\mathbf{V}^H \mathbf{H}^H \mathbf{H}\mathbf{V}\right)\right]$. As

$\mathbf{V}^H \mathbf{H}^H \mathbf{H}\mathbf{V} = \Lambda_M$ the proof of (17.82) ends.

17.5.2 Computation of the Optimal Power Allocation Coefficients a_i

Maximization with Constraints

From Fig. 17.5 and the prerequisites we see that (17.57) can be used for determining the total capacity, if we replace λ_i by $\lambda_i a_i$. Then the capacity with transmit power loading prefilter $\mathbf{A}$ in (17.77) is obtained as

$$C = \sum_{i=1}^{P} \log_2\left(1 + \frac{a_i \lambda_i E_S}{\sigma_n^2}\right) \tag{17.83}$$

which has to be maximized with respect to $a_1, \ldots, a_P$ and under the two constraints

$$\sum_{j=1}^{P} a_j = M \iff g(a_1, a_2, \ldots, a_P) = \sum_{j=1}^{P} a_j - M = 0 \tag{17.84}$$

and

$$a_j \geq 0 \; ; \; j = 1, 2, \ldots, P \tag{17.85}$$

Please note that we have restricted (17.76) to $P \leq M$, because there are only P eigenmodes and hence only P free coefficients a_i are left for the maximization. The remaining coefficients $a_i \; ; \; i = P + 1, \ldots, M$ are set equal to zero.

General Solution with Lagrange Method

The Lagrange method defines a new target function composed as a linear combination of the previous function C and the constraint $g(a_1, a_2, \ldots, a_P)$

$$J = C + L\,g = \max_{\{a_1, \ldots, a_P, L\}} \tag{17.86}$$

L is the Lagrange multiplier and is considered as a new free parameter. The constraint (17.85), which is an inequality, has to be treated separately in a second step. We recognize J as a concave function of the free parameters $a_1, \ldots, a_P, L$. Consequently, setting the partial derivatives of J with respect to the parameters equal to zero is a necessary and sufficient condition for the maximum

$$\frac{\partial J}{\partial a_m} = \frac{\partial C}{\partial a_m} + L \frac{\partial g}{\partial a_m} = 0 \; ; \; m = 1, 2, \ldots, P \tag{17.87}$$

We express $\log_2(x) = \eta \ln(x)$ with $\eta = \frac{1}{\ln(2)}$ and knowing that $\frac{\partial \ln(bx)}{\partial x} = b\frac{1}{x}$ we obtain from (17.87)

$$\eta \frac{1}{1 + \frac{a_m \lambda_m E_S}{\sigma_n^2}} \frac{\lambda_m E_S}{\sigma_n^2} + L = 0 \tag{17.88}$$

from which follows

$$a_m = -\frac{\eta}{L} - \frac{\sigma_n^2}{\lambda_m E_S} \; ; \; m = 1, 2, \ldots, P \tag{17.89}$$

Next we calculate

$$\frac{\partial J}{\partial L} = g = 0 \tag{17.90}$$

which yields the constraint (17.84)

$$\sum_{j=1}^{P} a_j = M \tag{17.91}$$

as is typically for the Lagrange method. Because L is a free parameter, we can redefine the Lagrange multiplier as $K = -\frac{\eta}{L}$ and consider K as a new parameter. Then (17.89) yields

$$a_m = K - \frac{\sigma_n^2}{\lambda_m E_S} \; ; \; m = 1, 2, \ldots, P \tag{17.92}$$

Equation (17.91) and (17.92) are $P + 1$ equations for the $P + 1$ unknowns $a_1, \ldots, a_P$, and K. However, the second constraint (17.85) still has to be met. Therefore we conclude from (17.92) the optimal solution in a first approach

$$a_m^{opt} = \left(K^{opt} - \frac{\sigma_n^2}{\lambda_m E_S} \right)_+ \; ; \; m = 1, 2, \ldots, P \tag{17.93}$$

with the function

$$(x)_+ = \begin{cases} x \; ; \; x > 0 \\ 0 \; ; \; x \leq 0 \end{cases} \tag{17.94}$$

and from (17.91) follows

$$\sum_{j=1}^{P} a_j^{opt} = M \tag{17.95}$$

Let

$$a_m^{opt} \begin{cases} > 0 \; ; \; m = 1, 2, \ldots, P_0 \\ = 0 \; ; \; m = P_0 + 1, \ldots, P \end{cases} \tag{17.96}$$

Then we conclude from (17.93)

$$a_m^{opt} = \begin{cases} K^{opt} - \frac{\sigma_n^2}{\lambda_m E_S} \; ; \; m = 1, 2, \ldots, P_0 \\ = 0 \qquad \; ; \; m = P_0 + 1, \ldots, P \end{cases} \tag{17.97}$$

Inserting (17.97) into (17.95) yields

$$\sum_{j=1}^{P_0} \left(K^{opt} - \frac{\sigma_n^2}{\lambda_j E_S} \right) = M \tag{17.98}$$

from which we obtain

$$K^{opt} = \frac{M}{P_0} + \frac{1}{P_0} \sum_{j=1}^{P_0} \frac{\sigma_n^2}{\lambda_j E_S} \tag{17.99}$$

In summary, (17.96), (17.97), and (17.99) represent the final conditions for the solution.

17.5.3 Graphical Interpretation of the Water Filling Solution

$\lambda_1 \geq \lambda_2 \geq \cdots \geq \lambda_P > 0$ shall be given. As depicted in Fig. 17.6, the terms $\frac{\sigma_n^2}{\lambda_m E_S} \; ; \; m = 1, 2, \ldots, P$ are modeled to built up the ground of a vessel. If we are pouring water with a volume of $\sum_{j=1}^{P} a_j$ into that vessel, the level will rise up to K^{opt} given in (17.99).

However, for the given shape of the ground, the total "water" M distributes in such a way that $\sum_{j=1}^{P_0} a_j^{opt} = M$ holds, because no more water is available. Consequently, $a_m^{opt} = 0$ for $m = P_0 + 1, \ldots, P$. From (17.97) we also conclude

$$a_m^{opt} + \frac{\sigma_n^2}{\lambda_m E_S} = K^{opt} \; ; \; m = 1, 2, \ldots, P_0 \tag{17.100}$$

Fig. 17.6 Graphical interpretation of the water filling algorithm to maximize capacity. Note $b = \frac{\sigma_n^2}{E_S}$

and K^{opt} is determining the final water level. From Fig. 17.6 we see for $m > P_0$ the ground of the vessel $\frac{\sigma_n^2}{\lambda_m E_S}$ exceeds the final water level, because no more water is available and the remaining a_m are zero. In general, the solution is iterative. With the optimal coefficients we obtain the maximal capacity from (17.83) as

$$C^{opt} = \sum_{i=1}^{P_0} \log_2 \left(1 + \frac{a_i^{opt} \lambda_i E_S}{\sigma_n^2} \right) \tag{17.101}$$

For a small b in Fig. 17.6 i.e. a high signal-to-noise ratio $\frac{E_S}{\sigma_n^2}$ all a_j^{opt} tend to have the same size and thus the water filling algorithm can not change the capacity significantly. A similar result is obtained, if all singular values $\sqrt{\lambda_i}$ are approximately equal. This is reflected in Fig. 17.7 as the result of a computer simulation using the "webdemo" [4]. The lower two curves represent a 2x2 MIMO channel with strong transmit and receive correlation ($\mathbf{R}_{tx} = \mathbf{R}_{rx} = \begin{pmatrix} 1 & \rho \\ \rho & 1 \end{pmatrix}$, $\rho = 0.95$). Apparently, the capacity gains vanish for increasing SNR $\frac{E_S}{\sigma_n^2}$.

17.5.4 Iterative Solution and Example

In the following the principle algorithm is given as a pseudo code.

Begin   Set $a_{P+1} < 0$
           For $m = P, P-1, ..., 1$ do
                if $a_{m+1} > 0$ set $K_m = K^{opt}$ else $K_m = \frac{1}{m} \left(M + \frac{\sigma_n^2}{E_S} \sum_{j=1}^{m} \frac{1}{\lambda_j} \right)$

$$a_m = K_m - \frac{\sigma_n^2}{\lambda_m E_S}$$

if $a_m > 0$ set $a_m^{opt} = a_m$ and $K_m = K^{opt}$

if $a_m \leq 0$ set $a_m^{opt} = 0$

End

End

Fig. 17.7 Capacity of a MIMO system with $M = 2$ transmit and $N = 2$ receive antennas under Gaussian noise, transmit and receive correlation $\rho = 0.95$. Upper two curves without (w/o) and lower two curves with (w/) transmit and receive correlation, with and without water filling (WF). *Source* Online platform "webdemo" [4]

Example 1

Given a MIMO system with $M = 8$ transmit antennas and $\frac{\sigma_n^2}{E_S} = 19$. The channel matrix **H** has the rank $P = 4$. A singular value decomposition has provided the eigenvalues $\lambda_1 = 19$, $\lambda_2 = \frac{19}{2}$, $\lambda_3 = \frac{19}{10}$, $\lambda_4 = 1$, $\lambda_5 = \lambda_6 = \lambda_7 = \lambda_8 = 0$. Find the optimal power allocation coefficients and the maximal capacity.

Solution: The iteration steps according to the pseudo code are as follows.

$m = 4$ $K_4 = 10$
          $a_4 = -9$, $a_4^{opt} = 0$
$m = 3$ $K_3 = 7$
          $a_3 = -3$, $a_3^{opt} = 0$
$m = 2$ $K_2 = \frac{11}{2}$
          $a_2 = \frac{7}{2}$ $a_2^{opt} = \frac{7}{2}$ $K_2 = K^{opt} = \frac{11}{2}$
$m = 1$ $a_1 = \frac{9}{2}$ $a_1^{opt} = \frac{9}{2}$

The remaining coefficients are $a_5 = a_6 = a_7 = a_8 = 0$. We also see that $P_0 = 2$.

Now we are going to calculate the resulting capacity in bit/channel use taking (17.83) with the result

$$C^{opt} \approx \log_2 \left(\frac{11}{2}\right) + \log_2 \left(\frac{11}{4}\right) \approx 2.46 + 1.46 = 3.92$$

Without optimal power loading the capacity (bit/channel use) is

$$C = \log_2 (2) + \log_2 \left(\frac{3}{2}\right) + \log_2 \left(\frac{11}{10}\right) + \log_2 \left(\frac{20}{19}\right) \approx 1.79$$

according to (17.57). In this example the channel capacity has doubled by employing the optimal transmit power loading.

Example 2

Given the results from Example 1, $E_S = 1$ and the transform matrix $\mathbf{V}$. Find the matrix $\mathbf{A}$ of the power allocation filter and the covariance matrix $\mathbf{R}_{\tilde{s}\tilde{s}}$ of the input signal $\tilde{s}$ of the eigenmode channel $\mathbf{D}$. Calculate the covariance matrix $\mathbf{R}_{ss}$ of the input signal s of the channel $\mathbf{H}$.

Solution: $\mathbf{A}$ can be calculated with (17.77) as the 8×8 matrix
$$\mathbf{A} = \mathbf{V} \operatorname{diag}\left(\sqrt{\frac{9}{2}}, \sqrt{\frac{7}{2}}, 0, 0, 0, 0, 0, 0\right) \text{ and } \mathbf{R}_{\tilde{s}\tilde{s}} \text{ follows from (17.75) as}$$
$$\mathbf{R}_{\tilde{s}\tilde{s}} = \operatorname{diag}\left(\frac{9}{2}, \frac{7}{2}, 0, 0, 0, 0, 0, 0\right).$$
The covariance matrix $\mathbf{R}_{ss}$ is given by (17.67) as $\mathbf{R}_{ss} = E_S \mathbf{A}\mathbf{A}^H$ from which

follows $\mathbf{R}_{ss} = \mathbf{V}\operatorname{diag}\left(\frac{9}{2}, \frac{7}{2}, 0, 0, 0, 0, 0, 0\right)\mathbf{V}^H$, which is in general not a diagonal

matrix. Thus, the output signal s of the prefilter is spatially correlated, as expected.

As a conclusion, the power allocation filter $\mathbf{A}$ is condensing the available total mean power $8E_S$ covered by the output signal vector s of the eight transmit antenna elements to just two eigenmodes. All other eigenmodes do not carry power. However, this does not mean that we can switch off any transmit antenna element.

Example 3

Given the matrices $\mathbf{V}$ and the diagonal matrix $\mathbf{F}$. Show with the following example that $\mathbf{G} = \mathbf{V}\mathbf{F}\mathbf{V}^H$ is a diagonal matrix only if $\mathbf{F} = \mathbf{I}_2$ is the identity matrix.

$$\mathbf{V} = \begin{pmatrix} 1 & j \\ j & 1 \end{pmatrix}, \quad \mathbf{F} = \begin{pmatrix} 1 & 0 \\ 0 & f \end{pmatrix}$$

Solution:
We obtain $\mathbf{G} = \mathbf{V}\mathbf{F}\mathbf{V}^H = \begin{pmatrix} 1+f & j(f-1) \\ -j(f-1) & 1+f \end{pmatrix}$ and recognize that $\mathbf{G}$ is a diagonal matrix only for $f = 1$ resulting in $\mathbf{F} = \mathbf{I}_2$.

17.6   Capacity of a Stochastic MIMO Channel

We have already pointed out that the capacity $C(k)$ considered so far is the instantaneous capacity, as it depends on time. Capacity is an important parameter for the quality of service and therefore the instantaneous capacity is of limited value for guaranteeing a satisfying operation of a system over a long time period. To this end two new measures have been introduced for characterizing quality of service of a MIMO system namely the "ergodic capacity" and the "outage capacity", which will be addressed briefly in the following.

17.6.1   Ergodic Channel Capacity

We consider $\mathbf{H}(k)$ as a stochastic process and define the ergodic capacity as the expectation of $C(k)$

$$C_{erg} = \mathbf{E}[C] = \int_0^\infty C p_C(C) dC \qquad (17.102)$$

where $p_C(C)$ is the probability density function of the capacity C. Please note that $C \geq 0$ and hence the integration starts at zero. $p_C(C)$ can be found from adequate channel models or measurements of the MIMO system, which is not considered here in more detail.

17.6.2   Outage Capacity

Knowing the probability density function $p_C(C)$ of the capacity C we can define the outage capacity C_{out} to be considered as the capacity threshold where the system starts to leave the guaranteed quality of service. This event will occur with the outage probability

$$P_{out} = \Pr[0 \leq C \leq C_{out}] = \int_0^{C_{out}} p_C(C) dC \qquad (17.103)$$

As an example, if the system is required to operate at full quality with a probability of $\geq 99.99\%$, the required outage probability has to be $P_{out} \leq 0.01\%$.

References

1. Shannon, C.: A mathematical theory of communication. Bell Syst. Tech. J. **27** (1948)
2. Telatar, I.: Capacity of multi-antenna Gaussian channels. Eur. Trans. Telecommun. **10** (1999)
3. Kiessling, M.: Statistical analysis and transmit prefiltering for MIMO wireless systems in correlated fading environments. Ph.D. thesis, University of Stuttgart, Institute of Telecommunications, Shaker Publisher (2004). ISBN 3-8322-3444-6
4. Meng, X., Wang, X.: Capacity limits of MIMO channels, webdemo. Technical report. Institute of Telecommunications, University of Stuttgart, Germany (2018). http://webdemo.inue.uni-stuttgart.de

Chapter 18
MIMO Systems with Precoding

18.1 Principle of MIMO Precoding

In Chap. 14 we have investigated the zero-forcing and the minimum mean squared error (MMSE) receiver, which are able to remove or at least minimize the inter-channel interference to the expense of a potential increase of the mean noise power at the receiver output. To maximize the channel capacity we have already investigated a prefilter in Chap. 17, which acts as a power allocation filter at the transmitter. Now we are going to consider prefilters also denoted as precoders to reduce inter-channel interference and thus move the receive filter in principle to the transmitter. As for single input single output wire-line and wireless systems, one motivation is to relocate the hardware complexity of the receiver to some extend to the transmitter [1–3]. This strategy is advantageous in the downlink scenario from the base station to the user, where the receivers are the individual user terminals, which then could be less complex. In most cases the resulting hardware increase of the base station transmitter can be afforded, because its cost is shared among the large number of the users. However, there is a significant burden, because precoding requires knowledge about the channel parameters at the transmitter side to be able to adjust the precoder. Consequently, the channel estimator, which is located at the receiver, has to sent appropriate channel parameters, e.g., the full channel matrix $\mathbf{H}$ to the transmitter via a feedback channel. This arrangement is thus denoted as closed loop scheme. Precoding is also used in the downlink of multi-user scenarios and described in Chap. 20. Precoders, which do not require channel knowledge, are called space-time encoders, are open loop schemes, and discussed in Chap. 19.

In Fig. 18.1 the principle block diagram of a MIMO downlink transmission with a prefilter matrix $\mathbf{A}(k)$ is shown. $\mathbf{c}(k)$ is the input signal vector containing the symbols from a QAM mapper.

The output of the precoder emits the signal vector $\mathbf{s}(k) \in \mathbb{C}^{M \times 1}$ using a MIMO antenna with M elements. The fading channel is modeled by the channel matrix $\mathbf{H}(k) \in \mathbb{C}^{N \times M}$, which owns a frequency-flat spectrum. $\mathbf{r}(k) \in \mathbb{C}^{N \times 1}$ is the receive vector

© Springer Nature Switzerland AG 2021
J. Speidel, *Introduction to Digital Communications*,
Signals and Communication Technology,
https://doi.org/10.1007/978-3-030-67357-4_18

Fig. 18.1 MIMO
transmission with precoding
matrix **A** at the transmitter

and $\mathbf{n}(k) \in \mathbb{C}^{N \times 1}$ the additive noise with zero mean. While the dimensions of $\mathbf{s}$, $\mathbf{H}$, $\mathbf{n}$, and $\mathbf{r}$ are the same as in the previous chapters, the selection of the dimension of $\mathbf{A}$ and $\mathbf{c}$ deserves consideration. If we would define $\mathbf{A}$ as a $M \times M$ matrix such as in Chap. 17 on MIMO capacity, it would be prohibitive to apply the pseudo inverse $\mathbf{H}^{+}$ or $\mathbf{H}^{++}$ (both $M \times N$) of $\mathbf{H}$, which could play the role as a prefilter in our further investigations. Therefore we define $\mathbf{A} \in \mathbb{C}^{M \times N}$ and as a consequence $\mathbf{c} \in \mathbb{C}^{N \times 1}$.

In the next section we will discuss a scheme, which just requires the precoding matrix, and the receive matrix $\mathbf{W}$ in Fig. 18.1 is not present. Alternatively, we investigate in Sect. 18.3 also some precoding schemes, which require an additional receive matrix. In the following we first consider the general block diagram in Fig. 18.1 and specialize later. We drop the discrete time variable k to simplify the notation and obtain for the transmit signal

$$\mathbf{s} = \mathbf{Ac} \tag{18.1}$$

and for the receive vector $\mathbf{r} \in \mathbb{C}^{N \times 1}$

$$\mathbf{r} = \mathbf{HAc} + \mathbf{n} \tag{18.2}$$

We recognize that the signal part of the receive signal is

$$\mathbf{r}_s = \mathbf{HAc} \tag{18.3}$$

where $\mathbf{HA}$ characterizes the inter-channel interference. Please recall that the inter-channel interference is completely removed, if $\mathbf{HA}$ is a diagonal matrix, $\mathbf{HA} = \mathbf{I}_N$, or a scalar. The noise part in the receive signal is just

$$\mathbf{r}_n = \mathbf{n} \tag{18.4}$$

and obviously untouched by the precoder. For the mean power of $\mathbf{r}_s$ we obtain

$$\mathbf{E}\left[\|\mathbf{r}_s\|^2\right] = \mathrm{tr}\left(\mathbf{R}_{cc}\,(\mathbf{HA})^{H}\,\mathbf{HA}\right) \tag{18.5}$$

with the covariance matrix of $\mathbf{c}$

$$\mathbf{R}_{cc} = \mathbf{E}\left[\mathbf{cc}^{H}\right] \in \mathbb{C}^{N \times N} \tag{18.6}$$

Equation (18.5) is easily shown with (18.3) as follows.

$$\mathbf{E}\left[\|\mathbf{r}_s\|^2\right] = \mathbf{E}\left[\text{tr}\left(\mathbf{r}_s\mathbf{r}_s^H\right)\right] = \text{tr}\left(\mathbf{HA}\,\mathbf{E}\left[\mathbf{cc}^H\right](\mathbf{HA})^H\right) = \text{tr}\left(\mathbf{HAR}_{cc}(\mathbf{HA})^H\right) = \text{tr}\left(\mathbf{R}_{cc}(\mathbf{HA})^H\,\mathbf{HA}\right),$$ where we have used for the last term the cyclic permutation rule for the trace of a matrix product, as outlined in the Appendix B.

With the mean noise power

$$\mathbf{E}\left[\|\mathbf{r}_n\|^2\right] = \mathbf{E}\left[\|\mathbf{n}\|^2\right] \tag{18.7}$$

we finally obtain the signal-to-noise ratio at the receiver

$$\gamma_r = \frac{\mathbf{E}\left[\|\mathbf{r}_s\|^2\right]}{\mathbf{E}\left[\|\mathbf{r}_n\|^2\right]} = \frac{\text{tr}\left(\mathbf{R}_{cc}(\mathbf{HA})^H\,\mathbf{HA}\right)}{\mathbf{E}\left[\|\mathbf{n}\|^2\right]} \tag{18.8}$$

For a system with receive filter $\mathbf{W}$ we get from Fig. 18.1 and with (18.2)

$$\mathbf{y} = \mathbf{Wr} = \mathbf{WHAc} + \mathbf{Wn} \tag{18.9}$$

with the signal part

$$\mathbf{y}_s = \mathbf{WHAc} \tag{18.10}$$

and the noise part

$$\mathbf{y}_n = \mathbf{Wn} \tag{18.11}$$

The mean power of $\mathbf{y}_s$ is easily obtained from (18.5) by just replacing $\mathbf{H}$ by $\mathbf{WH}$

$$\mathbf{E}\left[\|\mathbf{y}_s\|^2\right] = \text{tr}\left(\mathbf{R}_{cc}(\mathbf{WHA})^H\,\mathbf{WHA}\right) \tag{18.12}$$

The mean noise power is

$$\mathbf{E}\left[\|\mathbf{y}_n\|^2\right] = \text{tr}\left(\mathbf{R}_{nn}\mathbf{W}^H\mathbf{W}\right) \tag{18.13}$$

where we have used again the cyclic permutation rule. Consequently, for the signal-to-noise ratio at the receiver output follows

$$\gamma_y = \frac{\mathbf{E}\left[\|\mathbf{y}_s\|^2\right]}{\mathbf{E}\left[\|\mathbf{y}_n\|^2\right]} = \frac{\text{tr}\left(\mathbf{R}_{cc}(\mathbf{WHA})^H\,\mathbf{WHA}\right)}{\text{tr}\left(\mathbf{R}_{nn}\mathbf{W}^H\mathbf{W}\right)} \tag{18.14}$$

The mean power of the transmit signal $\mathbf{s}$ is also of interest and we obtain the result just by replacing in (18.12) $\mathbf{WHA}$ by $\mathbf{A}$ yielding

$$\mathbf{E}\left[\|\mathbf{s}\|^2\right] = \text{tr}\left(\mathbf{R}_{cc}\mathbf{A}^H\mathbf{A}\right) \tag{18.15}$$

We observe that $\mathbf{E}\left[\|\mathbf{s}\|^2\right]$ could be enhanced by the precoder $\mathbf{A}$ compared to the mean power

$$\mathbf{E}\left[\|\mathbf{c}\|^2\right] = \text{tr}\left(\mathbf{R}_{cc}\right) \tag{18.16}$$

of **c**. This may cause an overload of the channel input and even a violation of power limits for wireless transmitters. In this case the mean power of **c** must be reduced, which impacts the signal-to-noise ratio at the receiver accordingly. Another remedy to overcome this drawback is the use of the Tomlinson-Harashima scheme, which can limit the amplitudes of the components of the transmit signal vector $\mathbf{s} = \left(s_1 \cdots s_M \right)^T$ by applying modulo operations on the input symbols, [4–6].

18.2   Zero-Forcing and MMSE Precoding

A single prefilter matrix **A** shall be designed to reduce the inter-channel interference without the need for a receive filter. Thus, the matrix **W** in Fig. 18.1 is dropped and we focus on the receive signal **r**.

18.2.1   Zero-Forcing Precoder

In the following we check whether the pseudo inverse of the channel matrix **H** from Chap. 14 depending on the number of transmit and receive antennas, M and N, can remove inter-channel interference.

$$\mathbf{A} = \begin{cases} \mathbf{H}^+ = \left(\mathbf{H}^H \mathbf{H}\right)^{-1} \mathbf{H}^H \in \mathbb{C}^{M \times N} \; ; \; M \le N \\ \mathbf{H}^{++} = \mathbf{H}^H \left(\mathbf{H}\mathbf{H}^H\right)^{-1} \in \mathbb{C}^{M \times N} \; ; \; M \ge N \end{cases} \qquad (18.17)$$

The prerequisites for this investigation are the full rank of $\mathbf{H}^H \mathbf{H}$ and $\mathbf{H}\mathbf{H}^H$ i.e. $\text{rank}\left(\mathbf{H}^H \mathbf{H}\right) = M$, if $M \le N$ and $\text{rank}\left(\mathbf{H}\mathbf{H}^H\right) = N$ if $M \ge N$, respectively. Otherwise the inverse matrices in (18.17) do not exist. Please note that $\text{rank}\left(\mathbf{H}^H \mathbf{H}\right) = \text{rank}\left(\mathbf{H}\right)$ and $\text{rank}\left(\mathbf{H}\mathbf{H}^H\right) = \text{rank}\left(\mathbf{H}\right)$ hold in each of the two cases.

Then the inter-channel interference in the receive signal (18.2) is given by

$$\mathbf{H}\mathbf{A} = \begin{cases} \mathbf{H}\mathbf{H}^+ = \mathbf{H}\left(\mathbf{H}^H \mathbf{H}\right)^{-1} \mathbf{H}^H \ne \mathbf{I}_N \; ; \; M \le N \\ \mathbf{H}\mathbf{H}^{++} = \mathbf{I}_N \qquad\qquad\quad ; \; M \ge N \end{cases} \qquad (18.18)$$

and we substantiate that only the pseudo inverse $\mathbf{H}^{++}$ as a prefilter owns the ability to completely remove inter-channel interference in general and we get from (18.2) the receive signal

$$\mathbf{r} = \mathbf{c} + \mathbf{n} \; ; \; M \ge N \qquad (18.19)$$

which is composed of the transmit vector **c** just corrupted by the noise **n**. Therefore the prefilter with $\mathbf{H}^{++}$ is the most interesting one. However, it is restricted to systems with $M \ge N$.

Apparently the precoder matrix $\mathbf{A} = \mathbf{H}^{++}$ is the same as for the zero-forcing receiver investigated in Chap. 14. However, the receive signal (18.19) is significantly different, because the noise $\mathbf{n}$ is untouched by the precoder. Remember, the use of a zero-forcing receive matrix $\mathbf{W}$ without a precoder results in $\mathbf{y} = \mathbf{s} + \mathbf{W}\mathbf{n}$ with a potential enhancement of the noise by the receive filter.

The signal-to-noise ratio at the receiver of the precoding scheme with $\mathbf{H}^{++}$ directly follows from (18.19)

$$\gamma_r = \frac{\mathbf{E}\left[\|\mathbf{c}\|^2\right]}{\mathbf{E}\left[\|\mathbf{n}\|^2\right]} \; ; \; M \geq N \tag{18.20}$$

which can also be deduced from (18.8) with (18.16) and (18.18).

Another interesting observation, albeit critical, is that the mean power $\mathbf{E}\left[\|\mathbf{s}\|^2\right]$ of the transmit signal $\mathbf{s}$ could be increased compared to the mean power of $\mathbf{c}$. This is discussed shortly in the following. To this end we refer to (18.15) and first determine $\mathbf{A}^H\mathbf{A}$ with (18.17). For the sake of completeness we do not drop the case with the prefilter $\mathbf{H}^+$ in the following. It is straightforward to show that

$$\mathbf{A}^H\mathbf{A} = \begin{cases} \left(\mathbf{H}^+\right)^H \mathbf{H}^+ = \mathbf{H}\left(\mathbf{H}^H\mathbf{H}\right)^{-1}\left(\mathbf{H}^H\mathbf{H}\right)^{-1}\mathbf{H}^H \; ; \; M \leq N \\ \left(\mathbf{H}^{++}\right)^H \mathbf{H}^{++} = \left(\mathbf{H}\mathbf{H}^H\right)^{-1} \qquad\qquad\; ; \; M \geq N \end{cases} \tag{18.21}$$

holds yielding from (18.15)

$$\mathbf{E}\left[\|\mathbf{s}\|^2\right] = \begin{cases} \mathrm{tr}\left(\mathbf{R}_{cc}\mathbf{H}\left(\mathbf{H}^H\mathbf{H}\right)^{-1}\left(\mathbf{H}^H\mathbf{H}\right)^{-1}\mathbf{H}^H\right) \; ; \; M \leq N \\ \mathrm{tr}\left(\mathbf{R}_{cc}\left(\mathbf{H}\mathbf{H}^H\right)^{-1}\right) \qquad\qquad\quad\; ; \; M \geq N \end{cases} \tag{18.22}$$

For illustration we consider the following example.

Example 1
We are going to discuss the mean power of the transmit signal $\mathbf{s}$. In many applications $\mathbf{c}$ is uncorrelated with covariance matrix

$$\mathbf{R}_{cc} = E_S\mathbf{I}_N \tag{18.23}$$

and consequently its mean signal power is obtained from (18.16) as

$$\mathbf{E}\left[\|\mathbf{c}\|^2\right] = NE_S \tag{18.24}$$

From (18.22) then follows

$$\mathbf{E}\left[\|\mathbf{s}\|^2\right] = \begin{cases} E_S\mathrm{tr}\left(\left(\mathbf{H}^H\mathbf{H}\right)^{-1}\right) \; ; \; M \leq N \\ E_S\mathrm{tr}\left(\left(\mathbf{H}\mathbf{H}^H\right)^{-1}\right) \; ; \; M \geq N \end{cases} \tag{18.25}$$

where the cyclic permutation rule was applied for the case $M \leq N$. We assume that the matrices $\mathbf{H}^H \mathbf{H}$ and $\mathbf{H}\mathbf{H}^H$ have the full rank, say $P = \min\{M, N\}$. As the matrices are Hermiteian, their eigenvalues $\lambda_1, \lambda_2, \ldots, \lambda_P$ are positive and unequal to zero. Then from (18.25) follows with (18.24)[1]

$$E\left[\|\mathbf{s}\|^2\right] = \frac{E\left[\|\mathbf{c}\|^2\right]}{N} \sum_{i=1}^{P} \lambda_i^{-1} \qquad (18.26)$$

We clearly recognize that the mean power of $\mathbf{s}$ can be significantly larger than that of $\mathbf{c}$, if some eigenvalues λ_i exhibit small values. This is equivalent to say that the determinants of the two matrices $\mathbf{H}^H \mathbf{H}$ and $\mathbf{H}\mathbf{H}^H$ are small. Matrices with this property are called ill conditioned. The mean signal power of $\mathbf{s}$ will be enhanced in this case and may even cause an overload of the channel input. To make matters worse, if the matrices do not have full rank, $P < \min\{M, N\}$, then some eigenvalues are zero and the mean power of the transmit signal $\mathbf{s}$ even approaches infinity, theoretically. Finally, please note that not only the mean power but also the magnitude of the signal $\mathbf{s}$ at the output of the zero-forcing precoder can be enhanced in case of ill conditioned matrices.

18.2.2   MMSE Precoder

In this section we are looking for a precoder which minimizes the difference Δ between the original transmit signal $\mathbf{c}$ and the receive signal $\mathbf{r}$ in Fig. 18.1

$$\Delta = \mathbf{c} - \mathbf{r} \qquad (18.27)$$

in the mean squared error (MMSE) sense similar to an MMSE receiver discussed in Chap. 14. Thus, the minimization problem can be formulated as

$$J = E\left[\|\Delta\|^2\right] = \text{tr}\left(E\left[\Delta\Delta^H\right]\right) = \min_{\mathbf{A}} \qquad (18.28)$$

Below we prove that the squared error is

$$J = \text{tr}\left\{ \mathbf{R}_{cc} + \left(\mathbf{R}_{cn}^H - \mathbf{R}_{cc}\right)(\mathbf{HA})^H + \mathbf{HA}\left(\mathbf{R}_{cn} - \mathbf{R}_{cc}\right)\right\} \\ + \text{tr}\left\{ \mathbf{HA}\mathbf{R}_{cc}\,(\mathbf{HA})^H - \mathbf{R}_{cn} - \mathbf{R}_{cn}^H + \mathbf{R}_{nn}\right\} \qquad (18.29)$$

with the covariance matrices $\mathbf{R}_{cc}$ in (18.6) of the signal and $\mathbf{R}_{nn} = E\left[\mathbf{n}\mathbf{n}^H\right]$ of the noise as well as the cross-correlation matrices $\mathbf{R}_{cn} = E\left[\mathbf{c}\mathbf{n}^H\right]$ and $\mathbf{R}_{nc} = E\left[\mathbf{n}\mathbf{c}^H\right] = \mathbf{R}_{cn}^H$. The precoder matrix minimizing J turns out to be

---

[1] Note from Appendix B: For an $M \times M$ matrix $\mathbf{Q}$ with eigenvalues $\lambda_1, \lambda_2, \ldots, \lambda_M$ holds $\text{tr}\,(\mathbf{Q}) = \sum_{i=1}^{M} \lambda_i$, $\text{tr}\,(\mathbf{Q}^{-1}) = \sum_{i=1}^{M} \lambda_i^{-1}$ (if non zero eigenvalues), and $\det(\mathbf{Q}) = \lambda_1\lambda_2\cdots\lambda_M$.

$$\mathbf{A} = \left(\mathbf{H}^H \mathbf{H}\right)^{-1} \mathbf{H}^H \left(\mathbf{R}_{cc} - \mathbf{R}_{cn}^H\right) \mathbf{R}_{cc}^{-1} \ ; \ M \leq N \qquad (18.30)$$

Now we assume that $\mathbf{c}$ and $\mathbf{n}$ are uncorrelated yielding $\mathbf{R}_{cn} = \mathbf{E}\left[\mathbf{c}\right]\mathbf{E}\left[\mathbf{n}^H\right] = 0$, where the last term follows from the zero mean of the noise. Then the MMSE precoding matrix finally is

$$\mathbf{A} = \left(\mathbf{H}^H \mathbf{H}\right)^{-1} \mathbf{H}^H = \mathbf{H}^+ \ ; \ M \leq N \qquad (18.31)$$

and identical with the zero-forcing precoding matrix for $M \leq N$. Please note that the condition $M \leq N$ in (18.30) and (18.31) guarantees that $\left(\mathbf{H}^H \mathbf{H}\right)^{-1}$ exists, if $\mathbf{H}$ has full rank M.

However, from (18.18) we see that the inter-channel interference cannot be removed by an MMSE precoding matrix $\mathbf{A} = \mathbf{H}^+$. Below we determine the MMSE from (18.29) under the realistic assumption that $\mathbf{c}$ and $\mathbf{n}$ are uncorrelated, $\mathbf{R}_{cn} = \mathbf{E}\left[\mathbf{c}\right]\mathbf{E}\left[\mathbf{n}^H\right] = 0$, and $\mathbf{R}_{cc} = E_S \mathbf{I}_N$ as

$$J\left(\mathbf{H}^+\right) = J_{min} = \mathbf{E}\left[\|\mathbf{n}\|^2\right] \qquad (18.32)$$

Approach (18.28) has provided a precoder matrix only for the case $M \leq N$. One may speculate that

$$\mathbf{A} = \mathbf{H}^{++} = \mathbf{H}^H \left(\mathbf{H}\mathbf{H}^H\right)^{-1} \ ; \ M \geq N \qquad (18.33)$$

could be the solution for $M \geq N$. This is really true, because we show that the same MMSE as in (18.32) is achieved. This result is not surprising, as the prefilter $\mathbf{H}^{++}$ is completely removing the inter-channel interference according to (18.18), second line, resulting in the receive signal $\mathbf{r} = \mathbf{c} + \mathbf{n}$ yielding $\Delta = \mathbf{c} - \mathbf{r} = -\mathbf{n}$ and thus

$$J\left(\mathbf{H}^{++}\right) = \mathbf{E}\left[\|\mathbf{n}\|^2\right] = J_{min} \qquad (18.34)$$

In summary, we conclude that the precoding matrices for the zero-forcing and the MMSE precoder are identical. Consequently, the same holds for the signal-to-noise ratios at the receiver given by (18.20). Both MMSE precoders provide the same MMSE under the condition of an uncorrelated transmit signal $\mathbf{c}$, uncorrelated noise $\mathbf{n}$, and $\mathbf{R}_{cn} = 0$. Only the precoder matrix $\mathbf{H}^{++}$ can completely remove the inter-channel interference.

Example 2
Given a transmit signal $\mathbf{c}$ with covariance matrix $\mathbf{R}_{cc} = E_S \mathbf{I}_N$. Show that both MMSE precoders, $\mathbf{H}^+$ and $\mathbf{H}^{++}$, deliver the same signal-to-noise ratio $\gamma_r = \frac{\mathbf{E}[\|\mathbf{c}\|^2]}{\mathbf{E}[\|\mathbf{n}\|^2]}$ at the receiver.

Solution:
We just have to show that both precoders provide the same mean power of the signal $\mathbf{r}_s$ at the receiver, because the mean noise power always is $\mathbf{E}\left[\|\mathbf{n}\|^2\right]$. From (18.5)

follows $\mathbf{E}\left[\|\mathbf{r}_s\|^2\right] = E_S \text{tr}\left((\mathbf{HA})^H \, \mathbf{HA}\right)$. For $\mathbf{A} = \mathbf{H}^+$ we get with (18.18) and the cyclic permutation rule $\mathbf{E}\left[\|\mathbf{r}_s\|^2\right] = N E_S = \mathbf{E}\left[\|\mathbf{c}\|^2\right]$. The same result is obtained for $\mathbf{A} = \mathbf{H}^{++}$.

Proof of (18.29) and (18.30)

With (18.27) and (18.2) we obtain $\Delta\Delta^H = (\mathbf{c} - \mathbf{HAc} - \mathbf{n})\left(\mathbf{c}^H - \mathbf{c}^H \mathbf{A}^H \mathbf{H}^H - \mathbf{n}^H\right)$. After multiplication, the expected value follows as

$\mathbf{E}\left[\Delta\Delta^H\right] = \mathbf{R}_{cc} + \left(\mathbf{R}_{cn}^H - \mathbf{R}_{cc}\right)\mathbf{A}^H \mathbf{H}^H + \mathbf{HA}\left(\mathbf{R}_{cn} - \mathbf{R}_{cc}\right) + \mathbf{HAR}_{cc}\mathbf{A}^H \mathbf{H}^H - \mathbf{R}_{cn} - \mathbf{R}_{cn}^H + \mathbf{R}_{nn}$. Next we apply the trace operator yielding (18.29) and differentiate with respect to $\mathbf{A}^*$ using the differentiation rules summarized in Appendix B. Then follows $\frac{\partial J}{\partial \mathbf{A}^*} = \mathbf{H}^H\left(\mathbf{R}_{cn}^H - \mathbf{R}_{cc}\right) + \mathbf{H}^H \mathbf{HAR}_{cc}$. Setting this derivative equal to zero yields

$\mathbf{A} = \left(\mathbf{H}^H \mathbf{H}\right)^{-1} \mathbf{H}^H \left(\mathbf{R}_{cc} - \mathbf{R}_{cn}^H\right)\mathbf{R}_{cc}^{-1}$ and the proof is finished.

Proof of (18.32)

From the first line of (18.18) follows $\mathbf{HA} = \mathbf{HH}^+ = (\mathbf{HA})^H$. Then we obtain from (18.29) with the prerequisite $\mathbf{R}_{cn} = 0$

$J = \text{tr}\left(\mathbf{R}_{cc}\right) + \text{tr}\left(\mathbf{R}_{nn}\right) - 2\text{tr}\left(\mathbf{R}_{cc}\mathbf{HH}^+\right) + \text{tr}\left(\mathbf{R}_{cc}\mathbf{HH}^+\mathbf{HH}^+\right)$. With the prerequisite $\mathbf{R}_{cc} = E_S \mathbf{I}_N$ and the cyclic permutation rule follows

$\text{tr}\left(\mathbf{R}_{cc}\mathbf{HH}^+\right) = E_S \text{tr}\left(\mathbf{H}\left(\mathbf{H}^H \mathbf{H}\right)^{-1}\mathbf{H}^H\right) = E_S \text{tr}\left(\left(\mathbf{H}^H \mathbf{H}\right)^{-1}\mathbf{H}^H \mathbf{H}\right) = N E_S$.

In a similar way we get

$\text{tr}\left(\mathbf{R}_{cc}\mathbf{HH}^+\mathbf{HH}^+\right) = E_S \text{tr}\left(\mathbf{H}\left(\mathbf{H}^H \mathbf{H}\right)^{-1}\mathbf{H}^H \mathbf{H}\left(\mathbf{H}^H \mathbf{H}\right)^{-1}\mathbf{H}^H\right) = N E_S$.

Finally, we obtain $J = \text{tr}\left(\mathbf{R}_{nn}\right) = \mathbf{E}\left[\|\mathbf{n}\|^2\right]$ and the proof ends.

18.3   Precoding Based on Singular Value Decomposition

18.3.1   SVD-Based Precoder and Receiver

Precoder and Receiver Matrix

Using the theory of the eigenmode decomposition we now determine the precoder matrix as

$$\mathbf{A} = \mathbf{V} \tag{18.35}$$

where the unitary matrix $\mathbf{V} \in \mathbb{C}^{M \times M}$ stems from the right-hand side of the singular value decomposition of $\mathbf{H} \in \mathbb{C}^{N \times M}$

$$\mathbf{H} = \mathbf{UDV}^H \tag{18.36}$$

Then the receive signal (18.2) is

$$\mathbf{r} = \mathbf{HVc} + \mathbf{n} \tag{18.37}$$

Though the inter-channel interference $\mathbf{HV}$ is not completely reduced, we require a receive filter with matrix $\mathbf{W}$, which we select according to the left hand side matrix of (18.36)

$$\mathbf{W} = \mathbf{U}^H \in \mathbb{C}^{N \times N} \tag{18.38}$$

The output signal of the receiver is $\mathbf{y} = \mathbf{Wr}$ and with (18.37) and (18.38) follows

$$\mathbf{y} = \mathbf{U}^H \mathbf{HVc} + \mathbf{n}' \tag{18.39}$$

with the filtered noise

$$\mathbf{n}' = \mathbf{U}^H \mathbf{n} \tag{18.40}$$

Please note that the dimensions of the matrices and vectors partly differ from Sect. 18.1, namely $\mathbf{c} \in \mathbb{C}^{M \times 1}$, $\mathbf{A} = \mathbf{V} \in \mathbb{C}^{M \times M}$, and $\mathbf{y} \in \mathbb{C}^{N \times 1}$. On the first glance it seems not plausible that the input and output vectors, $\mathbf{c}$ and $\mathbf{y}$, of the system have different dimensions, M and N, respectively. We will see later that $|N - P|$ components of $\mathbf{y}$ will be zero, where P is the rank of $\mathbf{H}$.

Replacing $\mathbf{H}$ in (18.39) by (18.36) and knowing that $\mathbf{U}$ and $\mathbf{V}$ are unitary matrices we get the final result

$$\mathbf{y} = \mathbf{Dc} + \mathbf{n}' \tag{18.41}$$

with

$$\mathbf{D} = \left(\begin{array}{cccc|c} \sqrt{\lambda_1} & 0 & 0 & 0 & \\ 0 & \sqrt{\lambda_2} & 0 & 0 & \\ & & \ddots & & \mathbf{0} \\ 0 & 0 & 0 & \sqrt{\lambda_P} & \\ \hline & \mathbf{0} & & & \mathbf{0} \end{array} \right) \in \mathbb{R}^{N \times M} \tag{18.42}$$

$\sqrt{\lambda_i}$; $i = 1, 2, \ldots, P$ are the singular values of $\mathbf{H}$ and $P = \mathrm{rank}\,(\mathbf{H}) \leq \min\{M, N\}$ holds. The matrix $\mathbf{D}$ contains at the bottom $N - P$ lines with zeros and on the right-hand side $M - P$ columns with zeros . Therefore (18.41) boils down to

$$y_i = \sqrt{\lambda_i} c_i + n_i' \; ; \; i = 1, 2, \ldots, P \tag{18.43}$$

Obviously, to recover $c_1, c_2, \ldots, c_M$ from the replicas $y_1, y_2, \ldots, y_M$ the channel matrix must have the rank $P = M$. The remaining receiver output signals

$$y_i = n_i' \; ; \; i = M + 1, \ldots, N \tag{18.44}$$

just contain noise and have to be discarded by the receiver. We substantiate that the multi-user interference is completely removed, because the transmit signal c_i is just multiplied by a factor $\sqrt{\lambda_i}$.

Example 3

We consider a MIMO system with $M = 3$ transmit and $N = 4$ receive antennas. $\mathbf{H}$ shall have rank P. Then we can write (18.41) as

$$
\begin{pmatrix} y_1 \\ y_2 \\ y_3 \\ y_4 \end{pmatrix} = \begin{pmatrix} \sqrt{\lambda_1} & 0 & 0 \\ 0 & \sqrt{\lambda_2} & 0 \\ 0 & 0 & \sqrt{\lambda_3} \\ 0 & 0 & 0 \end{pmatrix} \begin{pmatrix} c_1 \\ c_2 \\ c_3 \end{pmatrix} + \begin{pmatrix} n_1' \\ n_2' \\ n_3' \\ n_4' \end{pmatrix}
\tag{18.45}
$$

The last row will be discarded, because y_4 is just noise. If the channel matrix has full rank $P = 3$, all symbols c_1, c_2, c_3 can be recovered. In case of $P = 2$ the singular value $\sqrt{\lambda_3} = 0$ and $y_3 = n_3'$ follow. Consequently, c_3 cannot be recovered. We also see that a weak eigenmode λ_i provides a low signal part $\sqrt{\lambda_i} c_i$.

Signal-to-Noise Ratio

To determine the signal-to-noise ratio of the SVD-based precoding scheme we consider in (18.39) the signal part

$$
\mathbf{y}_s = \mathbf{U}^H \mathbf{H} \mathbf{V} \mathbf{c}
\tag{18.46}
$$

and the noise part

$$
\mathbf{y}_n = \mathbf{n}' = \mathbf{U}^H \mathbf{n}
\tag{18.47}
$$

Then we obtain the mean signal power at the receiver

$$
\mathbf{E}\left[\|\mathbf{y}_s\|^2\right] = \mathbf{E}\left[\mathrm{tr}\left(\mathbf{y}_s \mathbf{y}_s^H\right)\right] = \mathrm{tr}\left(\mathbf{R}_{cc} \mathbf{V}^H \mathbf{H}^H \mathbf{H} \mathbf{V}\right)
\tag{18.48}
$$

Using (18.36) yields $\mathbf{E}\left[\|\mathbf{y}_s\|^2\right] = \mathrm{tr}\left(\mathbf{R}_{cc} \mathbf{D}^T \mathbf{D}\right)$. With

$$
\mathbf{D}^T \mathbf{D} = \Lambda_M = \mathrm{diag}\left(\lambda_1, \lambda_2, \ldots, \lambda_P, 0, \ldots, 0\right) \in \mathbb{R}^{M \times M}
\tag{18.49}
$$

follows the result

$$
\mathbf{E}\left[\|\mathbf{y}_s\|^2\right] = \mathrm{tr}\left(\mathbf{R}_{cc} \Lambda_M\right)
\tag{18.50}
$$

In case of a channel matrix $\mathbf{H}$ with $M \leq N$ and full rank, $P = M$ holds.

The mean power of the noise part $\mathbf{y}_n$ does not change, because in the Appendix B it is shown that a unitary matrix maintains the mean power. Consequently

$$
\mathbf{E}\left[\|\mathbf{y}_n\|^2\right] = \mathbf{E}\left[\|\mathbf{n}\|^2\right]
\tag{18.51}
$$

holds and for the signal-to-noise ratio at the receiver output follows with (18.48) and (18.50)

$$
\gamma_y = \frac{\mathbf{E}\left[\|\mathbf{y}_s\|^2\right]}{\mathbf{E}\left[\|\mathbf{y}_n\|^2\right]} = \frac{\mathrm{tr}\left(\mathbf{R}_{cc} \mathbf{V}^H \mathbf{H}^H \mathbf{H} \mathbf{V}\right)}{\mathbf{E}\left[\|\mathbf{n}\|^2\right]} = \frac{\mathrm{tr}\left(\mathbf{R}_{cc} \Lambda_M\right)}{\mathbf{E}\left[\|\mathbf{n}\|^2\right]}
\tag{18.52}
$$

With the same argument as for (18.51) we substantiate that the SVD-based precoder does not change the mean power of the transmit signal $\mathbf{s}$, hence

$$\mathbf{E}\left[\|\mathbf{s}\|^2\right] = \mathbf{E}\left[\|\mathbf{c}\|^2\right] = \mathrm{tr}\left(\mathbf{R}_{cc}\right) \qquad (18.53)$$

holds.

Example 4
In many applications the symbol vector $\mathbf{c} \in \mathbb{C}^{M \times 1}$ is uncorrelated with covariance matrix $\mathbf{R}_{cc} = E_S \mathbf{I}_M$. Then we obtain for the signal-to-noise ratio from (18.52) with the cyclic permutation rule

$$\gamma_y = \frac{\mathbf{E}\left[\|\mathbf{y}_s\|^2\right]}{\mathbf{E}\left[\|\mathbf{y}_n\|^2\right]} = \frac{E_S \mathrm{tr}\left(\mathbf{H}^H \mathbf{H}\right)}{\mathbf{E}\left[\|\mathbf{n}\|^2\right]} = \frac{E_S \mathrm{tr}\left(\Lambda_M\right)}{\mathbf{E}\left[\|\mathbf{n}\|^2\right]} = \frac{E_S \sum_{i=1}^{P} \lambda_i}{\mathbf{E}\left[\|\mathbf{n}\|^2\right]} \qquad (18.54)$$

Table 18.1 Comparison: Singular value decomposition (SVD) versus zero-forcing/MMSE based precoder for uncorrelated transmit signal $\mathbf{c}$ with $\mathbf{R}_{cc} = E_S \mathbf{I}_M$

Feature	SVD-based precoding	Zero-forcing (and MMSE) precoding
Precoder matrix	$\mathbf{A} - \mathbf{V}$	$\mathbf{A} = \mathbf{H}^{++} = \mathbf{H}^H \left(\mathbf{H}\mathbf{H}^H\right)^{-1}$ $M \geq N$
Receive matrix	$\mathbf{W} = \mathbf{U}^H$	Not required
Receive signal	$\mathbf{y} = \mathbf{D}\mathbf{c} + \mathbf{U}^H \mathbf{n}$	$\mathbf{r} = \mathbf{c} + \mathbf{n}$
Inter-channel interference	Completely removed	Completely removed
Signal-to-noise ratio	$\gamma_y = \frac{E_S \mathrm{tr}(\mathbf{H}^H \mathbf{H})}{\mathbf{E}[\|\mathbf{n}\|^2]}$	$\gamma_r = \frac{N E_S}{\mathbf{E}[\|\mathbf{n}\|^2]} \; ; \; M \geq N$
Mean power of transmit signal	$M E_S$	$E_S \mathrm{tr}\left(\left(\mathbf{H}\mathbf{H}^H\right)^{-1}\right) \; ; \; M \geq N$
Mean power of receive noise	$\mathbf{E}\left[\|\mathbf{n}\|^2\right]$	$\mathbf{E}\left[\|\mathbf{n}\|^2\right]$

18.3.2   Comparison of Zero-Forcing and SVD-Based Precoding

In Table 18.1 the main features of the zero-forcing and the SVD- based precoding for an uncorrelated transmit signal c are compared. As seen before, the zero-forcing and the MMSE precoder own the same matrices. Therefore we just mention the first one in the following. The zero-forcing precoder with $A = H^{++}$ can completely cancel the inter-channel interference, which is also true for the SVD-based precoding. Moreover, both schemes are not enhancing the noise at the receiver. The SVD-based method leaves the transmit mean power unchanged, whereas the zero-forcing precoder can cause an unfavorable enhancement of the transmit power. The zero-forcing precoding scheme does not require a receive filter.

References

1. Fischer, R.F.: Precoding and Signal Shaping for Digital Transmission. Wiley, New York (2002)
2. Joham, M., Utschick, W., Nossek, J.: Linear transmit processing in MIMO communications systems. IEEE Trans. Signal Process. (2005)
3. Vu, M., Paulraj, A.: MIMO wireless linear precoding. IEEE Signal Process. Mag. (2007)
4. Harashima, H., Miyakawa, H.: Matched-transmission technique for channels with intersymbol interference. IEEE Trans. Commun. (1972)
5. Tomlinson, M.: New automatic equalizer employing modulo arithmetic. Electron. Lett. 7 (1971)
6. Fischer, R., Windpassinger, C., Lampe, A., Huber, J.: Space-time transmission using Tomlinson-Harashima precoding. In: 4th ITG- Conference on Source and Channel Coding (2002)

Chapter 19
Principles of Space-Time Coding

19.1 Space-Time Block Coding

Figure 19.1 shows the principle block diagram of a MIMO transmitter with space-time encoding. The incoming bit sequence $b(n)$ is fed into the QAM mapper, which periodically maps κ consecutive bits to a QAM symbol $c(k')$, constituting a 2^κ-ary QAM. $b(n)$ may contain redundancy bits from a forward error correction encoder, [1, 2].

We focus on the description of the space-time encoder, which allocates to the input QAM symbols $c(k')$ dedicated redundancy symbols in the space-time domain to improve transmission quality. The operation of the encoder is block-wise. The sequence $c(k')$ of symbols from a QAM mapper shall occur at time instances $t = k'T'$; $k' \in \mathbb{Z}$ with symbol rate $v_S = \frac{1}{T'}$. The space-time encoder outputs a sequence of space-time symbol vectors, which appear at time instances $t = kT$; $k \in \mathbb{Z}$ and with rate $v_{ST} = \frac{1}{T}$. In the time frame $N_c T'$, which we also call block length, a QAM symbol vector

$$\mathbf{c}(k') = \big(c(k')\; c(k'+1) \cdots c(k'+N_c-1) \big)^T \tag{19.1}$$

is mapped to the output matrix

$$\mathbf{S}(k) = \big(\mathbf{s}(k)\; \mathbf{s}(k+1) \cdots \mathbf{s}(k+L-1) \big)$$

$$= \begin{pmatrix} s_1(k) & s_1(k+1) & \cdots & s_1(k+L-1) \\ \vdots & \vdots & \ddots & \vdots \\ s_j(k) & s_j(k+1) & \cdots & s_j(k+L-1) \\ \vdots & \vdots & \ddots & \vdots \\ s_M(k) & s_M(k+1) & \cdots & s_M(k+L-1) \end{pmatrix} \tag{19.2}$$

© Springer Nature Switzerland AG 2021
J. Speidel, *Introduction to Digital Communications*,
Signals and Communication Technology,
https://doi.org/10.1007/978-3-030-67357-4_19

Fig. 19.1 Principle block diagram of a MIMO system with space-time encoding and decoding

in time frame LT. In other words, N_c QAM symbols are mapped to L space-time symbol vectors

$$\mathbf{s}(k+l) = \left(s_1(k+l)\ s_2(k+l) \cdots s_M(k+l) \right)^T \ ;\ l = 0, 1, \ldots, L-1 \quad (19.3)$$

which are the column vectors of $\mathbf{S}(k)$. There are various denominations for $\mathbf{S}(k)$ either transmission matrix, space-time coding matrix, space-time codeword matrix, or space-time code block. The mapping of the vector $\mathbf{c}(k')$ to the matrix $\mathbf{S}(k)$ can be done in various ways, which is the subject of the space-time coding theory and will be addressed in the following sections in more detail. e.g., a linear space-time block encoder employs linear combining on the input symbols, such as addition and subtraction of the original symbols, their real or imaginary parts, or their conjugate complex values. For a (nonlinear) trellis space-time encoder the input-output relation is defined by a trellis diagram.

We recognize that in general there are different time basis at the input and at the output of the space-time encoder indicated by the discrete time variables k' and k, respectively. For synchronous block-wise operation between input and output,

$$N_c T' = LT \quad (19.4)$$

must hold. Please note that the definition of $\mathbf{S}(k)$ in (19.2) does not respect causality. Consequently from an implementation point of view the space-time symbol vectors $\mathbf{s}(k), \mathbf{s}(k+1), \ldots$ have to be delayed by $N_c T'$ time intervals, because in general the space-time encoder can start the mapping operation not before all N_c QAM symbols have been entered.

From (19.2) we see that row j of $\mathbf{S}(k)$ represents a sequence of L samples sent out by the transmit antenna j. On the other hand column $k+l$ is the space-time symbol vector emitted by all M transmit antenna elements at the same time instant $k+l$. An important parameter of any encoder is the code rate, which is defined as the ratio between the number of symbols at the input and the number of symbols at the output, both considered in the same time frame LT. In contrast to an encoder for temporal forward error correction on bit level the output of the space-time encoder is a vector, which is counted as one element despite of the fact that it is composed of M components. Hence, the spatial code rate of the space-time encoder is defined as

$$r_s = \frac{\text{number of input symbols in time frame } LT}{\text{number of output symbol vectors in time frame } LT} \qquad (19.5)$$

For synchronous block-wise operation specified by (19.4) the spatial code rate is

$$r_s = \frac{N_c}{L} \qquad (19.6)$$

As already mentioned, the time basis at the input and the output of the space-time encoder are in general different. Given the QAM symbol rate at the input as

$$v_S = \frac{1}{T'} \qquad (19.7)$$

we conclude from (19.4) and (19.6) for the space-time symbol vector rate

$$v_{ST} = \frac{1}{T} = \frac{L}{N_c} v_S = \frac{v_S}{r_s} \qquad (19.8)$$

We can determine the redundant symbols allocated by the space-time encoder in the time frame (19.4) recognizing that we have in total LM symbols at the output and N_c QAM symbols at the input yielding $LM - N_c = N_c \left(\frac{M}{r_s} - 1 \right)$ redundant output symbols. From this consideration follows that the space-time encoder allocates redundancy, if $LM > N_c$ or with (19.6)

$$r_s < M \qquad (19.9)$$

There is no redundancy if $LM = N_c$. From $L \in \mathbb{N}$ follows for this special case that N_c has to be an integer multiple of M and we obtain

$$r_s = M \qquad (19.10)$$

This is exactly the code rate of a serial-to-parallel converter, as will be outlined in Sect. 19.2. From the viewpoint of classical forward error correction for bit streams, where the maximal code rate is one, a code rate larger than one of the space-time encoder is surprising. Obviously, systems with more than one transmit antenna can demonstrate that property.

We conclude from (19.8), if $r_S = 1$ then the clock rates of the output and the input sequences are identical, $v_{ST} = v_S$. As a consequence, the bandwidth of the baseband of the space-time symbol sequence $\mathbf{s}(k)$ is not increased compared to the bandwidth of the QAM symbol sequence $c(k')$ although $N_c(M - 1)$ redundant symbols are contained in each output block of length LT. For $r_s < 1$ the speed of the output sequence and accordingly the bandwidth increase by factor $\frac{1}{r_s} = \frac{v_{ST}}{v_S}$.

For the sake of completeness we calculate the code rates of the building blocks of the transmitter in Fig. 19.1. Assume that in the time interval (19.4), the data source feeds κN_b information bits into the forward error correction encoder (not depicted in Fig. 19.1) and that we have at its output κN_c bits including $\kappa(N_c - N_b)$ redundancy bits. Furthermore, the QAM mapper shall assign one QAM symbol to a sequence of κ consecutive input bits, which results in a 2^κ-ary QAM scheme with κ bit/symbol. Then, the code rates are as follows:

- code rate of (temporal) forward error correction including temporal interleaving, $r_t = \frac{N_b}{N_c}$
- "code rate" of QAM mapper, $r_m = \kappa$
- overall code rate at the output of the space-time encoder, $r_{total} = \frac{\kappa N_b}{L} = \kappa \frac{N_b}{N_c} \frac{N_c}{L} = r_m r_t r_s$

As depicted in Fig. 19.1, at the receiving end a linear space-time decoder with N receive antennas can be employed, which performs in principle the reverse operation of the encoder. Alternatively, a nonlinear receiver with N receive antennas and a maximum likelihood detector is feasible. Then, under ideal conditions the original bit sequence is recovered, $\hat{b}(n) = b(n)$.

The receive signal can be determined as follows. The transmit signal vector at time instant $k + l$ is $s(k + l)$ according to (19.3). Then the receive vector is

$$\mathbf{r}(k + l) = \mathbf{H}(k + l)\mathbf{s}(k + l) + \mathbf{n}(k + l) \; ; \; l = 0, 1, \ldots, L - 1 \qquad (19.11)$$

where $\mathbf{H}(k + l)$ is the channel matrix and $\mathbf{n}(k + l)$ is the noise vector. Now, we can assemble all receive vectors in a receive matrix, which is the response to the transmit signal matrix $\mathbf{S}(k)$ in (19.2)

$$\big(\mathbf{r}(k) \cdots \mathbf{r}(k + L - 1)\big) =$$

$$= \big(\mathbf{H}(k)\mathbf{s}(k) \cdots \mathbf{H}(k + L - 1)\mathbf{s}(k + L - 1)\big) + \big(\mathbf{n}(k) \cdots \mathbf{n}(k + L - 1)\big)$$

$$(19.12)$$

If the channel exhibits only block fading, its channel matrix is approximately unchanged in the time interval of the block, say from time instant k to $k + L - 1$,

$$\mathbf{H}(k + l) \approx \mathbf{H}(k) \; ; \; l = 1, \ldots, L - 1 \qquad (19.13)$$

Then follows from (19.12) with (19.2) the matrix of the receive vectors approximately

$$\big(\mathbf{r}(k) \cdots \mathbf{r}(k + L - 1)\big) = \mathbf{H}(k)\mathbf{S}(k) + \big(\mathbf{n}(k) \cdots \mathbf{n}(k + L - 1)\big) \qquad (19.14)$$

In summary, we see that multiple transmit antennas enable the concatenation of symbols of the input symbol sequence $c(k')$ in the spatial and temporal domain, which justifies the name space-time coding. This method is a powerful alternative to binary forward error correction. As an example, it can be seen from (19.9) that a spatial code rate $r_s = 1$ adds redundancy to the output sequence for $M \geq 2$ without increasing the symbol vector rate v_{ST} in (19.8) of the output signal. The space-time encoder can be combined with a multi-antenna receiver to exploit additional receive diversity. In principle, the scheme can even be equipped with forward error correction as a layered space-time coding technique, as discussed in Sect. 19.5 or as a concatenated encoding scheme allowing iterative (Turbo) decoding at the receiver. Obviously, space-time coding is an open loop procedure, as it does not require channel information at the transmitter through feedback from the receiver. A comprehensive survey on space-time coding is given by [3]. In the next sections we will outline some examples of space-time coding schemes in more detail.

19.2 Spatial Multiplexing

Spatial multiplexing can be considered as a simple form of linear space-time encoding. As no redundancy is allocated to the information symbols $c(k')$, this scheme is often not ranked as a space-time encoder. The principle is shown in Fig. 19.2. At the transmitter a spatial demultiplexer is employed, which is just a serial-to-parallel converter.

The input QAM symbol sequence

$$c(k'), c(k' + 1), \ldots, c(k' + M - 1), \ldots \qquad (19.15)$$

with symbol rate $v_S = \frac{1}{T'}$ is entering a shift register. After MT' clock cycles (block length) the space-time symbol vector

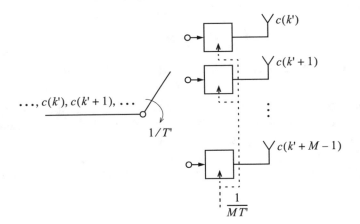

Fig. 19.2 Principle of a transmitter with spatial demultiplexer

$$\mathbf{s}(k) = \big(c(k') \; c(k'+1) \; \cdots \; c(k'+M-1)\big)^T \tag{19.16}$$

is available at the parallel outputs of the shift register and each component is forwarded to the respective antenna. In the following MT' clock cycles the next space-time symbol vector

$$\mathbf{s}(k+1) = \big(c(k'+M) \; c(k'+M+1) \; \cdots \; c(k'+2M-1)\big)^T \tag{19.17}$$

is generated and so forth. Please note that in contrast to the general scheme in Fig. 19.1 no processing other than storage and serial-to-parallel conversion is performed. Furthermore, the clock rate of the output space-time symbol vectors is $v_{ST} = \frac{1}{MT'} = \frac{v_S}{M}$ and thus by factor $\frac{1}{M}$ lower than the QAM symbol rate v_S at the input. We easily check that for $M = 1$ the clock rates are identical, as expected. Similarly as depicted in Fig. 19.1, at the receiving end a zero-forcing, MMSE, or maximum likelihood receiver with N receive antennas can be employed providing M output signals to the space-time decoder, which in this case is just a spatial multiplexer. It operates as a simple parallel-to-serial converter and outputs a serial symbol sequence, which is identical to (19.15), if no transmission errors are present. We can easily calculate the spatial code rate as follows: M QAM symbols $c(k')$ are entering in the time interval $MT' = LT$ (with $L = 1$) and one symbol vector is going out. Thus

$$r_s = M \tag{19.18}$$

With spatial multiplexing one original QAM symbol is transmitted by no more than one antenna. Consequently, there is no transmit diversity and we say the transmit diversity has order one. However, if the receiver is equipped with N receive antennas, they can provide receive diversity of order N, because each transmit symbol is received by all N receive antennas. The maximum diversity order of the total spatial multiplexing scheme then is N.

19.3   Orthogonal, Linear Space-Time Block Coding

We are now considering space-time encoders, which apply linear combining on the input symbols, such as addition or subtraction of the original symbols, their real or imaginary parts, or their conjugate complex values, respectively. Motivated by the Alamouti encoding scheme [4] described in the next section, the rows of the space-time coding matrix $\mathbf{S}$ turn out to be pairwise orthogonal. The same holds for the columns.

19.3.1 The Alamouti Encoder for MISO System with Two Transmit Antennas

The Transmission Scheme

The Alamouti space-time encoder was presented in [4] and is a simple but effective technique to achieve transmit diversity. It can be shown that this scheme is unique, as it employs the only 2×2 space-time block coding matrix with complex entries to achieve full spacial code rate $r_s = 1$ and full transmit diversity of the order two. In this section we consider a multiple input single output (MISO) transmission scheme with $M = 2$ transmit and $N = 1$ receive antennas. In the next section this approach is extended to a MIMO system with two receive antennas. The principal block diagram is depicted in Fig. 19.3. The multiple input single output (MISO) channel is described by the channel matrix, which is actually a row vector

$$\mathbf{H}(k) = \mathbf{h}^T(k) = \left(h_{11}(k) \ h_{12}(k) \right) \tag{19.19}$$

The Alamouti space-time encoding scheme operates under the assumption of block fading. This means that the channel coefficients do not change during a block time, here from time instants k to $k + 1$. Consequently

$$h_{11}(k) \approx h_{11}(k + 1) \ ; \ h_{12}(k) \approx h_{12}(k + 1) \ \Rightarrow \mathbf{h}^T(k) \approx \mathbf{h}^T(k + 1) \tag{19.20}$$

Furthermore, we assume that the channel coefficients are exactly known at the receiver. In reality they can only be estimated with some deviation from the actual values. However, the channel estimation scheme is not considered here and not shown in Fig. 19.3.

We will see that the time basis of the input and the output of the encoder are the same, thus $k' = k$. The space-time encoder maps the input QAM symbol vector

$$\mathbf{c}(k) = \left(c(k) \ c(k + 1) \right)^T \tag{19.21}$$

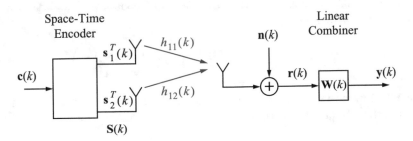

Fig. 19.3 Block diagram of a 2×1 multiple input single output (MISO) transmission scheme with Alamouti space-time encoder

to the space-time coding matrix with $L = 2$

$$S(k) = \begin{pmatrix} s_1(k) & s_1(k+1) \\ s_2(k) & s_2(k+1) \end{pmatrix} = \big(s(k) \ s(k+1) \big) \tag{19.22}$$

where we have introduced the space-time coding vectors at the output

$$s(k) = \big(s_1(k) \ s_2(k) \big)^T \ ; \ s(k+1) = \big(s_1(k+1) \ s_2(k+1) \big)^T \tag{19.23}$$

as column vectors of $S(k)$. In Fig. 19.3 the row vectors $s_j^T(k) = \big(s_j(k) \ s_j(k+1) \big)$ of $S(k)$ are shown, which indicate the output symbols of the antenna j for the time instants k and $k+1$, where $j = 1, 2$. The encoder responds to the input symbol $c(k)$ by the vector $s(k)$, and the receive signal is

$$r(k) = h^T(k)s(k) + n(k) \tag{19.24}$$

The next input symbol $c(k+1)$ generates the output vector $s(k+1)$ and the receive signal

$$r(k+1) = h^T(k+1)s(k+1) + n(k+1) \tag{19.25}$$

The block-wise processing can be simply described by introducing the following vector notation with (19.24) and (19.25)

$$\begin{pmatrix} r(k) \\ r(k+1) \end{pmatrix} = \begin{pmatrix} h^T(k) \\ h^T(k+1) \end{pmatrix} \big(s(k) \ s(k+1) \big) + \begin{pmatrix} n(k) \\ n(k+1) \end{pmatrix} \tag{19.26}$$

With (19.22) follows

$$\begin{pmatrix} r(k) \\ r(k+1) \end{pmatrix} = \begin{pmatrix} h^T(k) \\ h^T(k+1) \end{pmatrix} S(k) + \begin{pmatrix} n(k) \\ n(k+1) \end{pmatrix} \tag{19.27}$$

The space-time coding matrix is designed such that the signal vector $s(k)$ can be recovered at the receiver in the noise-free case by multiplying (19.27) from the left by an appropriate matrix. The Alamouti space-time coding matrix is defined for block fading channels (19.20) as

$$S(k) = \begin{pmatrix} c(k+1) & -c^*(k) \\ c(k) & c^*(k+1) \end{pmatrix} = \begin{pmatrix} s_1(k) & -s_2^*(k) \\ s_2(k) & s_1^*(k) \end{pmatrix} \tag{19.28}$$

where the last term is just a short hand notation. Obviously, the space-time encoder keeps the first output vector $s(k)$ and determines the second vector as

$$s(k+1) = \big(-s_2^*(k) \ s_1^*(k) \big)^T \tag{19.29}$$

Now we introduce (19.19), (19.20), and (19.28) in (19.27) and obtain after some simple manipulations

$$\underbrace{\begin{pmatrix} r(k) \\ r^*(k+1) \end{pmatrix}}_{\mathbf{r}(k)} = \underbrace{\begin{pmatrix} h_{11}(k) & h_{12}(k) \\ h_{12}^*(k) & -h_{11}^*(k) \end{pmatrix}}_{\mathbf{U}(k)} \underbrace{\begin{pmatrix} s_1(k) \\ s_2(k) \end{pmatrix}}_{\mathbf{s}(k)} + \underbrace{\begin{pmatrix} n(k) \\ n^*(k+1) \end{pmatrix}}_{\mathbf{n}(k)} \qquad (19.30)$$

Before we are going to investigate the receive matrix $\mathbf{W}(k)$ we look at some properties of $\mathbf{S}(k)$.

Properties of the Alamouti Space-Time Coding Matrix S(k)

In the following we drop k to simplify the notation.

- $\mathbf{S}$ has the property

$$\mathbf{SS}^H = \alpha \mathbf{I}_2 \qquad (19.31)$$

with

$$\alpha = |s_1|^2 + |s_2|^2 \qquad (19.32)$$

Consequently $\frac{1}{\sqrt{\alpha}}\mathbf{S}$ is a unitary matrix and the rows of $\mathbf{S}$ are orthogonal. The proof is straightforward. Similarly,

$$\mathbf{S}^H \mathbf{S} = \alpha \mathbf{I}_2 \qquad (19.33)$$

holds and the columns of $\mathbf{S}$ are orthogonal as well.
- $\mathbf{S}$ has full rank 2 and thus the Alamouti precoder can provide full transmit diversity of order two.

Determination of Receive Signal and Receiver Matrix

Obviously, $\mathbf{U}(k)$ in (19.30) has the property

$$\mathbf{U}^H \mathbf{U} = \beta \mathbf{I}_2 \qquad (19.34)$$

where

$$\beta(k) = |h_{11}(k)|^2 + |h_{12}k)|^2 \geq 0 \qquad (19.35)$$

Similar to (19.33) we conclude that $\frac{1}{\sqrt{\beta}}\mathbf{U}$ is a unitary matrix. To recover the signal vector $\mathbf{s}(k)$ the linear receiver in Fig. 19.3 is equipped with a matrix

$$\mathbf{W}(k) = \mathbf{U}^H(k) = \begin{pmatrix} h_{11}^*(k) & h_{12}(k) \\ h_{12}^*(k) & -h_{11}(k) \end{pmatrix} \qquad (19.36)$$

which yields the output signal vector

$$\mathbf{y}(k) = \begin{pmatrix} y(k) \\ y(k+1) \end{pmatrix} = \mathbf{U}^H(k) \begin{pmatrix} r(k) \\ r^*(k+1) \end{pmatrix}$$

and with (19.30) and (19.34) follows

$$\mathbf{y}(k) = \beta(k)\mathbf{s}(k) + \mathbf{U}^H(k)\mathbf{n}(k) \tag{19.37}$$

Obviously, the output of the receive filter is composed of the signal part $\beta(k)\mathbf{s}(k)$ and the noise part $\mathbf{n}'(k) = \mathbf{U}^H(k)\mathbf{n}(k)$.

The coefficient $\beta(k)$ demonstrates the expected transmit diversity. As we can see from (19.35), the signal part only fades out, if $h_{11}(k) = 0$ and $h_{12}(k) = 0$ at the same time. However, this event occurs with small probability. Consequently, the space-time coding scheme with two transmit antennas is very effective to achieve low symbol error rate in particular for situations, where the (temporal) forward error correction cannot help, because one link is in deep fade. The removal of the inter-channel interference in (19.37) by the receive matrix is also appreciated.

Unitary matrices preserve the mean power of signals, as is shown in the Appendix B. If the channel coefficients in (19.35) are normalized such that $\beta = 1$, then the mean powers of the signal and the noise part in the receive signal (19.37) are not enhanced by the receive matrix.

The signal $\mathbf{y}(k)$ in Fig. 19.3 is subject to a final decision, e.g., using the maximum likelihood detection algorithm. Please note that there is no need to feedback the channel coefficients to the transmitter, because the space-time encoding matrix is independent of the channel coefficients. This is different from closed loop systems, which use a transmit filter for allocating individual power to the antennas or a zero forcing precoder.

We also recognize that in the time interval $[k, k + 1]$ two QAM symbols $c(k)$ and $c(k + 1)$ are input and two space-time coding vectors are output resulting in the spatial code rate $r_s = 1$. Therefore also the clock rates at the input and at the output are identical, $v_{ST} = v_S$. The space-time encoder adds redundancy, because the two input symbols generate four output samples contained in the coding matrix $\mathbf{S}(k)$, which results in a redundancy of two symbols. As already pointed out, to calculate the spatial code rate for space-time coding we consider all transmit antennas as a single unit emitting a vector rather than counting the individual samples. In contrast, a code rate 1 indicates for conventional forward error correction that no redundancy is allocated to the information bits.

Please note that the definition of $\mathbf{S}(k)$ does not respect the causality of a real system. Causality can be simply obtained by introducing an adequate delay to ensure that all QAM symbols have entered the encoder before the output is generated. The Alamouti scheme has found frequent applications both for wireless and wire-line digital communications, [5]. In [6, 7] an application for digital transmission over more than two copper wires is described, which became part of the final standard to establish MIMO in-home power line communications.

19.3.2 The Alamouti Space-Time Encoder for a 2 × 2 MIMO System

The principle block diagram of the system is depicted in Fig. 19.4. The transmitter is unchanged and operates with two antennas. As the receiver employs also two antennas, the 2×2 MIMO channel matrix is

$$\mathbf{H}(k) = \begin{pmatrix} h_{11}(k) & h_{12}(k) \\ h_{21}(k) & h_{22}(k) \end{pmatrix} \tag{19.38}$$

Similar to (19.27) the signal at the receiver i is given by

$$\begin{pmatrix} r_i(k) \\ r_i(k+1) \end{pmatrix} = \begin{pmatrix} \mathbf{h}_i^T(k) \\ \mathbf{h}_i^T(k+1) \end{pmatrix} \mathbf{S}(k) + \begin{pmatrix} n_i(k) \\ n_i(k+1) \end{pmatrix} ; \quad i = 1, 2 \tag{19.39}$$

with the channel row vectors $\mathbf{h}_i^T(k) = \left(h_{i1}(k) \ h_{i2}(k) \right)$; $i = 1, 2$ and the same space-time coding matrix $\mathbf{S}(k)$ as given in (19.28). Obviously, the upper branch of the receiver was already investigated in Sect. 19.3.1 and the lower branch is quite similar. Therefore we allocate an index i to $\mathbf{r}, \mathbf{n}, \mathbf{U}, \mathbf{W}, \mathbf{y}$ and replace h_{11} by h_{i1}, and h_{12} by h_{i2}. Then follows from (19.30)

$$\underbrace{\begin{pmatrix} r_i(k) \\ r_i^*(k+1) \end{pmatrix}}_{\mathbf{r}_i(k)} = \underbrace{\begin{pmatrix} h_{i1}(k) & h_{i2}(k) \\ h_{i2}^*(k) & -h_{i1}^*(k) \end{pmatrix}}_{\mathbf{U}_i(k)} \underbrace{\begin{pmatrix} s_1(k) \\ s_2(k) \end{pmatrix}}_{s(k)} \mathbf{I} \underbrace{\begin{pmatrix} n_i(k) \\ n_i^*(k+1) \end{pmatrix}}_{\mathbf{n}_i(k)} ; \quad i = 1, 2$$

$$\tag{19.40}$$

where $i = 1$ and $i = 2$ indicate the upper and the lower branch in Fig. 19.4, respectively. Furthermore

$$\mathbf{U}_i^H \mathbf{U}_i = \beta_i \mathbf{I}_2 ; \quad i = 1, 2 \tag{19.41}$$

and

$$\beta_i(k) = |h_{i1}(k)|^2 + |h_{i2}(k)|^2 \geq 0 ; \quad i = 1, 2 \tag{19.42}$$

hold. Hence, $\frac{1}{\sqrt{\beta_i}} \mathbf{U}_i$ is a unitary matrix. The receiver matrices follow from (19.36) as

$$\mathbf{W}_i(k) = \mathbf{U}_i^H(k) = \begin{pmatrix} h_{i1}^*(k) & h_{i2}(k) \\ h_{i2}^*(k) & -h_{i1}(k) \end{pmatrix} ; \quad i = 1, 2 \tag{19.43}$$

The output of the receiver is

$$\mathbf{y}_i(k) = \begin{pmatrix} y_i(k) \\ y_i(k+1) \end{pmatrix} = \mathbf{U}_i^H(k) \begin{pmatrix} r_i(k) \\ r_i^*(k+1) \end{pmatrix} ; \quad i = 1, 2 \tag{19.44}$$

and with (19.40) it can be written as

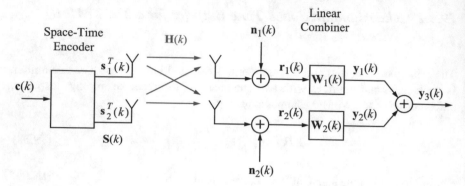

Fig. 19.4 Block diagram of 2 × 2 MIMO transmission scheme with Alamouti space-time encoder

$$\mathbf{y}_i(k) = \beta_i(k)\mathbf{s}(k) + \mathbf{U}_i^H(k)\mathbf{n}_i(k) \; ; \; i = 1, 2 \tag{19.45}$$

The receive signal of both branches exhibits a signal part $\beta_i(k)\mathbf{s}(k)$ and a noise part $\mathbf{n}_i'(k) = \mathbf{U}_i^H(k)\mathbf{n}_i(k) \; ; \; i = 1, 2$.

According to Fig. 19.4 we calculate the total receiver output signal block-wise as

$$\mathbf{y}_3(k) = \begin{pmatrix} y(k) \\ y(k+1) \end{pmatrix} = \mathbf{y}_1(k) + \mathbf{y}_2(k) = \begin{pmatrix} y_1(k) + y_2(k) \\ y_1(k+1) + y_2(k+1) \end{pmatrix} \tag{19.46}$$

With (19.45) and the Frobenius norm of the channel matrix $\mathbf{H}(k)$

$$\|\mathbf{H}(k)\|_F^2 = \sum_{i=1}^{2} \sum_{j=1}^{2} |h_{ij}(k)|^2 = \beta_1(k) + \beta_2(k) \tag{19.47}$$

the output vector eventually is

$$\mathbf{y}_3(k) = \|\mathbf{H}(k)\|_F^2 \, \mathbf{s}(k) + \mathbf{U}_1^H(k)\mathbf{n}_1(k) + \mathbf{U}_2^H(k)\mathbf{n}_2(k) \tag{19.48}$$

Consequently, the signal part in $\mathbf{y}_3(k)$ only vanishes, if all four channel coefficient are zero at the same time and the resilience against fading is significantly increased by using a second receive antenna. If the channel coefficients in (19.42) are normalized such that $\beta_1 = \beta_2 = 1$, then the noise part in the receive signal (19.48) is not enhanced by the receiver matrices.

Figure 19.5 shows the symbol error rate (SER) as a function of the signal-to-noise-ratio (SNR). An i.i.d. channel model with Rayleigh fading, frequency flat spectrum, and 4-PSK are used. Apparently, the 2 × 1 Alamouti precoding with one receive antenna (2nd curve from top) outperforms the SISO system (1st curve from top). As expected, a further improvement is achieved by using the Alamouti precoding with two receive antennas (2 × 2 Alamouti, 4th curve from top). For comparison an 1 ×

Fig. 19.5 Symbol error rate (SER) as a function of the signal-to-noise ratio (SNR) for the Alamouti space-time coding scheme with one and two receive antennas. A frequency flat channel with Rayleigh fading and 4-PSK are used. For comparison a 1×1 (SISO) and a 1×2 SIMO system without precoding are also shown

2 SIMO scheme (3rd curve from top) without precoding and two receive antennas is also given, providing respectable performance as well.

19.3.3 Orthogonal Space-Time Block Codes for More Than Two Transmit Antennas

This mathematically demanding subject goes back to [8]. We introduce the short hand notation for the input QAM symbols in (19.1) as

$$c(k' + l) = c_{l+1} \; ; \; l = 0, 1, \ldots, N_c - 1 \qquad (19.49)$$

For linear space-time block codes the entries of the $M \times L$ space-time coding matrix **S** are composed of $\pm c_l$ and $\pm c_l^*$ as with the Alamouti encoder and $\pm \mathrm{Re}\,[c_l]$, $\pm \mathrm{Im}\,[c_l]$ as well as linear combinations thereof, $l = 0, 1, \ldots, N_c - 1$. Major design targets are: The decoding should be simple, i.e. a linear decoder should employ a unitary matrix to avoid matrix inversion processing. Furthermore, the scheme should achieve full transmit diversity M and full receive diversity N resulting in a total maximum diversity order of MN.

Motivated by the 2×2 Alamouti space-time coding matrix, a generalization can be done using $M \times L$ space-time coding matrices $\mathbf{S}$ with orthogonal rows, which fulfill the condition

$$\mathbf{SS}^H = \alpha \mathbf{I}_M \tag{19.50}$$

with

$$\alpha = |c_1|^2 + |c_2|^2 + \cdots + |c_{N_c}|^2 \tag{19.51}$$

Hence, $\frac{1}{\sqrt{\alpha}}\mathbf{S}$ is unitary. In [8] the authors show that full transmit diversity M can be achieved with a spatial code rate $r_s = \frac{N_c}{L}$.

Some Examples of Space-Time Codes

Without proof we review some important space-time coding matrices with complex entries, which meet the above requirements [3, 8].

- Space-time coding matrix for $M = 2$ and $r_s = 1$ (Alamouti scheme)

$$\mathbf{S} = \begin{pmatrix} c_1 & -c_2^* \\ c_2 & c_1^* \end{pmatrix} = \begin{pmatrix} \mathbf{s}_1 & \mathbf{s}_2 \end{pmatrix} \tag{19.52}$$

Two input symbols c_1, c_2 are mapped to two output space-time symbol vectors $\mathbf{s}_1, \mathbf{s}_2$. As there are $M = 2$ transmit antennas, $\mathbf{S}$ has two rows. The input and output clock rates are equal, thus there is no bandwidth increase. The mapping by the space-time encoder cannot start until the last QAM symbol c_2 is cached at the input. Consequently, the output signal has a delay of two time intervals.

- Space-time coding matrix for $M = 3$ and $r_s = \frac{1}{2}$

$$\mathbf{S} = \begin{pmatrix} c_1 & -c_2 & -c_3 & -c_4 & c_1^* & -c_2^* & -c_3^* & -c_4^* \\ c_2 & c_1 & c_4 & -c_3 & c_2^* & c_1^* & c_4^* & -c_3^* \\ c_3 & -c_4 & c_1 & c_2 & c_3^* & -c_4^* & c_1^* & c_2^* \end{pmatrix}$$

$$= \begin{pmatrix} \mathbf{s}_1 & \mathbf{s}_2 & \mathbf{s}_3 & \mathbf{s}_4 & \mathbf{s}_5 & \mathbf{s}_6 & \mathbf{s}_7 & \mathbf{s}_8 \end{pmatrix} \tag{19.53}$$

Four input symbols $c_1, c_2, \ldots, c_4$ are mapped to eight output space-time symbol vectors $\mathbf{s}_1, \mathbf{s}_2, \ldots, \mathbf{s}_8$. The matrix has three rows, because of $M = 3$. The output speed is increased by factor 2, also the required bandwidth. The encoder has a delay of four input time intervals. The second half of $\mathbf{S}$ is the conjugate complex of the first half.

- Space-time coding matrix for $M = 4$ and $r_s = \frac{1}{2}$

$$\mathbf{S} = \begin{pmatrix} c_1 & -c_2 & -c_3 & -c_4 & c_1^* & -c_2^* & -c_3^* & -c_4^* \\ c_2 & c_1 & c_4 & -c_3 & c_2^* & c_1^* & c_4^* & -c_3^* \\ c_3 & -c_4 & c_1 & c_2 & c_3^* & -c_4^* & c_1^* & c_2^* \\ c_4 & c_3 & -c_2 & c_1 & c_4^* & c_3^* & -c_2^* & c_1^* \end{pmatrix}$$

$$= \begin{pmatrix} s_1 & s_2 & s_3 & s_4 & s_5 & s_6 & s_7 & s_8 \end{pmatrix} \qquad (19.54)$$

Four input symbols $c_1, c_2, \ldots, c_4$ are mapped to eight output space-time symbol vectors $s_1, s_2, \ldots, s_8$. The matrix has four rows because of $M = 4$. The output speed is increased by factor 2 and also the required bandwidth. The output signal has a delay of four input time intervals. The second half of $\mathbf{S}$ is the conjugate complex of the first half.

- Space-time coding matrix for $M = 3$ and $r_s = \frac{3}{4}$

$$\mathbf{S} = \begin{pmatrix} c_1 & -c_2^* & \frac{\sqrt{2}}{2}c_3^* & \frac{\sqrt{2}}{2}c_3^* \\ c_2 & c_1^* & \frac{\sqrt{2}}{2}c_3^* & -\frac{\sqrt{2}}{2}c_3^* \\ \frac{\sqrt{2}}{2}c_3 & \frac{\sqrt{2}}{2}c_3 & -\mathrm{Re}[c_1]+j\mathrm{Im}[c_2] & \mathrm{Re}[c_2]+j\mathrm{Im}[c_1] \end{pmatrix}$$

$$= \begin{pmatrix} s_1 & s_2 & s_3 & s_4 \end{pmatrix} \qquad (19.55)$$

Three input symbols c_1, c_2, c_3 are mapped to four output space-time symbol vectors $s_1, s_2, \ldots, s_4$. The matrix has three rows, because of $M = 3$. The output speed is increased by factor $\frac{4}{3}$, also the required bandwidth. The mapping cannot start until the last QAM symbol c_3 is cached at the input. Consequently, the output signal has a delay of three input time intervals.

- Alternative space-time coding matrix for $M = 3$ and $r_s = \frac{3}{4}$, [9].

$$\mathbf{S} = \begin{pmatrix} c_1 & c_2^* & c_3^* & 0 \\ -c_2 & c_1^* & 0 & -c_3^* \\ c_3 & 0 & c_1^* & c_2^* \end{pmatrix}$$

$$= \begin{pmatrix} s_1 & s_2 & s_3 & s_4 \end{pmatrix} \qquad (19.56)$$

Three input symbols c_1, c_2, c_3 are mapped to four output space-time symbol vectors $s_1, s_2, \ldots, s_4$. As $M = 3$ the matrix has three rows. The output speed is increased by factor $\frac{4}{3}$, also the required bandwidth. The output signal has a minimal delay of three input time intervals.

- There exist also space-time coding matrices for real symbols c_k only, which are feasible for amplitude shift keying or real valued phase shift keying. In case of $M = 2$, the Alamouti matrix (19.52) can be employed by setting $c_i^* = c_i$; $i = 1, 2$. For $M = 3$ the matrix (19.53) is applicable, if all conjugate complex elements are dropped resulting in a 3×4 matrix. Similarly for $M = 4$, (19.54) can provide the solution, if reduced to a 4×4 matrix by deleting the conjugate complex elements.
- For all given space-time coding matrices it is easy to check that the rows are pairwise orthogonal with property (19.50).

Layered space-time coding is an enhancement of spatial multiplexing. We will discuss such schemes in a separate Sect. 19.5, because they paved the way for the first applications of MIMO systems in wireless communications.

19.4 Principle of Space-Time Trellis Coding

Space-time trellis codes are an extension of conventional trellis codes [10] to MIMO
systems. They have been introduced by [11]. Since then a large number of codes
and their design have been widely explored and the improved coding gain and spectral efficiency have been demonstrated for fading channels. However, it was also
recognized that the decoder is much more complex compared to the orthogonal
space-time block codes. A survey is available in [3]. Here we are outlining the principle by showing an example. In general, the input-output relation is given by a
trellis diagram which is a state transition diagram enhanced by a discrete-time coor-
dinate k. As in the case of sequential detection of a data signal, a trellis decoder is
also used here employing the Viterbi algorithm. Figure 19.6 shows an example of
a space-time trellis encoder with 4-PSK, which provides transmit delay diversity.
We maintain the general term QAM symbol also for the PSK symbols. Consecutive
pairs of bits ($\kappa = 2$) of the input bit stream $b(n)$ are mapped to one QAM symbol
$c(k)$ by the mapper. The space-time encoder allocates to each $c(k)$ a space-time sym-
bol vector $\mathbf{s}(k) = \left(s_1(k)\ s_2(k) \right)^T$. In case of the transmit delay diversity $s_1(k) = c(k)$
and $s_2(k) = c(k-1)$ hold. As in the same time frame one QAM symbol is going
in and one space-time symbol vector is output, the spatial code rate is $r_s = 1$ and
consequently for the space-time coding matrix follows $\mathbf{S}(k) = \mathbf{s}(k)$.

We describe the space-time trellis encoder by means of a trellis diagram. To
this end we determine all independent state variables, which are associated with
the storage elements. The encoder in Fig. 19.6 owns only one state variable $s_2(k) =
c(k-1)$. With 4-QAM the symbol alphabet consists of $2^\kappa = 4$ distinct signal points,
which we denote as $\{a_1, a_2, a_3, a_4\}$ and which are shown in the complex domain in
Fig. 19.7 (left). The state variable can take on these 4 values. Consequently, the system
can stay in 4 different states, which can be denoted arbitrarily such as state 1, state 2,
state 3, and state 4. The state transitions are shown in the trellis diagram depicted in
Fig. 19.7 (right). Depending on the input signal the system can remain in the present
state or change into another state. At the bottom the time instants $k-1$ and k are
indicated enhancing the state transition diagram to a trellis diagram. The transitions
are labeled as $(c(k) \mid s_2(k))$, where the first element indicates the input QAM symbol
and the second element is the output symbol $s_2(k) = c(k-1)$. Horizontal arrows

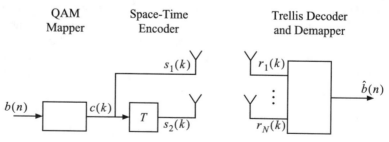

Fig. 19.6 Principle block diagram of a space-time trellis encoder and decoder for delay diversity

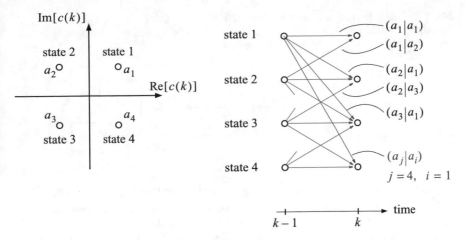

Fig. 19.7 Signal constellation diagram for 4-PSK (left), trellis diagram of space-time trellis encoder (right)

reveal that the system maintains the present state, if a new symbol comes in, whereas oblique arrows illustrate a change of the state from time instants $k - 1$ to k. Please note that not all transitions are sketched in Fig. 19.7 to avoid an overload of the graph. In our notation of the states and the state variables the transitions are indicated as $(a_j \mid a_i)$, which means, if the encoder is in state i and the new input signal is a_j, then the encoder moves to state j. The output signals are $s_2(k) = a_i$ and $s_1(k) = a_j$. As already mentioned, the receiver acts as a trellis decoder. Its complexity grows exponentially with the number of states, in our example 2^κ.

19.5 Layered Space-Time Architecture

The layered space-time architecture is based on the demultiplexing of the incoming bit stream into parallel layers, each feeding a transmit antenna. The bit stream can undergo forward error correction and mapping to symbols before or after the demultiplexing operation. In essence, layered space-time coding employs spatial multiplexing as described in 19.2 and is categorized in vertical, horizontal, and diagonal layered space-time coding, but with some ambiguity, [3, 12].

19.5.1 Vertical Layered Space-Time Coding

This scheme originates from the Bell Laboratories [13] and is also called Bell Labs Layered Space-Time (BLAST). The principle block diagram is straightforward and

Fig. 19.8 Block diagram of the original vertical layered space-time encoder (V-BLAST)

Forward Error Correction   De-Multiplexer   QAM Mapper

depicted in Fig. 19.8. The main building block is a demultiplexer $1 : M$ as described in Sect. 19.2, which divides the incoming bit sequence $b(n)$ into M parallel output streams. Their bit rates are reduced by factor M compared to the input $b(n)$. The QAM mapper allocates κ−tuples of consecutive bits to QAM symbols, which are directly the output signals $s_1(k), \ldots, s_M(k)$ composing the space-time symbol vector $\mathbf{s}(k)$. In view of the vertical shape of this column vector, which allocates the QAM symbols to each transmit antenna from top to bottom the scheme is denoted as vertical layered space-time coding or V-BLAST. First versions have even abstained from temporal encoding and have employed the same QAM signal constellations for each layer [13]. If the information bit sequence $b'(n')$ is input to a forward error correction encoder including a temporal interleaver and the code rate between $b'(n')$ and $b(n)$ is r_t, then the overall code rate of the layered space-time transmitter results in

$$r_{total} = r_t \kappa M \tag{19.57}$$

Obviously, r_{total} can be larger than M. For example, if we take 16-QAM ($\kappa = 4$) and a typical temporal code rate of $r_t = \frac{1}{2}$ for wireless communications the overall code rate is $r_{total} = 2M$. A vertical layered space-time encoder without temporal forward error correction ($r_t = 1$) is just a spatial demultiplexer and thus exhibits no transmit diversity. We then say the transmit diversity order is one. If temporal encoding is uses, i.e. $r_t < 1$, the transmit diversity order can exceed one and consequently the total diversity order of the overall system will be larger than the number of receive antennas N.

In Fig. 19.9 an equivalent structure of a V-BLAST encoder is shown. It results from the scheme in Fig. 19.8, if the QAM mapping is moved to the input of the demultiplexer. Figure 19.10 shows a receiver for a MIMO signal with vertical space-time encoding. The first building block can be a linear zero-forcing or a MMSE receiver. The M output signals $y_1, y_2, \ldots, y_M$ enter a spatial multiplexer $M : 1$, which operates as a parallel-to-serial converter providing estimates $\hat{c}(k')$ of the transmit QAM symbols. After demapping the estimates $\hat{b}(n)$ of the encoded bit stream are available, which are fed into a forward error correction decoder yielding the estimates $\hat{b}'(n')$. Obviously, the output signal of the spatial multiplexer is processed as in case of a digital single input single output (SISO) receiver. Consequently, all advanced techniques can be applied for improved demapping and decoding, such as iterative

Forward   QAM   De-
Error   Mapper   Multiplexer
Correction

Fig. 19.9 Block diagram of an alternative vertical layered space-time encoder

ZF or MMSE   Multiplexer   QAM   Forward
Receiver   Demapper   Error
                                             Correction
                                             Decoder

Loops for iterative (Turbo) decoding

Fig. 19.10 Block diagram of a linear receiver for vertical layered space-time decoding

(Turbo) decoding or soft demapping [14] using the Turbo principle indicated by the dashed feedback lines in Fig. 19.10. Alternatively, the linear receiver including the spatial multiplexer can be enhanced by an iterative a-posterior detector for the receive vector $\left(r_1 \, r_2 \, \cdots \, r_N \right)^T$. Then the complexity of the receiver is significantly increased. The original V-BLAST receiver employed the "BLAST algorithm" [13, 15] with ordered successive interference cancellation (OSIC).

19.5.2 Horizontal Layered Space-Time Coding

As outlined in the previous section, the vertical layered space-time coder operates with only one temporal encoder, if at all. Alternatively, the scheme in Fig. 19.8 can be equipped with dedicated forward error correction encoders and QAM mappers for each layer at the output of the $1 : M$ demultiplexer. In view of the horizontal arrangement and the fact that coding and mapping are independent from one layer to the other the resulting scheme in Fig. 19.11 is referred to as horizontal layered space-time coding or horizontal Bell Labs layered space-time (H-BLAST). Please note

Fig. 19.11 Block diagram of horizontal layered space-time encoding. The spatial interleaver is in operation only for diagonal layered space-time coding

that the spatial interleaver indicated by dashed lines is only present for the diagonal layered space-time encoder described in the next section. Obviously, the temporal encoder and the QAM mapper are M-fold compared to Fig. 19.9, if M transmit antennas are used. This is an increase of complexity. However, the processing of the signals is at clock rates, which are by a factor M lower compared to the input $b'(m)$ of the demultiplexer. The overall code rate can be easily calculated and is the same as for the vertical layered space-time coding scheme given by (19.57). As the signals in the different layers are independent and because each antenna sends no other signal than that of its own layer, there is no transmit diversity available on symbol level. Hence, the transmit diversity order of H-BLAST is one. The signals coming from each of the M transmit antennas are received by all N receive antennas resulting in a receive diversity order of N. Consequently, we conclude that the diversity order of the horizontal layered space-time coding scheme on a symbol basis can not be larger than N. Besides the diversity order, also the coding gain of a transmission scheme is of importance for achieving superior transmission quality. The coding gain depends on the selected forward error correction code and the performance of the decoding algorithm.

Figure 19.12 shows the principle block diagram of a linear receiver for horizontal layered space-time decoding. After a linear zero-forcing or MMSE receiver the M output signals $y_1, y_2, \ldots, y_M$ can be processed separately using conventional single input single output decoding principles. As mentioned before with the vertical layered system, also maximum likelihood joint decoding of the components of the receive vector $(r_1\ r_2\ \cdots\ r_N)^T$ is an alternative here, however, with much higher complexity.

19.5.3   Diagonal Layered Space-Time Coding

This scheme builds upon the horizontal layered space-time system. By inserting a spatial interleaver, as shown by dashed lines in Fig. 19.11, symbols of different layers can be interleaved crosswise. This cross layer operation has motivated the nomencla-

Fig. 19.12 Block diagram of a linear receiver for horizontal layered space-time decoding. The spatial deinterleaver is in operation only for diagonal layered space-time coding

ture diagonal layered space-time coding or diagonal (D-) BLAST. In [13] a stream rotator with periodic antenna allocation is preferred as an interleaver. If the code words are made large enough, bits or even codewords of each layer are transmitted over all M antennas and the transmit diversity order approaches M. An appropriate receiver with N antennas then can get full diversity order MN of the overall system. The total code rate is the same as in (19.57). The receiver is more complex than that of the horizontal layered system owing to the spatial interleaving. V-, H-, and D-BLAST schemes have been one of the first MIMO systems introduced in the second half of the 1990's starting with zero-forcing or MMSE receivers combined with successive interference cancellation.

19.5.4 Iterative Receivers for Layered Space-Time Systems

Iterative detection using the Turbo principle can significantly increase the performance of a receiver. Turbo codes and their iterative decoding (also referred to as Turbo decoding) have been first introduced by [16] in the framework of two concatenated codes for forward error correction. A comprehensive survey on the Turbo principle is given in [18]. A serially concatenated Turbo coding scheme consists of the cascade of a first encoder (outer encoder), an interleaver, and a second encoder (inner encoder). Ideally, the interleaver generates an input signal for the inner encoder, which is statistically independent of the output signal of the outer encoder. Thus, in the ideal case the two encoders operate with input bit streams, which do not hang together statistically. Similarly at the receiver the Turbo decoder is equipped with a cascade or a parallel arrangement of two decoders called inner and outer decoder separated by a deinterleaver. The two decoders do not operate isolated and mutually exchange reliability information on the intermediate decoding results through a feedback loop, which has motivated the term "Turbo". The reliability information exchange is done as extrinsic and intrinsic information on the basis of log-likelihood values [17]. To decode one bit the decoding process is operated several times also referred to as decoding iterations. After each iteration the decoding result has improved, i.e. the mean bit error rate has decreased. Depending on the convergence conditions the iteration cycle is finally abandoned and the output of the outer decoder, which is an a posterior probability (APP) decoder, is subject to hard decision and provides the final decoded bit. For APP decoding the BCJR algorithm invented by [19] is optimal. Owing to its high computational complexity several sub-optimal solutions have been proposed, such as the soft-output Viterbi algorithm [20]. For code design and convergence properties the EXIT chart has proven to be a powerful method [21–23].

An effective yet not very complex method is a MIMO receiver with soft demapping, as indicated in Fig. 19.10. Only the middle feedback loop depicted by dashed lines is employed. The multiplexed samples $\hat{c}(k')$ are entering the demapper. A conventional demapper would perform a hard decision and would output the associated binary codewords from a look-up table. To apply soft demapping at the receiver the transmitter is equipped with a forward error correction encoder, which is named as outer encoder and the mapper plays the part of the inner encoder. Likewise for soft demapping at the receiver the forward error correction decoder operates as outer decoder and the demapper acts as the inner decoder. In contrast to a conventional demapper, the output of the soft demapper provides log-likelihood values [17]. Through a recursive loop the outer decoder provides a-prior information for the soft demapper. After several iterations the output signal of the outer decoder undergoes a hard decision resulting in the output bits $\hat{b}'(n')$. With an increasing number of iterations the Turbo cliff builds up, which is characterized by a steep decent of the mean bit error probability as a function of the signal-to-noise ratio.

References

1. Richardson, T., Urbanke, R.: Modern Coding Theory. Cambridge University Press, Cambridge (2008)
2. Moon, T.K.: Error Correction Coding - Mathematical Methods and Algorithms. Wiley Interscience, Hoboken (2005)
3. Vucetic, B., Yuan, J.: Space-Time Coding. Wiley, Hoboken (2003)
4. Alamouti, S.: A simple channel diversity technique for wireless communications. IEEE J. Sel. Areas Commun. **16** (1998)
5. Physical channels and modulation, Technical Specifications. TS 36.211, V11.5.0. Technical report, 3GPP (2012)
6. Schneider, D., Speidel, J., Stadelmeier, L., Schill, D., Schwager, A.: MIMO for inhome power line communications. In: International Conference on Source and Channel Coding (SCC), ITG Fachberichte (2008)
7. Berger, L.T., Schwager, A., Pagani, P., Schneider, D.: MIMO power line communications - Narrow and broadband standards. CRC Press, EMC and advanced processing (2014)
8. Tarokh, V., Jafarkhani, H., Calderbank, A.: Space-time block codes from orthogonal designs. IEEE Trans. Inf. Theory **45** (1999)
9. Hochwald, B., Marzetta, T.L., Papadias, C.B.: A transmitter diversity scheme for wideband CDMA systems based on space-time spreading. IEEE J. Sel. Areas Commun. **19** (2001)
10. Biglieri, E., Divsalar, D., McLane, P., Simon, M.: Introduction to Trellis-coded Modulation with Applications. Macmillan, New York (1991)

11. Tarokh, V., Seshadri, N., Calderbank, A.: Space-time codes for high data rate wireless communication: performance criterion and code construction. IEEE Trans. Inform. Theory **44** (1998)

12. Paulraj, A., Nabar, R., Gore, D.: Introduction to Space-Time Wireless Communications. Cambridge University Press, Cambridge (2003)

13. Foschini, G.: Layered space-time architecture for wireless communication in a fading environment when using multi-element antennas. Bell Syst. Tech. J. (1996)

14. ten Brink, S., Speidel, J., Yan, R.-H.: Iterativeerative demapping for QPSK modulation. Electron. Lett. **34** (1998)

15. Foschini, G., Gans, M.: On limits of wireless communications in a fading environment when using multiple antennas. IEEE Wirel. Pers. Commun. **6** (1998)

16. Berrou, C., Glavieux, A., Thitimajshima, P.: Near Shannon limit error-correcting coding and decoding. In: International Conference on Communications ICC (1993)

17. Hagenauer, J., Offer, E., Papke, L.: Iterative decoding of binary block and convolutional codes. IEEE Trans. Inform. Theory **42** (1996)

18. Hagenauer, J.: The turbo principle: tutorial introduction and state of the art. In: Proceedings of 1st International Symposium on Turbo Codes (1997)

19. Bahl, L., Cocke, J., Jelinek, F., Raviv, J.: Optimal decoding of linear codes for minimizing symbol error rate. IEEE Trans. Inform. Theory **20** (1974)

20. Hagenauer, J., Hoeher, P.: A viterbi algorithm with soft-decision outputs and its applications. In: IEEE International Conference on Global Communications (GLOBECOM) (1989)

21. ten Brink, S.: Convergence of iterative decoding. Electron. Lett. **35** (1999)

22. ten Brink, S.: Designing iterative decoding schemes with the extrinsic information transfer chart. AEÜ Int. J. Electron. Commun. **54** (2000)

23. ten Brink, S.: Design of concatenated coding schemes based on iterative decoding convergence. Ph.D. thesis, University of Stuttgart, Institute of Telecommunications, Shaker Publication, ISBN 3-8322-0684-1 (2001)

Chapter 20
Principles of Multi-user MIMO Transmission

20.1 Introduction

Hitherto, we have considered the MIMO transmission between a base station and one user, which we call single-user MIMO transmission. In a communications network the base station has to serve a large number of users, e.g., in an in-house area with a wireless local area networks (WLAN) according to the standard IEEE 802.11. Also in outdoor scenarios a multitude of users has to be addressed with the cellular networks of the type 3G (year 2004), 4G (year 2010), and 5G (year 2020) with data rates of about 8 Mbit/s, 100 Mbit/s, and up to 1 Gbit/s, respectively. In this Chapter we investigate methods for data communication between the base station and the users. Each transmitter and receiver shall be equipped with multiple antennas to benefit from the MIMO principle, which we call multi-user MIMO (MU MIMO) transmission. Also in the case that each user equipment has only one antenna, the term MIMO is used, because the antennas of all users taken together are considered as multiple input or multiple output. We differentiate between the directions from the base station to the users called the downlink and between the link from the users to the base station, which is referred to as uplink. Conventionally, without the MIMO principle the base station allocates certain time slots or frequency bands to the users. With multiple antennas a multi-user MIMO scheme can serve all users at the same time and in the same frequency band, hence providing a higher efficiency. While the WLAN standards IEEE 802.11 a, b, g, and n do not support multi-user MIMO techniques, the later versions AC-WLAN or AC Wave 2 own these benefits. However, the transmission of different signals in the same frequency band and in the same time slots in the downlink gives rise to interference at the user terminals. Of course, a similar situation will occur in the uplink, when many users are going to address the same base station. Hence, an important question for the system designer is how to overcome or at least minimize the impact of the interference. Several methods for multi-user MIMO systems are described in [1, 2] as a survey. In the next sections we outline some selected principles in more detail. On top of these methods

© Springer Nature Switzerland AG 2021
J. Speidel, *Introduction to Digital Communications*,
Signals and Communication Technology,
https://doi.org/10.1007/978-3-030-67357-4_20

protocols of the higher system layers, such as the data link and the network layer can reduce residual interference by proper scheduling of the user access. However, this is beyond the scope of this Chapter.

20.2 Precoding for Multi-user MIMO Downlink Transmission

20.2.1 Precoding by "Channel Inversion"

As the base station transmits signals to the various users in the same frequency band and at the same time, we have to make sure that each user gets the individually devoted signal without interference from the other users. Hence, for each user terminal the *multi-user interference* also called *multi-access interference* has to be minimized. Several schemes operate with precoding at the transmitter to keep the complexity at the the cost sensitive user equipment low. In Chap. 18 linear precoding is discussed in quite some detail as an alternative to a linear receiver for the reduction of the inter-channel interference for single-user MIMO systems. The prefilter for multi-user applications, which will be outlined in the following, shall provide a reduction of the multi-user interference at all receive terminals. We will see that even the inter-channel interference can be canceled by some of these methods.

The principle block diagram is depicted in Fig. 20.1. Apparently, the downlink transmission can be regarded as a broadcast or point-to-multi-point mode, in which the base station transmits signals to all users. Hence, the combined channel with matrix $\mathbf{H}(k)$ is also called broadcast channel. Let us start with the receiver, where U user terminals also denoted as user equipment and indicated by the index $u = 1, 2, ..., U$ have to be served.

The user terminal u shall be equipped with N_u receive antennas and its receive signal is denoted as $\mathbf{r}_u(k) \in \mathbb{C}^{N_u \times 1}$. In the following we drop the discrete-time variable k to simplify the notation. The combined receive column vector can be written as

$$\mathbf{r} = \begin{pmatrix} \mathbf{r}_1 \\ \vdots \\ \mathbf{r}_u \\ \vdots \\ \mathbf{r}_U \end{pmatrix} \in \mathbb{C}^{N \times 1} \tag{20.1}$$

and the total number of receive antennas for all users shall be

$$N = \sum_{u=1}^{U} N_u \tag{20.2}$$

Fig. 20.1 Block diagram of a multi-user MIMO downlink transmission between a base station and various user terminals

Similarly, the additive noise vector for the user terminal u is $\mathbf{n}_u \in \mathbb{C}^{N_u \times 1}$ and the combined noise vector is defined as

$$\mathbf{n} = \begin{pmatrix} \mathbf{n}_1 \\ \vdots \\ \mathbf{n}_u \\ \vdots \\ \mathbf{n}_U \end{pmatrix} \in \mathbb{C}^{N \times 1} \tag{20.3}$$

At the transmitter side the base station has structured its MIMO antennas into U groups, where the group u owns M_u antenna elements and the associated transmit signal vector is $\mathbf{s}_u \in \mathbb{C}^{M_u \times 1}$, $u = 1, 2, ..., U$. Then the total number of transmit antenna elements at the base station is

$$M = \sum_{u=1}^{U} M_u \tag{20.4}$$

We assume a fading channel with frequency-flat spectrum defined by an $N \times M$ matrix $\mathbf{H} \in \mathbb{C}^{N \times M}$ and introduce the combined vector of all transmit signals

$$\mathbf{s} = \begin{pmatrix} s_1 \\ \vdots \\ s_u \\ \vdots \\ s_U \end{pmatrix} \in \mathbb{C}^{M \times 1} \tag{20.5}$$

The input signals to the precoder are the QAM symbols devoted to the various users and are also structured as vectors $\mathbf{c}_u \in \mathbb{C}^{N_u \times 1}$, $u = 1, 2, ..., U$. Hence, $\mathbf{c}_u$ represents the vector of N_u streams of QAM symbols entering the precoder. Then the combined input signal vector of the precoder can be defined as

$$\mathbf{c} = \begin{pmatrix} \mathbf{c}_1 \\ \vdots \\ \mathbf{c}_u \\ \vdots \\ \mathbf{c}_U \end{pmatrix} \in \mathbb{C}^{N \times 1} \tag{20.6}$$

Consequently, as portrait in Fig. 20.1 the precoding matrix $\mathbf{T}$ must have the dimension $M \times N$ with in general complex entries. $\mathbf{T} \in \mathbb{C}^{M \times N}$ maps the input signal vector $\mathbf{c}$ to the output vector $\mathbf{s}$ yielding

$$\mathbf{s} = \mathbf{T}\mathbf{c} \tag{20.7}$$

Please note, at the moment no individual mapping of an input $\mathbf{c}_u$ to an output s_j is required. We obtain for the receive signal vector

$$\mathbf{r} = \mathbf{H}\mathbf{s} + \mathbf{n} \tag{20.8}$$

Plugging (20.7) into (20.8) yields

$$\mathbf{r} = \mathbf{H}\mathbf{T}\mathbf{c} + \mathbf{n} \tag{20.9}$$

As described in Sect. 18.2, the zero-forcing and the MMSE precoder with matrix

$$\mathbf{T} = \mathbf{H}^{++} = \mathbf{H}^H \left(\mathbf{H}\mathbf{H}^H\right)^{-1} \; ; \; M \geq N \tag{20.10}$$

completely removes the interference, where $\mathbf{H}^{++}$ is the pseudo inverse matrix of $\mathbf{H}$. Please note that the channel matrix $\mathbf{H}$ must have full rank N otherwise the inverse in (20.10) does not exist. Hence, the total number M of transmit antennas at the base station must be equal to or larger than the sum N of the receive antennas of all user equipment

$$\sum_{u=1}^{U} M_u \geq \sum_{u=1}^{U} N_u \tag{20.11}$$

With

$$\mathbf{HT} = \mathbf{HH}^{++} = \mathbf{I}_N \qquad (20.12)$$

follows the receive signal

$$\mathbf{r} = \mathbf{I}_N\mathbf{c} + \mathbf{n} = \mathbf{c} + \mathbf{n} \qquad (20.13)$$

and the multi-user interference as well as the inter-channel interference are completely canceled. The right hand side of (20.13) can be interpreted as the diagonalization of the system of equations (20.9), where the diagonal matrix is just the identity matrix $\mathbf{I}_N$. From (20.13) the individual signal for the user u can be decomposed with the help of (20.1), (20.3), and (20.6) as

$$\mathbf{r}_u = \mathbf{c}_u + \mathbf{n}_u \; ; \; u = 1, 2, ..., U \qquad (20.14)$$

showing that the noise $\mathbf{n}_u$ remains as the only impairment for the receive signal. (20.14) requires that the dimension of $\mathbf{c}_u$ equals the dimension of $\mathbf{r}_u$. Consequently, the number N_u of signal components constituting the transmit symbol vector $\mathbf{c}_u \in \mathbb{C}^{N_u \times 1}$ determines the number N_u of receive antennas for the user u, which is a prerequisite of our derivation. Apparently, the base station devotes N_u parallel symbol streams to user u. Furthermore, we recognize that no individual mapping of an input $\mathbf{c}_u$ to an output $\mathbf{s}_j$ is required at the transmitter.

Now, consider an application, in which only one user u is part of the system. Then $M = M_u$ and $N = N_u$ and from the condition $M \geq N$ in (20.11) follows

$$M_u \geq N_u \qquad (20.15)$$

This scenario can occur for any terminal. Therefore (20.15) must hold for all users $u = 1, 2, ..., U$.

As shown in Fig. 20.1, the last stage is the detector at the user terminals, in many cases a maximum likelihood detector, which takes $\mathbf{r}_u$ as the input and decides for the most likely QAM symbol vector $\hat{\mathbf{c}}_u$. Of course, if the channel matrix is a square matrix $\mathbf{H} \in \mathbb{C}^{N \times N}$ with full rank N the pseudo inverse matrix $\mathbf{H}^{++}$ turns into the inverse matrix resulting in $\mathbf{T} = \mathbf{H}^{-1}$. This is the reason why the described method for multi-user interference reduction is also called *channel inversion*.

Furthermore, for all precoding techniques discussed in this Chapter the transmitter must have the full knowledge about the channel matrix $\mathbf{H}$. Hence, channel estimation at the receiver has to be performed and the entries of the estimated matrix are then sent as channel state information (CSI) to the transmitter via a separate feedback loop. Therefore, this technique is called *closed loop precoding*. Depending on the number of users and antennas per user terminal the dimension of $\mathbf{H}$ can be reasonably large and the calculation of the inverse or pseudo inverse becomes expensive. In the next section a transmitter consisting of an individual precoder per user is discussed.

Example 1

Consider the multi-user downlink with $U = 3$ user terminals. User 1 will be provided with two and the other two users with one symbol stream. The channel matrix $\mathbf{H}$ is estimated at the receivers and in a small time interval

$$\mathbf{H} = \begin{pmatrix} 1 & 0.2 & -0.1 & 0.1 & 0 \\ -0.2 & 1 & 0.1 & 0 & 0.1 \\ 0.1 & -0.1 & 1 & 0.2 & -0.1 \\ 0 & -0.1 & -0.2 & 1 & 0.2 \end{pmatrix} \qquad (20.16)$$

is given. For simplicity we assume real matrix entries. Precoding with a matrix $\mathbf{T}$ shall be applied to minimize the multi-user interference. Find $\mathbf{T}$ using the method of "channel inversion".

Solution:

From the channel matrix we conclude $M = 5$ and $N = 4$. Furthermore, $N_1 = 2$ and $N_2 = N_3 = 1$ are given. Thus, in Fig. 20.1 $\mathbf{c}_1$ and $\mathbf{r}_1$ are 2×1 vectors whereas c_2, c_3, r_2, and r_3 are scalars. Apparently, $\mathbf{H}$ is non-quadratic and hence no inverse matrix exists. Therefore we calculate the precoding matrix as the pseudo inverse (20.10) of $\mathbf{H}$. The result is

$$\mathbf{T} = \mathbf{H}^H \left(\mathbf{H}\mathbf{H}^H \right)^{-1} = \begin{pmatrix} 0.95 & -0.19 & 0.09 & -0.10 \\ 0.20 & 0.94 & -0.07 & -0.02 \\ -0.08 & 0.11 & 0.94 & -0.16 \\ 0.01 & 0.09 & 0.20 & 0.93 \\ -0.02 & 0.11 & -0.09 & 0.19 \end{pmatrix} \; ; \; \mathbf{H}\mathbf{T} = \begin{pmatrix} 1 & 0 & 10^{-4} & 0 \\ 0 & 1 & 0 & 10^{-4} \\ 0 & 0 & 1,0001 & 0 \\ 0 & 10^{-4} & 0 & 1 \end{pmatrix}$$

$$(20.17)$$

Apparently, $\mathbf{H}\mathbf{T}$ is a good approximation of the identity matrix $\mathbf{I}_4$. Then follows from (20.9) with $\mathbf{c}_1 = \left(c_{11} \ c_{12} \right)^T$, $\mathbf{r}_1 = \left(r_{11} \ r_{12} \right)^T$, and $\mathbf{n}_1 = \left(n_{11} \ n_{12} \right)^T$

$$\begin{pmatrix} r_{11} \\ r_{12} \\ r_2 \\ r_3 \end{pmatrix} = \begin{pmatrix} c_{11} \\ c_{12} \\ c_2 \\ c_3 \end{pmatrix} + \begin{pmatrix} n_{11} \\ n_{12} \\ n_2 \\ n_3 \end{pmatrix} \qquad (20.18)$$

where the elements with double indices belong to user 1. It is well appreciated that the multi-user as well as the inter channel interference are completely removed and the multi-user downlink turns into three independent single-user links with receive signals r_1, r_2, and r_3.

20.2.2 Precoding with Block Diagonalization

Input–Output Relation of the Downlink

An interesting closed form method to cancel the multi-user interference at each user terminal in the downlink is the design of a precoding matrix $\mathbf{T}$ in such a way that the interference term $\mathbf{HT}$ in (20.9) becomes a block-diagonal matrix. According to linear algebra a matrix $\mathbf{G}$ is said to be block diagonal, if

$$\mathbf{G} = \text{diag } (\mathbf{G}_1 \mathbf{G}_2 \dots \mathbf{G}_U) \tag{20.19}$$

holds, where the $\mathbf{G}_i$ are matrices with smaller dimensions than the dimension of $\mathbf{G}$. We consider the block diagram of the multi-user MIMO downlink in Fig. 20.2. The receiver side with the U user terminals is the same as in Fig. 20.1. However, at the transmitter the base station employs individual precoders with matrices $\mathbf{T}_u \in \mathbb{C}^{M \times \zeta_u}$, $u = 1, 2, ..., U$.

The input signals at the base station are $\mathbf{c}_u \in \mathbb{C}^{\zeta_u \times 1}$, $u = 1, 2, ..., U$, where $\mathbf{c}_u$ represents the column vector composed of ζ_u QAM symbols. Obviously,

$$\zeta_u \leq N_u , \quad u = 1, 2, ..., U \tag{20.20}$$

guarantees that all transmit symbols for user u are covered by the receive antennas of this user. The precoder output vector is $\mathbf{x}_u \in \mathbb{C}^{M \times 1}$

$$\mathbf{x}_u = \mathbf{T}_u \mathbf{c}_u ; \quad u = 1, 2, ..., U \tag{20.21}$$

Fig. 20.2 Multi-user MIMO downlink transmission between a base station with individual precoders and various user terminals

where M is the number of transmit antennas at the base station. The U precoder outputs are added component by component yielding the transmit signal vector

$$\mathbf{s} = \sum_{u=1}^{U} \mathbf{x}_u = \sum_{u=1}^{U} \mathbf{T}_u \mathbf{c}_u \tag{20.22}$$

with

$$\mathbf{s} = \left(s_1 \cdots s_M\right)^T \ \epsilon \ \mathbb{C}^{M \times 1} \tag{20.23}$$

Apparently, all $\mathbf{x}_u$ must have the same number M of components, otherwise they cannot be added. As a consequence, $\mathbf{T}_u \ \epsilon \ \mathbb{C}^{M \times \zeta_u}$, $u = 1, 2, ..., U$ holds. Furthermore, we can define the combined precoding matrix

$$\mathbf{T} = \left(\mathbf{T}_1 \cdots \mathbf{T}_u \cdots \mathbf{T}_U\right) \ \epsilon \ \mathbb{C}^{M \times \zeta} \tag{20.24}$$

which contains U vertical slices and

$$\zeta = \sum_{u=1}^{U} \zeta_u \leq N \tag{20.25}$$

holds. The inequality follows from (20.2) and (20.20). Between the transmitter and the receiver the broadcast channel with the combined matrix $\mathbf{H}(k)$ is present again. However, individual sub-matrices $\mathbf{H}_u \ \epsilon \ \mathbb{C}^{N_u \times M}$ are effective characterizing the channels from the M transmit antenna outputs of the base station to the N_u inputs of the user terminal u, $u = 1, 2, ..., U$. Hence, we define the combined channel matrix

$$\mathbf{H} = \begin{pmatrix} \mathbf{H}_1 \\ \vdots \\ \mathbf{H}_u \\ \vdots \\ \mathbf{H}_U \end{pmatrix} \tag{20.26}$$

and recognize that the sub-matrices $\mathbf{H}_u$ are horizontal slices of $\mathbf{H}$. Next, we determine the receive signals. The signal vector for user u is

$$\mathbf{r}_u = \mathbf{H}_u \mathbf{s} + \mathbf{n}_u \tag{20.27}$$

and with (20.22) follows

$$\mathbf{r}_u = \mathbf{H}_u \mathbf{T}_u \mathbf{c}_u + \mathbf{H}_u \sum_{\substack{i=1 \\ i \neq u}}^{U} \mathbf{T}_i \mathbf{c}_i + \mathbf{n}_u \ ; \ u = 1, 2, ..., U \tag{20.28}$$

where $\mathbf{n}_u$ is the additive noise at the user terminal u. We have decomposed $\mathbf{r}_u$ already into the desired receive signal $\mathbf{H}_u\mathbf{T}_u\mathbf{c}_u$ and the multi-user interference for user u

$$\mathbf{H}_u \sum_{\substack{i=1 \\ i \neq u}}^{U} \mathbf{T}_i\mathbf{c}_i \; ; \; u = 1, 2, ..., U \tag{20.29}$$

With the combined vectors (20.1), (20.6) and the matrices (20.24), (20.26) follows the matrix notation

$$\mathbf{r} = \mathbf{H}\mathbf{T}\mathbf{c} + \mathbf{n} = \begin{pmatrix} \mathbf{H}_1 \\ \vdots \\ \mathbf{H}_u \\ \vdots \\ \mathbf{H}_U \end{pmatrix} \begin{pmatrix} \mathbf{T}_1 \cdots \mathbf{T}_u \cdots \mathbf{T}_U \end{pmatrix} \mathbf{c} + \mathbf{n} \tag{20.30}$$

We can execute $\mathbf{H}\mathbf{T}$ just like the multiplication of a column vector with a row vector using the block matrices as entries and obtain

$$\begin{pmatrix} \mathbf{r}_1 \\ \vdots \\ \mathbf{r}_u \\ \vdots \\ \mathbf{r}_U \end{pmatrix} = \begin{pmatrix} \mathbf{H}_1\mathbf{T}_1 & \cdots & \mathbf{H}_1\mathbf{T}_u & \cdots & \mathbf{H}_1\mathbf{T}_U \\ & \ddots & & \ddots & \\ \mathbf{H}_u\mathbf{T}_1 & \cdots & \mathbf{H}_u\mathbf{T}_u & \cdots & \mathbf{H}_u\mathbf{T}_U \\ & \ddots & & \ddots & \\ \mathbf{H}_U\mathbf{T}_1 & \cdots & \mathbf{H}_U\mathbf{T}_u & \cdots & \mathbf{H}_U\mathbf{T}_U \end{pmatrix} \begin{pmatrix} \mathbf{c}_1 \\ \vdots \\ \mathbf{c}_u \\ \vdots \\ \mathbf{c}_U \end{pmatrix} + \begin{pmatrix} \mathbf{n}_1 \\ \vdots \\ \mathbf{n}_u \\ \vdots \\ \mathbf{n}_U \end{pmatrix} \tag{20.31}$$

Apparently, $\mathbf{H}\mathbf{T} \in \mathbb{C}^{U \times U}$ is a block matrix composed of sub-matrices or blocks $\mathbf{H}_i\mathbf{T}_j \in \mathbb{C}^{N_i \times \zeta_j}$.

Cancellation of the Multi-user Interference

Regarding (20.31) the multi-user interference is given by the blocks $\mathbf{H}_i\mathbf{T}_j$ with unequal indices $i \neq j$. If we can find precoding matrices, which fulfill the condition

$$\mathbf{H}_i\mathbf{T}_j = \mathbf{0} \; ; \; i, j = 1, 2, ..., U \; ; \; i \neq j \tag{20.32}$$

where $\mathbf{0}$ is the null matrix, then (20.31) results in

$$\begin{pmatrix} \mathbf{r}_1 \\ \vdots \\ \mathbf{r}_u \\ \vdots \\ \mathbf{r}_U \end{pmatrix} = \begin{pmatrix} \mathbf{H}_1\mathbf{T}_1 & \cdots & \mathbf{0} & \cdots & \mathbf{0} \\ & \ddots & & \ddots & \\ \mathbf{0} & \cdots & \mathbf{H}_u\mathbf{T}_u & \cdots & \mathbf{0} \\ & \ddots & & \ddots & \\ \mathbf{0} & \cdots & \mathbf{0} & \cdots & \mathbf{H}_U\mathbf{T}_U \end{pmatrix} \begin{pmatrix} \mathbf{c}_1 \\ \vdots \\ \mathbf{c}_u \\ \vdots \\ \mathbf{c}_U \end{pmatrix} + \begin{pmatrix} \mathbf{n}_1 \\ \vdots \\ \mathbf{n}_u \\ \vdots \\ \mathbf{n}_U \end{pmatrix} \tag{20.33}$$

Fig. 20.3 Multi-user MIMO downlink decoupled into U parallel single-user links with block diagonalization. The precoding matrices are given by $\mathbf{T}_u = \mathbf{V}_u \mathbf{A}_u$, $u = 1, 2, ..., U$ in (20.41)

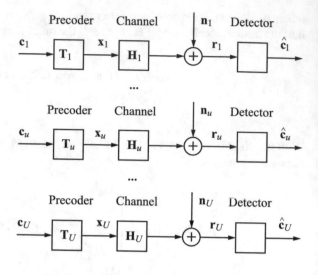

and $\mathbf{HT}$ becomes a block diagonal matrix of the type (20.19). Of course, for the precoding matrices $\mathbf{T}_j \neq \mathbf{0} \, \forall \, j$ must hold, otherwise the transmit signals will be zero. The condition (20.32) turns the multi-user downlink into a parallel structure of decoupled single-user MIMO links

$$\mathbf{r}_u = \mathbf{H}_u \mathbf{T}_u \mathbf{c}_u + \mathbf{n}_u \, , \quad u = 1, 2, ..., U \tag{20.34}$$

as depicted in Fig. 20.3. Then

$$\mathbf{r} = \text{diag}\,(\mathbf{H}_1\mathbf{T}_1, ..., \mathbf{H}_u\mathbf{T}_u, ..., \mathbf{H}_U\mathbf{T}_U)\,\mathbf{c} + \mathbf{n} \tag{20.35}$$

looks similar to an eigenmode transmission scheme. The interference defined in (20.29) is zero for arbitrary transmit symbol vectors $\mathbf{c}_1, ..., \mathbf{c}_U$. The interesting question now is how the precoder matrices can be found.

Determining the Precoder Matrices for Block Diagonalization

In principle, we have to look for a linear mapping with a matrix $\mathbf{T}$, which transforms $\mathbf{H}$ into the block diagonal form. For the multi-user MIMO downlink the solution was first described in [1, 3–5]. We restrict ourselves to the principal computational steps. The precoding matrix $\mathbf{T}_u$ for user u is the solution of the system of equations (20.32) and formally given by

$$\mathbf{T}_u = \arg \begin{bmatrix} \mathbf{H}_1\mathbf{T}_u = \mathbf{0} \\ \vdots \\ \mathbf{H}_{u-1}\mathbf{T}_u = \mathbf{0} \\ \mathbf{H}_{u+1}\mathbf{T}_u = \mathbf{0} \\ \vdots \\ \mathbf{H}_U\mathbf{T}_u = \mathbf{0} \end{bmatrix} \tag{20.36}$$

From $\mathbf{H}_i \mathbf{T}_u = \mathbf{0}$, $i = 1, 2, ..., U$, $i \neq u$ follows that each column vector of $\mathbf{T}_u$ must be orthogonal to the corresponding row vectors of all $\mathbf{H}_i$, $i \neq u$. The solution is obtained from a singular value decomposition of the matrix $\mathbf{H}_u^-$, which contains all sub-channel matrices except $\mathbf{H}_u$

$$
\mathbf{H}_u^- = \begin{pmatrix} \mathbf{H}_1 \\ \vdots \\ \mathbf{H}_{u-1} \\ \mathbf{H}_{u+1} \\ \vdots \\ \mathbf{H}_U \end{pmatrix} = \mathbf{U}_u \mathbf{D}_u \left(\tilde{\mathbf{V}}_u \ \mathbf{V}_u \right)^H \ ; \ \mathbf{D}_u = \begin{pmatrix} \Lambda_u^{\frac{1}{2}} & \mathbf{0} \\ \mathbf{0} & \mathbf{0} \end{pmatrix} \tag{20.37}
$$

With $\mathbf{H}_u \in \mathbb{C}^{N_u \times M}$ we conclude that $\mathbf{H}_u^- \in \mathbb{C}^{(N-N_u) \times M}$ and $\mathbf{D}_u \in \mathbb{R}^{(N-N_u) \times M}$ hold. Suppose rank $\left[\mathbf{H}_u^- \right] = L_u$, then the diagonal matrix $\Lambda_u^{\frac{1}{2}}$ contains the L_u nonzero singular values of $\mathbf{H}_u^-$ and $\mathbf{V}_u$ owns the dimension $M \times (M - L_u)$. The matrix $\left(\tilde{\mathbf{V}}_u \ \mathbf{V}_u \right)$ contains the right hand singular vectors of $\mathbf{H}_u^-$, where $\mathbf{V}_u$ assembles the $M - L_u$ vectors, which form an orthogonal basis for the null space of $\mathbf{H}_u^-$. The column vectors of $\mathbf{T}_u$ must be located in this null space. Consequently, the precoder can be determined by

$$
\mathbf{T}_u = \mathbf{V}'_u \tag{20.38}
$$

composed of the first ζ_u eigenvectors of $\mathbf{V}_u$. The described procedure has to be executed for all $\mathbf{T}_u$, $u = 1, 2, ..., U$. The receive vector for the user u is obtained by inserting (20.38) into (20.34) resulting in

$$
\mathbf{r}_u = \mathbf{H}_u \mathbf{V}'_u \mathbf{c}_u + \mathbf{n}_u \ , \ u = 1, 2, ..., U \tag{20.39}
$$

The described solution is valid for the following sufficient condition on the number of transmit and receive antennas, [5]

$$
M > \max [\sum_{\substack{i = 1 \\ i \neq u}}^{U} N_i \ ; \ u = 1, 2, ..., U] \tag{20.40}
$$

To obtain the symbol vector $\mathbf{c}_u$ from (20.39) various methods known from single-user MIMO transmission can be applied, such as the linear precoding techniques with a matrix $\mathbf{A}_u$ described in Chap. 18 yielding the total precoder matrix

$$
\mathbf{T}_u = \mathbf{V}'_u \mathbf{A}_u \ , \ u = 1, 2, ..., U \tag{20.41}
$$

The decoupling of the single-user links is not affected, because (20.32) guarantees also $\mathbf{H}_i \mathbf{T}_j \mathbf{A}_j = \mathbf{0}$, $i \neq j$. Then we obtain the receive signal vector of user u from (20.34)

$$\mathbf{r}_u = \mathbf{H}_u \mathbf{V}'_u \mathbf{A}_u \mathbf{c}_u + \mathbf{n}_u \ , \quad u = 1, 2, ..., U \tag{20.42}$$

Example 3 addresses all numerical steps in quite some detail. The matrix $\mathbf{A}_u$ can be determined in different ways also to maximize the system capacity by transmit power loading using the water filling algorithm.

Finally, it should be mentioned that the channel inversion method described in the previous section can be applied as well for the multi-user MIMO scheme depicted in Fig. 20.2. The precoder is then defined by the matrix $\mathbf{T}$ in (20.10), where the channel matrix is given by (20.26).

20.2.3 Alternative Multi-user MIMO Precoding

As we have seen, block diagonalization can completely remove the multi-user interference. However, the system capacity is not taken into account with this procedure. Therefore, alternative approaches try to find a good compromise between low interference and high capacity. This could be done by just minimizing the multi-user interference rather than canceling it. Then the resulting matrix in (20.33) still shows some nonzero off-diagonal elements. Hence, the U single-user links are not anymore decoupled precisely and show some "leakage" to the other links. Several approaches have been investigated, such as precoding with minimum mean squared interference or successive interference cancellation to minimize this leakage. In addition to the precoder the receiver terminals can also be equipped with a filter and both filters are adapted in a closed control loop between the base station and the user terminal. These methods also have the potential to relax the stringent conditions (20.11) and (20.40), which require an increase of the total number of transmit antennas M, if the number of users and/or user antennas is elevated in the system.

As outlined in Chap. 18, zero-forcing precoding is designed on the basis of low noise at the receiver. Therefore precoders have been investigated, which use the matrix $\mathbf{H}^H \left(\mathbf{HH}^H + \alpha \mathbf{I}_N \right)^{-1}$ known from the single-user minimum mean squared error (MMSE) receiver, and this modification is called "regularized channel inversion". The parameter α is used to maximize the signal-to-interference-plus-noise ratio at the receiver, [1].

In Chap. 18 we have discussed that the output of a linear precoder can generate signals $\mathbf{s}$ with large magnitudes, in particular if the channel matrix is ill conditioned. This was recognized as a general drawback of the linear precoding methods. A remedy is achieved by the introduction of nonlinear operations, such as the dirty paper precoder or the Tomlinson–Harashima precoder, which is able to reduce the transmit signal level by modulo operations on the input symbols, e.g., [1, 6]. A challenge for all precoding techniques is also the precise channel estimation combined with an effective provision of significant channel parameters for the transmitter via a feedback channel or adequate reciprocity conditions of the downlink and uplink channels.

20.3 Beamforming for Multi-user Downlink

Multi-user transmit beamforming can be considered as a special case of the previously discussed multi-user precoding in the downlink. The block diagram in Fig. 20.2 still holds in principle with some changes. The precoder matrix is reduced to just a beamforming vector

$$\mathbf{T}_u = \mathbf{t}_u \in \mathbb{C}^{M \times 1} \; ; \; u = 1, 2, ..., U \tag{20.43}$$

The base station is still equipped with M transmit antennas, but only one symbol stream per user is allocated. Thus, $\zeta_u = 1$ holds and the symbol vector per user u is

$$\mathbf{c}_u = c_u \; ; \; u = 1, 2, ..., U \tag{20.44}$$

containing only one symbol. Although the first multi-user beamforming schemes started with only one receive antenna per user, each user terminal can still maintain N_u antennas for a more general case, where the total number is given by (20.2). The individual channel from the M transmit antennas to the user terminal u is $\mathbf{H}_u$ and (20.26) holds for the total channel matrix $\mathbf{H}$. From (20.22) follows with (20.43) and (20.44) the transmit signal vector

$$\mathbf{s} = \sum_{u=1}^{U} \mathbf{t}_u c_u \tag{20.45}$$

which represents the superposition of all symbols c_u of the users weighted by the different precoding vectors. The receive signal of user u is obtained from (20.28)

$$\mathbf{r}_u = \mathbf{H}_u \mathbf{t}_u c_u + \mathbf{H}_u \sum_{\substack{i=1 \\ i \neq u}}^{U} \mathbf{t}_i c_i + \mathbf{n}_u \; ; \; u = 1, 2, ..., U \tag{20.46}$$

where the first term on the right hand side is the desired signal $\mathbf{H}_u \mathbf{t}_u c_u$ and the second one characterizes the multi-user interference. With the same methods as discussed in Sect. 20.2.2 the multi-user interference can be completely removed. Again, the individual channel matrices $\mathbf{H}_u$ must be known at the base station via a feedback link from each user terminal, where the channel matrix is estimated.

   Actually, beamforming with vectors was the first precoding technique and was later extended to full precoding matrices. Consequently, the comments given in the previous sections apply. In particular the condition (20.40) on the number of antennas and users still holds for perfect cancellation. Again, to overcome this stringent requirement schemes have been investigated for which the interference is only minimized rather than canceled using different minimization criteria, such as the minimal squared interference power or the maximization of the signal-to-leakage-power ratio, [7, 8]. The beamforming method allows an illustrative explanation, which is discussed in the following example.

Example 2

Consider a multi-user downlink with beamforming vectors $t_u \in \mathbb{C}^{M \times 1}$; $u = 1, 2, ..., U$. The user terminals are equipped with N_u receive antennas, $u = 1, 2, ..., U$. The noise at the receivers shall be zero. Determine the transmit signal vector, if only one user u is active and all other user symbols are zero, $c_j = 0 \forall j \neq u$. Find the receive signal of user u.

Solution:

From (20.45) follows $s = t_u c_u$. The user u receives the signal vector $r_u = H_u t_u c_u$, which is obtained from (20.46). As is well known, a matrix changes the magnitude and the direction of a vector after multiplication. On the other hand, by selection of t_u the direction of the receive vector r_u can be adjusted for a given channel matrix H_u and we are able to steer the beam of the electro-magnetic wave in the direction of the user u. As a consequence, the impact for the remaining users is significantly reduced. If a second user signal c_j is active, the corresponding beamforming vector should be aligned in such a way that the receive vector r_j is almost orthogonal to r_u, which results in minimal interference. In summary, beamforming can increase the transmission quality in a multi-user downlink.

Example 3

With this exercise we investigate the block diagonalization described in Sect. 20.2.2 in quite some detail. Consider the multi-user downlink scheme in Fig. 20.2 and Example 1 with the broadcast channel matrix (20.16). User 1 shall be served with two QAM symbols per symbol vector. Thus, $\zeta_1 = 2$ and owing to (20.20) at least $N_1 = 2$ receive antennas are required. For user 2 and 3 one symbol per symbol vector has to be transmitted, hence $\zeta_2 = \zeta_3 = 1$ and $N_2 = N_3 = 1$ are given, respectively. The base station is equipped with $M = 5$ antennas. The channel matrix in (20.16) is decomposed into horizontal stripes as in (20.26), which indicate the individual channels from the M transmit to the dedicated receive antennas of each user terminal

$$H_1 = \begin{pmatrix} 1 & 0.2 & -0.1 & 0.1 & 0 \\ -0.2 & 1 & 0.1 & 0 & 0.1 \end{pmatrix} \tag{20.47}$$

$$H_2 = \begin{pmatrix} 0.1 & -0.1 & 1 & 0.2 & -0.1 \end{pmatrix} \tag{20.48}$$

$$H_3 = \begin{pmatrix} 0 & -0.1 & -0.2 & 1 & 0.2 \end{pmatrix} \tag{20.49}$$

The multi-user interference shall be canceled by individual precoders with the matrices T_1, T_2, and T_3 using the method of block diagonalization.

(a) Find the precoder matrices.
(b) Determine the receive signals. The noise shall be neglected.
(c) Reduce the inter-channel interference, if any, with an additional zero-forcing precoder.

Solution:

(a)

Calculation of $\mathbf{T}_1$

According to (20.37) we determine

$$\mathbf{H}_1^- = \begin{pmatrix} \mathbf{H}_2 \\ \mathbf{H}_3 \end{pmatrix} = \begin{pmatrix} 0.1 & -0.1 & 1 & 0.2 & -0.1 \\ 0 & -0.1 & -0.2 & 1 & 0.2 \end{pmatrix} \tag{20.50}$$

For the singular value decomposition a computer program is used yielding

$$\mathbf{H}_1^- = \mathbf{U}_1 \underbrace{\begin{pmatrix} 1.05 & 0 & 0\,0\,0 \\ 0 & 1.03 & 0\,0\,0 \end{pmatrix}}_{\mathbf{D}_1} \underbrace{\begin{pmatrix} \triangle & \triangle & -0.98 & -0.11 & 0.11 \\ \triangle & \triangle & -0.97 & 0.97 & 0.18 \\ \triangle & \triangle & 0.10 & 0.06 & 0.13 \\ \triangle & \triangle & -0.01 & 0.14 & -0.15 \\ \triangle & \triangle & 0.02 & -0.16 & 0.96 \end{pmatrix}^H}_{\left(\tilde{\mathbf{V}}_1\ \mathbf{V}_1\right)^H} \tag{20.51}$$

$\mathbf{D}_1$ incorporates two nonzero singular values. Therefore $\mathbf{H}_1^-$ has full rank $L_1 = 2$. The $M - L_1 = 3$ vectors on the right of $\left(\tilde{\mathbf{V}}_1\ \mathbf{V}_1\right)$ determine the matrix $\mathbf{V}_1$. Entries indicated by $\triangle$ belong to $\tilde{\mathbf{V}}_1$ and are of no interest. Also the matrix $\mathbf{U}_1$ is not needed in detail. Because $\zeta_1 = 2$, we only require two of the three vectors in $\mathbf{V}_1$ taken from the left to determine $\mathbf{T}_1$,

$$\mathbf{T}_1 = \begin{pmatrix} -0.98 & -0.11 \\ -0.97 & 0.97 \\ 0.10 & 0.06 \\ -0.01 & 0.14 \\ 0.02 & -0.16 \end{pmatrix} \tag{20.52}$$

We easily calculate the interference terms

$$\mathbf{H}_1\mathbf{T}_1 = \begin{pmatrix} -1.19 & 0.09 \\ -0.76 & 0.98 \end{pmatrix} \ ; \ \mathbf{H}_2\mathbf{T}_1 = \begin{pmatrix} 0.1 & 0 \end{pmatrix} \ ; \ \mathbf{H}_3\mathbf{T}_1 = \begin{pmatrix} 0.07 & 0 \end{pmatrix} \tag{20.53}$$

Calculation of $\mathbf{T}_2$

$$\mathbf{H}_2^- = \begin{pmatrix} \mathbf{H}_1 \\ \mathbf{H}_3 \end{pmatrix} = \begin{pmatrix} 1 & 0.2 & -0.1 & 0.1 & 0 \\ -0.2 & 1 & 0.1 & 0 & 0.1 \\ 0 & -0.1 & -0.2 & 1 & 0.2 \end{pmatrix} \tag{20.54}$$

$$\mathbf{H}_2^- = \mathbf{U}_2 \underbrace{\begin{pmatrix} 1.11 & 0 & 0 & 0\,0 \\ 0 & 1.02 & 0 & 0\,0 \\ 0 & 0 & 0.97 & 0\,0 \end{pmatrix}}_{\mathbf{D}_2} \underbrace{\begin{pmatrix} \triangle & \triangle & \triangle & 0.09 & 0.06 \\ \triangle & \triangle & \triangle & -0.06 & -0.10 \\ \triangle & \triangle & \triangle & 0.96 & 0.20 \\ \triangle & \triangle & \triangle & 0.22 & -0.16 \\ \triangle & \triangle & \triangle & -0.17 & 0.96 \end{pmatrix}^{H}}_{\left(\tilde{\mathbf{V}}_2 \ \mathbf{V}_2\right)^{H}} \tag{20.55}$$

Three singular values are unequal to zero. Therefore $\mathbf{H}_2^-$ has full rank $L_2 = 3$. Consequently, the $M - L_2 = 2$ vectors on the right of $\left(\tilde{\mathbf{V}}_2 \ \mathbf{V}_2\right)$ determine the matrix $\mathbf{V}_2$. Because $\zeta_2 = 1$, we just require one vector of $\mathbf{V}_2$ taken from the left to compose $\mathbf{T}_2$,

$$\mathbf{T}_2 = \begin{pmatrix} 0.09 \\ -0.06 \\ 0.96 \\ 0.22 \\ -0.17 \end{pmatrix} \tag{20.56}$$

Then we obtain

$$\mathbf{H}_1 \mathbf{T}_2 = \begin{pmatrix} 0 \\ 0 \end{pmatrix} \ ; \ \mathbf{H}_2 \mathbf{T}_2 = 1.04 \,; \ \mathbf{H}_3 \mathbf{T}_2 = 0 \tag{20.57}$$

Calculation of $\mathbf{T}_3$

$$\mathbf{H}_3^- = \begin{pmatrix} \mathbf{H}_1 \\ \mathbf{H}_2 \end{pmatrix} = \begin{pmatrix} 1 & 0.2 & -0.1 & 0.1 & 0 \\ -0.2 & 1 & 0.1 & 0 & 0.1 \\ 0.1 & -0.1 & 1 & 0.2 & -0.1 \end{pmatrix} \tag{20.58}$$

$$\mathbf{H}_3^- = \mathbf{U}_3 \underbrace{\begin{pmatrix} 1.05 & 0 & 0 & 0\,0 \\ 0 & 1.03 & 0 & 0\,0 \\ 0 & 0 & 1.02 & 0\,0 \end{pmatrix}}_{\mathbf{D}_3} \underbrace{\begin{pmatrix} \triangle & \triangle & \triangle & -0.11 & 0.03 \\ \triangle & \triangle & \triangle & -0.01 & -0.10 \\ \triangle & \triangle & \triangle & -0.18 & 0.08 \\ \triangle & \triangle & \triangle & 0.98 & 0.01 \\ \triangle & \triangle & \triangle & 0.01 & 0.99 \end{pmatrix}^{H}}_{\left(\tilde{\mathbf{V}}_3 \ \mathbf{V}_3\right)^{H}} \tag{20.59}$$

Three singular values are unequal to zero. Hence, $\mathbf{H}_3^-$ has full rank $L_3 = 3$. The $M - L_3 = 2$ vectors on the right of $\left(\tilde{\mathbf{V}}_3 \ \mathbf{V}_3\right)$ determine the matrix $\mathbf{V}_3$. Because $\zeta_3 = 1$, we just require one vector of $\mathbf{V}_3$ taken from the left and get

$$\mathbf{T}_3 = \begin{pmatrix} -0.11 \\ -0.01 \\ -0.18 \\ 0.98 \\ 0.01 \end{pmatrix} \tag{20.60}$$

We easily calculate

$$\mathbf{H}_1\mathbf{T}_3 = \begin{pmatrix} 0 \\ 0 \end{pmatrix} \; ; \; \mathbf{H}_2\mathbf{T}_3 = 0 \; ; \; \mathbf{H}_3\mathbf{T}_3 = 1.02 \tag{20.61}$$

(b)
Receive vectors
We insert the numerical results into (20.28) and obtain

$$\begin{pmatrix} \mathbf{r}_1 \\ r_2 \\ r_3 \end{pmatrix} = \left(\begin{pmatrix} -1.19\ 0.09 \\ -0.76\ 0.98 \end{pmatrix} \begin{pmatrix} 0 \\ 0 \end{pmatrix} \begin{pmatrix} 0 \\ 0 \end{pmatrix} \right) \begin{pmatrix} \mathbf{c}_1 \\ c_2 \\ c_3 \end{pmatrix} \tag{20.62}$$

We recognize the block diagonal structure of the combined system matrix. For r_2 and r_3 marginal multi-user interference is present owing to numerical errors. Nevertheless, we clearly see that the precoders $\mathbf{T}_1$, $\mathbf{T}_2$, and $\mathbf{T}_3$ turn the multi-user downlink into three single-user links, which are almost completely decoupled.
(c)
The receive signal $\mathbf{r}_1$ still contains inter channel interference caused by the matrix

$$\mathbf{G}_{11} = \begin{pmatrix} -1.19\ 0.09 \\ -0.76\ 0.98 \end{pmatrix} \tag{20.63}$$

As $\mathbf{G}_{11}$ is a square matrix with full rank 2, we can employ the inverse as an additional precoder in the zero-forcing sense

$$\mathbf{A}_1 = \mathbf{G}_{11}^{-1} = \begin{pmatrix} -0.89\ 0.08 \\ -0.69\ 1.08 \end{pmatrix} \tag{20.64}$$

and the combined precoding matrix for link 1 then is

$$\mathbf{T}_1^{ZF} = \mathbf{T}_1\mathbf{A}_1 = \begin{pmatrix} 0.95 & -0.2 \\ 0.19 & 0.97 \\ -0.13 & 0.07 \\ -0.09 & 0.15 \\ 0.09 & -0,17 \end{pmatrix} \tag{20.65}$$

The new multi-user interference terms associated with $\mathbf{T}_1^{ZF}$ then are

$$\mathbf{H}_1\mathbf{T}_1^{ZF} = \begin{pmatrix} 1\ 0 \\ 0\ 1 \end{pmatrix} \; ; \; \mathbf{H}_2\mathbf{T}_1^{ZF} = (0.09\ 0) \; ; \; \mathbf{H}_3\mathbf{T}_1^{ZF} = (0\ 0) \tag{20.66}$$

Furthermore, we use $\frac{1}{1.04}\mathbf{T}_2$ instead of $\mathbf{T}_2$ as well as $\frac{1}{1.02}\mathbf{T}_3$ instead of $\mathbf{T}_3$ and obtain from (20.62) finally

$$\begin{pmatrix} r_1 \\ r_2 \\ r_3 \end{pmatrix} = \left(\begin{pmatrix} 1 & 0 \\ 0 & 1 \\ (0.09\ 0) \\ (0\ 0) \end{pmatrix} \begin{pmatrix} 0 \\ 0 \\ 1 \\ 0 \end{pmatrix} \begin{pmatrix} 0 \\ 0 \\ 0 \\ 1 \end{pmatrix} \right) \begin{pmatrix} c_1 \\ c_2 \\ c_3 \end{pmatrix} \tag{20.67}$$

showing that just a minor distortion of 0.09 remains in r_2.

20.4 Principles of Multi-user MIMO Uplink Transmission

20.4.1 System Model of the Uplink

In this section we consider the uplink transmission from U user terminals to the base station. The block diagram is depicted in Fig. 20.4. The scenario is similar to Fig. 20.2, however, the transmission signals flow in the reverse direction from the right to the left. The user terminal u is equipped with M_u transmit antennas and allocates the QAM symbol vector $\mathbf{c}_u \in \mathbb{C}^{M_u \times 1}$ to the transmit signal vector $\mathbf{s}_u \in \mathbb{C}^{M_u \times 1}$, where $u = 1, 2, ..., U$. The combined transmit signal vector $\mathbf{s}$ is defined in (20.5).

The uplink channel is characterized by the channel matrix

$$\mathbf{H} = \begin{pmatrix} \mathbf{H}_1 & \cdots & \mathbf{H}_u & \cdots & \mathbf{H}_U \end{pmatrix} \in \mathbb{C}^{N \times M} \tag{20.68}$$

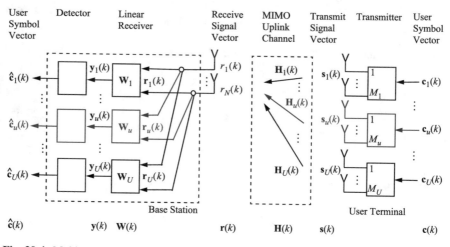

Fig. 20.4 Multi-user MIMO uplink transmission between individual user terminals and the base station

which is separated into horizontal blocks $\mathbf{H}_u \in \mathbb{C}^{N \times M_u}$, where $\mathbf{H}_u$ determines the uplink channel from the output of the M_u terminal antennas of the user u to the N receive antennas of the base station. Please note that we use the same notation for $\mathbf{H}$ and $\mathbf{H}_u$ as in the case of the downlink, although different in general. The total number of transmit antennas of all the users is M and defined in (20.4). The receive signal vector $\mathbf{r} \in \mathbb{C}^{N \times 1}$ is given by

$$\mathbf{r} = \begin{pmatrix} r_1 & r_2 & \cdots & r_N \end{pmatrix}^T \tag{20.69}$$

At the base station we apply linear processing using a bank of receive filters $\mathbf{W}_1, ..., \mathbf{W}_U$, where one filter is associated to each user. Thus, we abandon a pre-filter at each user terminal to keep the amount of the hardware, the software and the power consumption of cost-conscious mobile terminals low. A higher complexity of the base station is adequate owing to the fact that its cost is shared among all users in the network. $\mathbf{r}$ is also the input of the receive filters with matrix $\mathbf{W}_u \in \mathbb{C}^{M_u \times N}$ generating the output signal vector

$$\mathbf{y}_u \in \mathbb{C}^{M_u \times 1} \tag{20.70}$$

where $u = 1, 2, ..., U$. Each receive filter is followed by a signal detector, which outputs an estimate $\hat{\mathbf{c}}_u \in \mathbb{C}^{M_u \times 1}$ of the transmit symbol vector $\mathbf{c}_u$.

20.4.2 Receive Signal at the Base Station

From Fig. 20.4 we determine the receive signal as

$$\mathbf{r} = \mathbf{H}_u \mathbf{s}_u + \sum_{\substack{i = 1 \\ i \neq u}}^{U} \mathbf{H}_i \mathbf{s}_i + \mathbf{n} \tag{20.71}$$

and recognize the desired signal $\mathbf{H}_u \mathbf{s}_u$ for user u, which is corrupted by the multi-user interference given by the second term on the right hand side and the noise $\mathbf{n}$ at the base station. Please note that according to Fig. 20.4 $\mathbf{r} = \mathbf{r}_1 = \mathbf{r}_2 = \cdots = \mathbf{r}_U$ holds. The similarity with the downlink in (20.28) needs no explanation. In principle, the base station has to perform a multi-user detection.

20.4.3 Zero-Forcing Receiver for Multi-user Uplink Interference Reduction

The filter output signal for user u at the base station is

$$\mathbf{y}_u = \mathbf{W}_u \mathbf{r} = \mathbf{W}_u \mathbf{H}_u \mathbf{s}_u + \mathbf{W}_u \sum_{\substack{i=1 \\ i \neq u}}^{U} \mathbf{H}_i \mathbf{s}_i + \mathbf{W}_u \mathbf{n} \qquad (20.72)$$

The multi-user interference can be written with matrix notation as

$$\mathbf{W}_u \sum_{\substack{i=1 \\ i \neq u}}^{U} \mathbf{H}_i \mathbf{s}_i = \mathbf{W}_u \mathbf{B}_u \mathbf{b}_u \qquad (20.73)$$

with

$$\mathbf{B}_u = (\mathbf{H}_1 \cdots \mathbf{H}_{u-1} \mathbf{H}_{u+1} \cdots \mathbf{H}_U) \ ; \ \mathbf{b}_u = (\mathbf{s}_1 \cdots \mathbf{s}_{u-1} \mathbf{s}_{u+1} \cdots \mathbf{s}_U)^T \qquad (20.74)$$

Below we prove that the zero-forcing receive filter

$$\mathbf{W}_u^{ZF} = \mathbf{H}_u^H \left(\mathbf{I}_N - \mathbf{B}_u \mathbf{B}_u^+ \right) \qquad (20.75)$$

can completely remove the multi-user interference independently of $\mathbf{s}$, where

$$\mathbf{B}_u^+ = \left(\mathbf{B}_u^H \mathbf{B}_u \right)^{-1} \mathbf{B}_u^H \qquad (20.76)$$

is the pseudo inverse matrix of $\mathbf{B}_u$. Please note, $\mathbf{B}_u^H \mathbf{B}_u$ must have full rank, otherwise its inverse does not exist. Under this prerequisite

$$\mathbf{W}_u^{ZF} \mathbf{B}_u = \mathbf{H}_u^H \left(\mathbf{I}_N - \mathbf{B}_u \mathbf{B}_u^+ \right) \mathbf{B}_u = \mathbf{0} \qquad (20.77)$$

holds and we obtain the filter output signal

$$\mathbf{y}_u = \mathbf{W}_u^{ZF} \mathbf{H}_u \mathbf{s}_u + \mathbf{W}_u^{ZF} \mathbf{n} \ ; \ u = 1, 2, ..., U \qquad (20.78)$$

Apparently, the multi-user MIMO uplink turns into U decoupled single-user MIMO links defined by (20.78). $\mathbf{y}_u$ still suffers some inter-channel interference given by $\mathbf{W}_u^{ZF} \mathbf{H}_u$. This impairment can be reduced by conventional single-user design methods, such as an additional zero-forcing or MMSE filter, which follows the $\mathbf{W}_u^{ZF}$ filter and both can be combined to one unit. The last stage in the uplink processing in Fig. 20.4 is the decision device, which will mostly be a single-user maximum likelihood detector.

Motivated by the fact that the multi-user MIMO uplink can be turned into a parallel arrangement of single-user connections opens up several alternative methods, which try to find a good compromise between interference and noise reduction, in principle. Hence, a target is to maximize the signal-to-interference-plus-noise ratio at the receiver. Linear methods furnish a minimum mean squared error (MMSE). Then a maximum likelihood detection per user can follow. Other approaches employ additional precoders at the transmitter, which are adapted jointly with the receive filters. Depending on the statistics of the noise and the interference an optimal solution is the joint multi-user maximum likelihood detection, yet expensive owing to the complexity, which grows exponentially as a function of the system parameters.

Proof of (20.77)
With (20.75) and (20.76) follows from the left hand side of (20.77)
$$\mathbf{W}_u^{ZF} \mathbf{B}_u = \mathbf{H}_u^H \left(\mathbf{B}_u - \mathbf{B}_u \left(\mathbf{B}_u^H \mathbf{B}_u \right)^{-1} \mathbf{B}_u^H \mathbf{B}_u \right) = \mathbf{H}_u^H \left(\mathbf{B}_u - \mathbf{B}_u \right) = \mathbf{0}$$ and the proof is finished.

20.5 Outlook: Massive MIMO for Multi-user Applications

The ever increasing demand for high speed communications has motivated the research on methods to further increase the system capacity or spectral efficiency measured in bit/s per Hz bandwidth of a wireless network. A cell is composed of a base station with M antenna elements and U user equipment with in total N antenna elements. Marzetta [9] investigated theoretically the effect of a drastic increase of the number of base station antennas, $M \to \infty$, yet his findings are also applicable for $M < \infty$. He showed the large potential, which was then verified by computer simulation and with first test beds, [10–13]. With the term *massive MIMO* or *large MIMO* we characterize a multi-user MIMO scheme with a very large number of antennas at the base station and in total at the user equipment side, typically more than one hundred. Existing antenna masts for the 4G cellular networks carry already four MIMO antennas with about twenty antenna element arrays each. There are also ideas to integrate a large number of planar antennas into the facade of buildings or into the wallpaper in the in-house area. Research and development are ongoing in this field. Of course, the hardware complexity as well as the cost play an important part. In the following we just give a survey on the principles.

First, we consider the downlink scenario in Fig. 20.2 with M transmit antennas at the base station and U user equipment. The total number N of antennas at the user side is defined in (20.2). To determine the system capacity we can apply (17.57) in Sect. 17.2 and find $C = \sum_{i=1}^{P} \log_2 \left(1 + \frac{\lambda_i E_S}{\sigma_n^2} \right)$, where $P \leq \min\{M, N\}$ is the rank of the channel matrix in (20.26). Assume $M > N$ and a channel matrix $\mathbf{H}$ with full rank N. Then follows $P = N$ and the system capacity for the cell

$$C = \sum_{i=1}^{N} \log_2 \left(1 + \frac{\lambda_i E_S}{\sigma_n^2} \right) \tag{20.79}$$

increases with the total number N of user equipment antennas and thus with the number U of users. We observe that M must ascend as well owing to $M > N$. Hence, with a massive MIMO antenna at the base station we can serve a large number of users and achieve a high system capacity. If the base station operates with beamforming, narrow beams between the base station and the individual user equipment can be generated showing very small overlap. This also accounts for a higher energy efficiency and transmission quality.

The situation is similar in the case of an uplink transmission depicted in principle in Fig. 20.4. Please note that M and N change their role according to our definitions, where the total number M of antennas at the user side is then given by (20.4) and N is the number of base station antennas. "Asymptotic favorable transmission" has been observed meaning that $\frac{1}{N}\mathbf{h}_u^H \mathbf{h}_j$, $u \neq j$ tends to zero for very large N in the case of a Rayleigh fading and a line of sight channel. $\mathbf{h}_u = \mathbf{H}_u \in \mathbb{C}^{N \times 1}$ are the vectors of the channel matrix $\mathbf{H}$ in (20.68), if only one antenna per user equipment is present. The property $\frac{1}{N}\mathbf{h}_u^H \mathbf{h}_j \approx 0$ can be used by a maximum ratio combiner at the base station with the matrix $\mathbf{W}_u = \mathbf{h}_u^H$, $u = 1, 2, ..., U$ to minimize the multi-user interference.

In the research and development field on massive MIMO for multi-user systems work is ongoing to circumvent some obstacles, such as pilot contamination, hardware complexity and channel hardening. Pilot contamination can be present in a multi-cell system, in which the pilot signal of a user is reused in neighboring cells for other users, in cases where not sufficient orthogonal pilot signals are present. Then the users with the same pilots cannot be differentiated and thus cause impairments.

The large number of antenna elements increases the hardware complexity. Special designs for antennas and radio-frequency amplifiers are subject to intensive work. As a workaround, antenna switching techniques can be introduced and only the instantaneously active antenna is connected to an amplifier. This will reduce the number of amplifiers compared to the number of antennas.

Furthermore, asymptotic channel hardening is an observed effect, when the channel gain gets approximately constant for an increased number N of base station antennas,

$$\frac{\|\mathbf{h}_u\|^2}{\mathbf{E}\left[\|\mathbf{h}_u\|^2\right]} \to 1 \text{ for } N \to \infty \tag{20.80}$$

Thus, individual channels between the base station and the user equipment tend to become deterministic. However, many solutions for all these obstacles are in progress so that massive MIMO will become an integral part of a variety of wireless networks, [14].

References

1. Spencer, Q.H., Peel, C.B., Swindlehurst, A.L., Haardt, M.: An introduction to the multi-user MIMO downlink. IEEE Commun. Mag. (2004)
2. Khalid, F., Speidel, J.: Advances in MIMO techniques for mobile communications - a survey. Int. J. Commun., Netw. Syst. Sci. *3* (2010)
3. Choi, L.-U., Murch, R.D.: A downlink decomposition transmit preprocessing technique for multi-user MIMO systems. Proc. IST Mob. Wirel. Telecommun. Summit (2002)
4. Spencer, Q., Haardt, M.: Capacity and downlink transmission algorithms for a multi-user MIMO channel. In: Proceedings of 36th Asilomar Conference on Signals, Systems, and Computers (2002)
5. Choi, L.-U., Murch, R.D.: A transmit preprocessing technique for multiuser MIMO systems using a decomposition approach. IEEE Trans. Wirel. Commun. (2004)
6. Fischer, R., Windpassinger, C., Lampe, A., Huber, J.: Space-time transmission using Tomlinson-Harashima precoding. In: 4th ITG- Conference on Source and Channel Coding (2002)
7. Tarighat, A., Sadek, M., Sayed, A.: A multi-user beamforming scheme for downlink MIMO channels based on maximizing signal-to-leakage ratios. In: Proceedings IEEE International Conference on Accoustic, Speech, and Signal Processing (ICASSP) (2005)
8. Bjoernson, E., Bengtsson, M., Ottersten, B.: Optimal multi-user transmit beamforming: a difficult problem with a simple solution structure. IEEE Signal Proc. Mag. (2014)
9. Marzetta, T.L.: Noncooperative cellular wireless with unlimited numbers of base station antennas. IEEE Trans. Wirel. Commun. (2010)
10. Hoydis, J., ten Brink, S., Debbah, M.: Massive MIMO in the UL/DL of cellular networks: How many antennas do we need? IEEE J. Select. Areas Commun. (2013)
11. Larsson, E.G., Edfors, O., Tufvesson, F., Marzetta, T.L.: Massive MIMO for next generation wireless systems. IEEE Commun. Mag. **52** (2014)
12. Marzetta, T., Larsson, E.G., Yang, H., Ngo, H.Q.: Fundamentals of Massive MIMO. Cambridge University Press (2016)
13. Bjoernson, E., Larsson, E.G., Marzetta, T.L.: Massive MIMO: Ten myths and one critical question. IEEE Commun. Mag. (2016)
14. Bjoernson, E., Hoydis, J., Sanguinetti, L.: Massive MIMO Networks: Spectral, Energy, and Hardware Efficiency. Now Publishers (2017)

Appendix A
Some Fundamentals of Random Variables and Stochastic Processes

In the following we give a brief overview on random variables and stochastic processes. It should be understood as a summary of the most important findings, which are needed in digital communications and signal processing. For some lemmas the derivations and proofs are outlined. Beyond that the reader is referred to dedicated textbooks such as [1, 2].

A.1 Continuous Random Variables

We start with some basics on random variables. Let X be a real-valued random variable and x an event.

A.1.1 Probability Density Function and Probability

$p(x)$ is denoted as probability density function (in short density function or density) with the property $p(x) \geq 0$ and $\int_{-\infty}^{\infty} p(x)dx = 1$. We call $F(x) = \int_{-\infty}^{x} p(u)du$ the (cumulative) distribution function. $F(b)$ is the probability of the event that the random variable is located in the interval $-\infty < X < b$ and we write with the probability operator P,

$$P[X < b] = P[-\infty < X < b] = \int_{-\infty}^{b} p(x)dx \qquad (A.1)$$

from which we conclude

$$P[a < X < b] = \int_{a}^{b} p(x)dx = F(b) - F(a) \qquad (A.2)$$

© Springer Nature Switzerland AG 2021
J. Speidel, *Introduction to Digital Communications*,
Signals and Communication Technology,
https://doi.org/10.1007/978-3-030-67357-4

Please note, if the density function $p(x)$ contains no Dirac impulses at the boarders a and b, then also the equal sign holds for the interval. With $b \to \infty$ we obtain

$$P[X > a] = P[a < X < \infty] = \int_a^\infty p(x)dx \tag{A.3}$$

A.1.2 Two Random Variables

Joint Probability Density Function

For the two random variables X_1 and X_2 we define the (two-dimensional) joint probability density function $p_{12}(x_1, x_2)$. The density functions of the individual random variables X_i are $p_i(x_i)$, $i = 1, 2$ and called marginal probability density functions. They are calculated as

$$p_1(x_1) = \int_{-\infty}^\infty p_{12}(x_1, x_2)dx_2 \; ; \; p_2(x_2) = \int_{-\infty}^\infty p_{12}(x_1, x_2)dx_1 \tag{A.4}$$

Conditional Probability Density Function

We define $p_{1/2}(x_1 \mid X_2 = x_2)$ or with short hand notation $p_{1/2}(x_1 \mid x_2)$ as the conditional probability density function of X_1 under the condition $X_2 = x_2$.

$$p_{1/2}(x_1 \mid x_2) = \frac{p_{12}(x_1, x_2)}{p_2(x_2)} \tag{A.5}$$

Conditional Probabilities

With the conditional densities we can calculate conditional probabilities

$$P[X_1 < x_1 \mid X_2 = x_2] = \int_{-\infty}^{x_1} p_{1/2}(u_1 \mid x_2) \, du_1 \tag{A.6}$$

Bayes Theorem

The Bayes theorem relates the two conditional densities as

$$p_{1/2}(x_1 \mid x_2) = \frac{p_{2/1}(x_2 \mid x_1) \, p_1(x_1)}{p_2(x_2)} \;, \; p_2(x_2) \neq 0 \tag{A.7}$$

Statistical Independence of Random Variables

We call two random variables statistically independent, if and only if $p_{1/2}(x_1 \mid x_2) = p_1(x_1)$ is independent of x_2. Then follows from (A.5)

$$p_{12}(x_1, x_2) = p_1(x_1)p_2(x_2) \tag{A.8}$$

A.2 Statistical Parameters for Random Variables

A.2.1 Expected Value

The expected value (or in short expectation) of a real-valued random variable X with density function $p(x)$ is defined as

$$\mathbf{E}[X] = \int_{-\infty}^{\infty} x p(x) dx = m_x \tag{A.9}$$

$\mathbf{E}[X]$ is also called first moment or mean value of X. We see that $\mathbf{E}[...]$ is a linear operator.

A.2.2 Function of a Random Variable, nth Moment

Let $g(...)$ be a function of the random variable X yielding the new random variable $Y = g(X)$. Then

$$\mathbf{E}[Y] = \mathbf{E}\left[g(X)\right] = \int_{-\infty}^{\infty} g(x)p(x)dx \tag{A.10}$$

holds. On that basis we can define the nth *moment* of X as

$$\mathbf{E}[X^n] = \int_{-\infty}^{\infty} x^n p(x)dx \tag{A.11}$$

and in particular for $n = 2$ we obtain $\mathbf{E}[X^2]$, which is the quadratic mean and physically the *mean power* of the random variable X.

The nth *central moment* is defined as

$$\mathbf{E}\left[(X - m_x)^n\right] = \int_{-\infty}^{\infty} (x - m_x)^n \, p(x)dx \tag{A.12}$$

which yields for $n = 2$ the *variance* of X

$$\text{var}[X] = \sigma_x^2 = \mathbf{E}\left[(X - m_x)^2\right] \tag{A.13}$$

σ_x is known as *standard deviation*. It is straightforward to show that

$$\sigma_x^2 = \mathbf{E}\left[X^2\right] - m_x^2 \tag{A.14}$$

is true. If the random variable has zero mean, then the variance equals the mean power.

A.2.3 Covariance and Correlation of Two Random Variables

For two real-valued random variables X_i with $\mathbf{E}[X_i] = m_i$, the marginal density functions $p_i(x_i)$; $i = 1, 2$, and the joint density function $p_{12}(x_1, x_2)$ we define the *joint central moment* of order (k, n) as

$$\mathbf{E}\left[(X_1 - m_1)^k (X_2 - m_2)^n\right] = \int_{-\infty}^{\infty} \int_{-\infty}^{\infty} (x_1 - m_1)^k (x_2 - m_2)^n \, p_{12}(x_1, x_2) dx_1 dx_2 \tag{A.15}$$

from which follows for $k = n = 1$ the *covariance* between X_1 and X_2 as

$$\mu_{12} = \mathbf{E}[(X_1 - m_1)(X_2 - m_2)] = \int_{-\infty}^{\infty} \int_{-\infty}^{\infty} (x_1 - m_1)(x_2 - m_2) \, p_{12}(x_1, x_2) dx_1 dx_2 \tag{A.16}$$

It is straightforward to show that

$$\mu_{12} = \mathbf{E}[X_1 X_2] - m_1 m_2 \tag{A.17}$$

holds.

The *correlation coefficient* is defined as

$$\rho_{12} = \frac{\mu_{12}}{\sigma_1 \sigma_2} = \frac{\mathbf{E}[X_1 X_2] - m_1 m_2}{\sigma_1 \sigma_2} \tag{A.18}$$

Two random variables X_1 and X_2 are called *uncorrelated,* if and only if the covariance is zero

$$\mu_{12} = 0 \tag{A.19}$$

which yields $\rho_{12} = 0$[1] and finally

$$\mathbf{E}[X_1 X_2] = m_1 m_2 \tag{A.20}$$

---

[1] This is the reason why ρ_{12} is called correlation coefficient.

Consequently, if we would like to see whether two random variables are uncorrelated, we have to check, if their covariance is zero.

The *Correlation* between X_1 and X_2 is defined as

$$c_{12} = \mathbf{E}[X_1 X_2] = \int_{-\infty}^{\infty} \int_{-\infty}^{\infty} x_1 x_2 p_{12}(x_1, x_2) dx_1 dx_2 \qquad (A.21)$$

We call two random variables *orthogonal*, if

$$\mathbf{E}[X_1 X_2] = 0 \qquad (A.22)$$

Let X_1 and X_2 be two random variables each with zero mean $m_1 = m_2 = 0$. If they are uncorrelated, then their covariance (A.19) is zero and from (A.20) follows for the correlation $c_{12} = \mathbf{E}[X_1 X_2] = 0$.

Let X_1 and X_2 be *statistically independent* random variables. Then (A.8) holds and from (A.21) follows

$$\mathbf{E}[X_1 X_2] = \int_{-\infty}^{\infty} \left[\int_{-\infty}^{\infty} x_1 p_1(x_1) dx_1 \right] x_2 p_2(x_2) dx_2 = \mathbf{E}[X_1]\mathbf{E}[X_2] = m_1 m_2$$
$$(A.23)$$

which yields $\rho_{12} = 0$. Hence, we conclude that statistically independent random variables are also uncorrelated. In general, the reverse is not true.

A.3 Stochastic Processes

A.3.1 *Definition of a Stochastic Process*

For engineers a stochastic process is best explained with the help of a physical experiment. Consider a large number $N \to \infty$ of identical resistors, each resistor i generating a random noise voltage $X_i(t)$ as a function of time t, where $i = 1, 2, ..., N$. The stochastic process $X(t) = \{X_1(t), X_2(t), ..., X_N(t)\}$ represents the family also called ensemble of all voltages and $X_i(t)$ is the ith sample function or ith realization of the process. All sample functions ($i = 1, 2, ..., N$) belonging to the process have the same statistical parameters, such as probability density function, autocorrelation etc. To characterize the stochastic process statistical parameters can be defined in two ways, namely along the time axis of a dedicated sample function $X_i(t)$ or over all sample functions of $X(t)$ at a fixed time instant t_ν. Then $X(t_\nu)$ is a continuous random variable. In our measuring campaign we can further look at the stochastic process $X(t)$ at different time instants $t_1 < t_2 < t_3 < ... < t_M$ yielding a sequence of random variables

$$X(t_1), X(t_2), .., X(t_M) \qquad (A.24)$$

and the stochastic process $X(t)$ can be regarded for each fixed time instant t as a random variable. Consequently, the corresponding definitions for random variables can be applied to describe the statistical parameters of the stochastic process. The set of random variables in (A.24) is characterized by its joint probability density function

$$p_{1M}\left(x_{t_1}, x_{t_2}, ..., x_{t_M}\right) \tag{A.25}$$

where $x_{t_\nu} = x(t_\nu)$ is the short hand notation of an event of the random variable $X(t_\nu)$; $\nu = 1, 2, ..., M$. In general, the probability density function depends on the time instances $t_1, t_2, ..., t_M$, if the process is non-stationary, see Sect. A.3.4.

A.3.2 Autocovariance, Auto-, and Cross-Correlation Function

Single Stochastic Process $X(t)$

The *autocovariance function* $\mu_{xx}(t_1, t_2)$ of a stochastic process $X(t)$ is defined similar to the covariance (A.16) of two random variables $X(t_1)$ and $X(t_2)$

$$\mu_{xx}(t_1, t_2) = \mathbf{E}\left[(X(t_1) - m_x(t_1))(X(t_2) - m_x(t_2))\right] =$$
$$= \int_{-\infty}^{\infty}\int_{-\infty}^{\infty}\left(x_{t_1} - m_x(t_1)\right)\left(x_{t_2} - m_x(t_2)\right) p_{12}\left(x_{t_1}, x_{t_2}\right) dx_{t_1} dx_{t_2} \tag{A.26}$$

where $p_{12}\left(x_{t_1}, x_{t_2}\right)$ is the joint probability density function of $X(t_1)$ and $X(t_2)$.

The *expected values* are $m_x(t_i) = \mathbf{E}[X(t_i)]$; $i = 1, 2$. It is straightforward to show that

$$\mu_{xx}(t_1, t_2) = \mathbf{E}[X(t_1)X(t_2)] - m_x(t_1)m_x(t_2) \tag{A.27}$$

The *autocorrelation function* of the process $X(t)$ is defined similar to (A.21) as

$$R_{xx}(t_1, t_2) = \mathbf{E}[X(t_1)X(t_2)] = \int_{-\infty}^{\infty}\int_{-\infty}^{\infty} x_{t_1}x_{t_2}p_{12}\left(x_{t_1}, x_{t_2}\right) dx_{t_1} dx_{t_2} \tag{A.28}$$

Two Stochastic Processes $X(t)$ and $Y(t)$

We consider two stochastic processes $X(t)$ and $Y(t)$ with the corresponding random variables

$$X(t_i) ; \quad i = 1, 2, ..., M_x ; \quad t_1 < t_2 < t_3... \tag{A.29}$$

$$Y(\tilde{t}_j) ; \quad j = 1, 2, ..., M_y ; \quad \tilde{t}_1 < \tilde{t}_2 < \tilde{t}_3... \tag{A.30}$$

The joint set of random variables is characterized by the joint probability density function

$$p_{xy}\left(x_{t_1}, x_{t_2}, ..., x_{t_{M_x}}; y_{\tilde{t}_1}, y_{\tilde{t}_2}, ..., y_{\tilde{t}_{M_y}}\right) \qquad (A.31)$$

where $x_{t_\nu} = x(t_\nu)$ and $y_{\tilde{t}_\nu} = y(\tilde{t}_\nu)$. Then the *cross-correlation function* can be defined as

$$R_{xy}(t_1, t_2) = \mathbf{E}[X(t_1)Y(t_2)] = \int_{-\infty}^{\infty}\int_{-\infty}^{\infty} x_{t_1} y_{t_2} p_{xy}\left(x_{t_1}, y_{t_2}\right) dx_{t_1} dy_{t_2} \qquad (A.32)$$

where we have renamed $\tilde{t}_2$ as t_2.

A.3.3 Time-Domain Parameters and Ergodicity

As already alluded, we can define parameters for a stochastic process $X(t)$ along the time axis of a dedicated sample function $X_i(t)$ or over the ensemble at a fixed time instant t_ν, yielding $X(t_\nu)$. As a consequence, we differentiate between time-domain averages (or moments) on one hand and ensemble values also called expected values on the other hand. On the basis of a sample function $X_i(t)$, in general complex, we get the following time-domain parameters:

Mean Value

$$\bar{x} = \lim_{T_0 \to \infty} \frac{1}{2T_0} \int_{-T_0}^{T_0} x_i(t)dt \qquad (A.33)$$

Autocovariance Function

$$c_{xx}(\tau) = \lim_{T_0 \to \infty} \frac{1}{2T_0} \int_{-T_0}^{T_0} (x_i(t) - \bar{x})^* (x_i(t+\tau) - \bar{x}) dt \qquad (A.34)$$

Autocorrelation Function

$$R_{xx}(\tau) = \lim_{T_0 \to \infty} \frac{1}{2T_0} \int_{-T_0}^{T_0} x_i^*(t)x_i(t+\tau)dt \qquad (A.35)$$

For $\tau = 0$ follows the **Mean Power**

$$R_{xx}(0) = \lim_{T_0 \to \infty} \frac{1}{2T_0} \int_{-T_0}^{T_0} |x_i(t)|^2 dt \qquad (A.36)$$

Ergodicity

A wide sense stationary stochastic process (see Sect. A.3.4) is called ergodic, if all statistical parameters calculated on the basis of the ensemble and with respect to time of any sample function $X_j(t)$ are identical. Thus, an ergodic process can be statistically described by just one realization. In engineering ergodicity is often assumed as a hypothesis, because the experimental proof in many cases is difficult, although important. In the following we exclusively consider ergodic processes and focus on the ensemble values.

A.3.4 Stationary Stochastic Process

Strict Sense Stationary (SSS) Stochastic Process

- A stochastic process $X(t)$ is "strict sense stationary" (SSS), if $X(t)$ and $X(t + a) \, \forall a$ have the same statistical parameters. In other words their statistics do not depend on time. This holds for all probability density functions such as (A.25) and all M,

$$p_{1M}\left(x(t_1), x(t_2), .., x(t_M)\right) = p_{1M}\left(x(t_1 + a), x(t_2 + a), .., x(t_M + a)\right) \quad \text{(A.37)}$$

- Two stochastic processes $X(t)$ and $Y(t)$ are jointly strict sense stationary, if the joint statistics of $X(t)$ and $Y(t)$ are equal to the joint statistics of $X(t + a)$ and $Y(t + a)$, $\forall a$, respectively.
- A complex-valued stochastic process $Z(t) = X(t) + \mathrm{j}Y(t)$ is strict sense stationary, if this condition holds jointly for the real and imaginary part.

Wide Sense Stationary (WSS) Stochastic Process

The conditions for a "wide sense stationary" (WSS) process are much weaker than for a strict sense stationary process, as they just impose conditions on the first and second order moments. Higher moments are not touched. From (A.37) follows that $p_1\left(x_{t_1}\right) = p_1\left(x_{t_1+a}\right) \, \forall a$ with $x_{t_\nu} = x(t_\nu)$ as before. Consequently, the expected value is constant. Furthermore (A.37) results in $p_{12}\left(x_{t_1}, x_{t_2}\right) = p_{12}\left(x_{t_1+a}, x_{t_2+a}\right) \, \forall a$ and thus the density function, the second moments and the autocorrelation function depend only on a time difference $t_2 - t_1 = \tau$.
Definition:

- A stochastic process $X(t)$ is wide sense stationary, if its expected value is constant

$$\mathbf{E}[X] = m_x = const. \quad \text{(A.38)}$$

- and if its autocorrelation function just depends on a time difference $\tau = t_2 - t_1$. Then follows from (A.28) by using $t_2 - t_1 = \tau$ and replacing t_1 by the fixed time instant t

$$R_{xx}(t_1, t_2) = R_{xx}(t, t + \tau) = \mathbf{E}\left[X(t)X(t + \tau)\right] = R_{xx}(\tau) \quad \text{(A.39)}$$

where $R_{xx}(\tau)$ is a short hand notation.

For a complex stochastic process we define

$$R_{xx}(\tau) = \mathbf{E}\left[X^*(t)X(t+\tau)\right] \tag{A.40}$$

In general the autocorrelation function exhibits the following properties

$$R_{xx}(-\tau) = R_{xx}^*(\tau); \ R_{xx}(0) \geq |R_{xx}(\tau)| \tag{A.41}$$

$R_{xx}(0) = \mathbf{E}[|X(t)|^2]$ is always a real value and is called the mean power of $X(t)$. We also see that a SSS stochastic process is also WSS.

A.3.5 Uncorrelated WSS Stochastic Processes

A Single Process $X(t)$

To check whether a WSS stochastic process $X(t)$ is uncorrelated, we have to extend the definition of the covariance of a random variable in (A.16) to a process.

The *autocovariance function* of the WSS process $X(t)$ is thus given by

$$C_{xx}(t, t+\tau) = C_{xx}(\tau) = \mathbf{E}\left[(X^*(t) - m_x^*)(X(t+\tau) - m_x)\right] = R_{xx}(\tau) - |m_x|^2 \tag{A.42}$$

where $m_x = \mathbf{E}[X(t)]$ is the expected value of $X(t)$.

The autocovariance function $C_{xx}(\tau)$ specifies the expected value of the product of the two random variables $X^*(t) - m_x^*$ and $X(t+\tau) - m_x$ for any given time shift $\tau \neq 0$. Similar to (A.19) we can say that the two random variables are uncorrelated, if $C_{xx}(\tau) = 0$. However, we have to exclude $\tau = 0$, because in this case both random variables just differ in the sign of the imaginary part and of course are strongly correlated. Consequently, we can formulate the following meaningful definition:

A WSS stochastic process $X(t)$ is *uncorrelated*, if its autocovariance function meets the condition

$$C_{xx}(\tau) \begin{cases} \neq 0; \ \tau = 0 \\ = 0; \ \tau \neq 0 \end{cases} \tag{A.43}$$

For the *autocorrelation function* of an *uncorrelated* process then follows with (A.43) and $R_{xx}(0) = \mathbf{E}\left[|X(t)|^2\right]$

$$R_{xx}(\tau) = \begin{cases} \mathbf{E}\left[|X(t)|^2\right]; \ \tau = 0 \\ |m_x|^2 \quad ; \ \tau \neq 0 \end{cases} \tag{A.44}$$

Now, consider a WSS process with *zero mean*, $m_x = 0$. Then we find from (A.44) that this process is *uncorrelated*, if

$$R_{xx}(\tau) = 0 \ \forall \tau \neq 0 \tag{A.45}$$

Two Processes $X(t)$ and $Y(t)$

The statistical interrelation between two WSS processes $X(t)$ and $Y(t)$ is defined by the *cross-covariance function* similar to (A.16)

$$C_{xy}(\tau) = \mathbf{E}\left[(X(t + \tau) - m_x)\left(Y^*(t) - m_y^*\right)\right] =$$

$$= R_{xy}(\tau) - m_x m_y^* \tag{A.46}$$

with $m_x = \mathbf{E}[X(t)]$, $m_y = \mathbf{E}[Y(t)]$, and the *cross-correlation function*

$$R_{xy}(\tau) = \mathbf{E}\left[X(t + \tau)Y^*(t)\right] \tag{A.47}$$

Using similar arguments as before with $C_{xx}(\tau)$, but no exception for $\tau = 0$ is required here, we define:

Two (WSS) processes $X(t)$ and $Y(t)$ are *uncorrelated* if

$$C_{xy}(\tau) = 0 \ \forall \tau \tag{A.48}$$

Then follows from (A.46)

$$R_{xy}(\tau) = m_x m_y^* \ \forall \tau \tag{A.49}$$

If at least one stochastic process has zero mean, then the processes are referred to as *orthogonal*

$$R_{xy}(\tau) = 0 \ \forall \tau \tag{A.50}$$

It is straightforward to show that the cross-correlation function has the following symmetry property

$$R_{xy}(\tau) = R_{yx}^*(-\tau) \tag{A.51}$$

A.3.6 Statistically Independent Processes

Two stochastic processes $X(t)$ and $Y(t)$ are statistically independent, if and only if for any choice of t_i and $\tilde{t}_j$ as well as M_x and M_y

$$p_{xy}\left(x_{t_1}, x_{t_2}, ..., x_{t_{M_x}}; y_{\tilde{t}_1}, y_{\tilde{t}_2}, ..., y_{\tilde{t}_{M_y}}\right) = p_x\left(x_{t_1}, x_{t_2}, ..., x_{t_{M_x}}\right) p_y\left(y_{\tilde{t}_1}, y_{\tilde{t}_2}, ..., y_{\tilde{t}_{M_y}}\right) \tag{A.52}$$

holds, where $p_x\left(x_{t_1}, x_{t_2}, ..., x_{t_{M_x}}\right)$ and $p_y\left(y_{\tilde{t}_1}, y_{\tilde{t}_2}, ..., y_{\tilde{t}_{M_y}}\right)$ are the joint density functions of $x_{t_1}, x_{t_2}, ..., x_{t_{M_x}}$ and $y_{\tilde{t}_1}, y_{\tilde{t}_2}, ..., y_{\tilde{t}_{M_y}}$, respectively. From the statistical independence follows that the two processes are uncorrelated, but this does not hold reversely.

In summary, two WSS stochastic processes $X(t)$ and $Y(t)$ with joint probability density function $p_{xy}(x, y)$ are *statistically independent*, if and only if

$$p_{xy}(x, y) = p_x(x)p_y(y) \tag{A.53}$$

where $p_x(x)$ and $p_y(y)$ are the marginal probability density functions of the stochastic processes $X(t)$ and $Y(t)$, respectively.

A.4 Stochastic Processes and Linear Time-Invariant Systems

A.4.1 Input–Output Relation of Linear System in Time Domain

Let $h(t)$ be the (deterministic) impulse response of a linear time-invariant system. At its input the sample function $x(t)$ of a WSS stochastic process $X(t)$ is active. Throughout the following we always consider stationary processes. Then the output process $Y(t)$ with sample function $y(t)$ is also stationary [1] and given by the convolution

$$y(t) = x(t) * h(t) = \int_{-\infty}^{\infty} x(u)h(t-u)du \tag{A.54}$$

However, as the stochastic signals cannot be expressed by a mathematical formula we are not in a position to explore this equation further and have to find a statistical description using autocorrelation functions and power spectral densities.

A.4.2 Wiener–Lee Theorem for Input–Output Autocorrelation Functions

Let $R_{xx}(\tau)$ and $R_{yy}(\tau)$ be the autocorrelation functions of the input and output stochastic process $x(t)$ and $y(t)$, respectively. We calculate the autocorrelation function $R_{hh}(\tau)$ of the deterministic (and thus also ergodic) impulse response $h(t)$ according to (A.35)

$$R_{hh}(\tau) = \mathbf{E}\left[h^*(t)h(t+\tau)\right] = \int_{-\infty}^{\infty} h^*(t)h(t+\tau)dt \tag{A.55}$$

where we have dropped $\lim_{T_0 \to \infty} \frac{1}{2T_0}$, because $h(t)$ is a deterministic signal with finite energy $R_{hh}(0) = \int_{-\infty}^{\infty} |h(t)|^2 \, dt$. The integral for $R_{hh}(\tau)$ can be considered as the convolution between $h(\tau)$ and $h^*(-\tau)$

$$R_{hh}(\tau) = h(\tau) * h^*(-\tau) \tag{A.56}$$

The Wiener–Lee theorem describes the relation between the input and the output autocorrelation function of a linear time-invariant system as follows

$$R_{yy}(\tau) = R_{hh}(\tau) * R_{xx}(\tau) = h(\tau) * h^*(-\tau) * R_{xx}(\tau) \tag{A.57}$$

A.4.3 Wiener–Khintchine Theorem for Power Spectral Density

For communications engineers spectra of signals are important to get an idea about the required bandwidth. For deterministic signals with finite energy the Fourier spectrum exists according to the sufficient Dirichlet condition. However, a random signal $x(t)$ has infinite energy, because in general

$$\lim_{T_0 \to \infty} \int_{-T_0}^{T_0} |x(t)|^2 \, dt \to \infty \tag{A.58}$$

On the other hand an ergodic stochastic process $X(t)$ exhibits finite mean power which is

$$\mathbf{E}\left[|X|^2\right] = R_{xx}(0) = \lim_{T_0 \to \infty} \frac{1}{2T_0} \int_{-T_0}^{T_0} |x(t)|^2 \, dt < \infty \tag{A.59}$$

The Wiener–Khintchine theorem provides the power spectral density $S_{xx}(f)$ of $X(t)$ by means of the Fourier transform of the autocorrelation function of $X(t)$,

$$R_{xx}(\tau) \rightarrowtail S_{xx}(f) = \int_{-\infty}^{\infty} R_{xx}(\tau) e^{-j2\pi f \tau} \, d\tau \tag{A.60}$$

From the symmetry $R_{xx}(-\tau) = R_{xx}^*(\tau)$ follows the property that $S_{xx}(f)$ is real and moreover

$$S_{xx}(f) \geq 0 \tag{A.61}$$

holds. With $h(t) \rightarrowtail H(f)$ and $h^*(-t) \rightarrowtail H^*(f)$ follows from (A.56) with the Fourier transform

$$R_{hh}(\tau) \rightarrowtail S_{hh}(f) = |H(f)|^2 \tag{A.62}$$

and with $R_{yy}(\tau) \longmapsto S_{yy}(f)$ we obtain the power spectral density of the output process $Y(t)$ with (A.57)

$$S_{yy}(f) = |H(f)|^2 S_{xx}(f) \tag{A.63}$$

With the inverse Fourier transform we obtain the from (A.60)

$$R_{xx}(\tau) = \int_{-\infty}^{\infty} S_{xx}(f)e^{j2\pi f\tau}df \tag{A.64}$$

and for $\tau = 0$ the mean power of $X(t)$

$$R_{xx}(0) = \int_{-\infty}^{\infty} S_{xx}(f)df \tag{A.65}$$

Example:
White noise $X(t)$ is defined by its constant power spectral density,

$$S_{xx}(f) = a = const. \; \forall \; f \tag{A.66}$$

Consequently, the autocorrelation function is

$$R_{xx}(\tau) = a\delta(\tau) \tag{A.67}$$

and we see that $R_{xx}(\tau) = 0 \; \forall \tau \neq 0$. Thus, all pairs of random variables $X(t)$ and $X(t + \tau)$ are uncorrelated for $\tau \neq 0$. We observe that a is also the mean power of $X(t)$.

A.5 Modulation and Demodulation of a Stationary Stochastic Process

A.5.1 Modulation

We consider a WSS stationary process $X(t)$, which shall be modulated with the carrier $e^{j2\pi f_0 t}$. Then we obtain

$$Y(t) = X(t)e^{j2\pi f_0 t} \tag{A.68}$$

$X(t)$ shall have the autocorrelation function $R_{xx}(\tau) = \mathbf{E}[X^*(t)X(t + \tau)]$. For the autocorrelation function of the process $Y(t)$ we obtain

$$R_{yy}(\tau) = \mathbf{E}\left[Y^*(t)Y(t + \tau)\right] = \mathbf{E}\left[X(t + \tau)X^*(t)e^{j2\pi f_0 \tau}\right] \tag{A.69}$$

Noting that $e^{j2\pi f_0\tau}$ is deterministic yields the final result

$$R_{yy}(\tau) = R_{xx}(\tau)e^{j2\pi f_0\tau} \tag{A.70}$$

This shows that the modulation of a stationary stochastic process translates into the modulation of its autocorrelation function. With the frequency shifting property of the Fourier transform we obtain

$$R_{yy}(\tau) \rightarrowtail S_{yy}(f) = S_{xx}(f - f_0) \tag{A.71}$$

outlining that the modulation of $X(t)$ results in a frequency shift of its power spectral density $S_{xx}(f)$ by the carrier frequency f_0.

A.5.2   Demodulation

We consider the modulated stationary process $Y(t)$ and apply the synchronous demodulation with the carrier $e^{-j2\pi f_0 t}$ resulting in the demodulated stochastic process $Z(t)$,

$$Z(t) = Y(t)e^{-j2\pi f_0 t} = X(t) \tag{A.72}$$

Consequently, we obtain

$$R_{zz}(\tau) = R_{yy}(\tau)e^{-j2\pi f_0\tau} = R_{xx}(\tau) \tag{A.73}$$

and

$$S_{zz}(f) = S_{yy}(f + f_0) = S_{xx}(f) \tag{A.74}$$

We also see that modulation and demodulation does not change the mean power of the processes, because with $\mathbf{E}[|X|^2] = R_{xx}(0)$, $\mathbf{E}[|Y|^2] = R_{yy}(0)$, and $\mathbf{E}[|Z|^2] = R_{zz}(0)$ follows

$$\mathbf{E}\left[|X|^2\right] = \mathbf{E}\left[|Y|^2\right] = \mathbf{E}\left[|Z|^2\right] \tag{A.75}$$

A.6   Stationary, Real-Valued Bandpass Process

As is well known, e.g. [3], any real-valued bandpass signal can be written in general as

$$n(t) = x(t)\cos(2\pi f_0 t) - y(t)\sin(2\pi f_0 t) \tag{A.76}$$

where $x(t)$ and $y(t)$ are real-valued lowpass signals with cut-off frequency f_c. This model shall be adopted to a stochastic bandpass process $N(t)$ with power spectral density $S_{nn}(f)$ and with the passband in the range of $f_0 - f_c \le |f| \le f_0 + f_c$, where

$f_0 > f_c$ is the center frequency. $X(t)$ and $Y(t)$ shall be WSS lowpass processes with the power spectral densities

$$S_{xx}(f) \; ; \; S_{yy}(f) \quad \begin{cases} \neq 0 \; ; \; |f| \leq f_c \\ = 0 \; ; \quad \text{else} \end{cases} \tag{A.77}$$

A.6.1　Condition for Stationarity

We would like to know under which conditions this bandpass process $N(t)$ is wide sense stationary. Therefore we have to check whether $\mathbf{E}[N(t)] = const.$ holds and whether the autocorrelation function

$$R_{nn}(\tau) = \mathbf{E}\left[N(t)N(t+\tau)\right] \tag{A.78}$$

is independent of t.

Expected Value

With (A.76) follows

$$\mathbf{E}[N(t)] = \mathbf{E}[X(t)]\cos(2\pi f_0 t) - \mathbf{E}[Y(t)]\sin(2\pi f_0 t) \tag{A.79}$$

$\mathbf{E}[N(t)] = const. \; \forall t$ holds, if

$$\mathbf{E}[X(t)] = \mathbf{E}[Y(t)] = 0 \tag{A.80}$$

Consequently from (A.79) also follows

$$\mathbf{E}[N(t)] = 0 \tag{A.81}$$

Autocorrelation Function

Next, please consider (A.78). By applying basic trigonometric formulas and using the fact, that terms with sin(), cos(), and the arguments $2\pi f_0 t$ and $2\pi f_0(t+\tau)$ are non-random and therefore can be taken out from the expectation operator, we obtain finally with (A.76)

$$\begin{aligned} R_{nn}(\tau) = \quad & \tfrac{1}{2}\left[R_{xx}(\tau) + R_{yy}(\tau)\right]\cos(2\pi f_0 \tau) + \\ & + \tfrac{1}{2}\left[R_{xx}(\tau) - R_{yy}(\tau)\right]\cos(4\pi f_0 t + 2\pi f_0 \tau) - \\ & - \tfrac{1}{2}\left[R_{xy}(\tau) - R_{yx}(\tau)\right]\sin(2\pi f_0 \tau) - \\ & - \tfrac{1}{2}\left[R_{xy}(\tau) + R_{yx}(\tau)\right]\sin(4\pi f_0 t + 2\pi f_0 \tau) \end{aligned} \tag{A.82}$$

To get $R_{nn}(\tau)$ independent of t the following conditions must hold

$$R_{xx}(\tau) = R_{yy}(\tau) \tag{A.83}$$

and

$$R_{xy}(\tau) = -R_{yx}(\tau) \tag{A.84}$$

Then we obtain from (A.82)

$$R_{nn}(\tau) = R_{xx}(\tau)\cos(2\pi f_0\tau) - R_{xy}(\tau)\sin(2\pi f_0\tau) \tag{A.85}$$

Knowing that an autocorrelation function provides the mean power of the process for $\tau = 0$ we conclude from (A.85) and (A.83)

$$\mathbf{E}\left[|N(t)|^2\right] = \mathbf{E}\left[|X(t)|^2\right] = \mathbf{E}\left[|Y(t)|^2\right] \tag{A.86}$$

Furthermore, we can find another property by applying (A.51) on (A.84) yielding

$$R_{yx}(-\tau) = -R_{yx}(\tau) \tag{A.87}$$

which indicates that $R_{yx}(\tau)$ is an odd function. Consequently, $R_{yx}(0) = 0$ must be true and the property (A.84) yields

$$R_{xy}(0) = 0 \tag{A.88}$$

This means that the random variables $X(t)$ and $Y(t)$ for any given t are not correlated. Please note that $R_{xy}(0) = 0$ does not require $X(t + \tau)$ and $Y(t)$ to be uncorrelated for any τ. However, if the zero mean processes $X(t)$ and $Y(t)$ are assumed to be uncorrelated for any τ, $R_{xy}(\tau) = 0 \,\forall \tau$ holds and from (A.85) follows

$$R_{nn}(\tau) = R_{xx}(\tau)\cos(2\pi f_0\tau) \tag{A.89}$$

Using the Wiener–Khintchine theorem (A.60) and the frequency shifting property of the Fourier transform we obtain from (A.89) the power spectral density of the process $N(t)$

$$S_{nn}(f) = \frac{1}{2}\left[S_{xx}(f - f_0) + S_{xx}(f + f_0)\right] \tag{A.90}$$

which clearly exhibits a bandpass shape.

A.6.2   Summary on Stationary Bandpass Process

A WSS bandpass process $N(t)$ exhibits zero mean and is composed of the in-phase component $X(t)$ and the quadrature component $Y(t)$, which are zero mean WSS lowpass processes. Moreover, $N(t)$, $X(t)$, and $Y(t)$ have the same mean power.

For the cross-correlation holds $R_{xy}(0) = E[X(t)Y(t)] = 0$, which means that the random variables $X(t)$ and $Y(t)$ are uncorrelated for any fixed t.

A.6.3 Complex Envelope of a Bandpass Process

It is straightforward to show that (A.76) can be written as

$$N(t) = \mathrm{Re}\left[Z(t)e^{j2\pi f_0 t}\right] \tag{A.91}$$

where

$$Z(t) = X(t) + jY(t) \tag{A.92}$$

is called the complex envelope. If $X(t)$ and $Y(t)$ are WSS lowpass processes, then $Z(t)$ is a WSS complex lowpass process. It is easy to show that for the autocorrelation function of $Z(t)$ follows with (A.83) and (A.84)

$$R_{zz}(\tau) = E\left[Z^*(t)Z(t+\tau)\right] = 2\left[R_{xx}(\tau) - jR_{xy}(\tau)\right] \tag{A.93}$$

A.7 Two-Dimensional Gaussian Random Process

A.7.1 Joint Gaussian Probability Density Function

We consider now two real-valued SSS Gaussian processes $X(t)$ and $Y(t)$. For any fixed t they represent random variables [1] with the Gaussian joint probability density function

$$p_{xy}(x, y) = \frac{1}{2\pi\sigma_x\sigma_y\sqrt{1-\rho^2}}e^{-\frac{(x-m_x)^2\sigma_y^2 - 2(x-m_x)(y-m_y)\rho\sigma_x\sigma_y + (y-m_y)^2\sigma_x^2}{2\sigma_x^2\sigma_y^2(1-\rho^2)}} \tag{A.94}$$

with the mean values m_x and m_y,
the variances

$$\sigma_x^2 = E[X^2] - m_x^2 \; ; \; \sigma_y^2 = E[Y^2] - m_y^2 \tag{A.95}$$

the normalized cross-covariance (correlation coefficient)

$$\rho = \frac{R_{xy}(0) - m_x m_y}{\sigma_x\sigma_y} \tag{A.96}$$

and the marginal probability density function of $X(t)$

$$p_x(x) = \frac{1}{\sqrt{2\pi}\sigma_x} e^{-\frac{(x-m_x)^2}{2\sigma_x^2}} \tag{A.97}$$

as well as of $Y(t)$

$$p_y(y) = \frac{1}{\sqrt{2\pi}\sigma_y} e^{-\frac{(y-m_y)^2}{2\sigma_y^2}} \tag{A.98}$$

A.7.2 Uncorrelated Gaussian Random Processes

Let $X(t)$ and $Y(t)$ be two WSS, real and uncorrelated Gaussian processes. Then $R_{xy}(\tau) = m_x m_y$ holds according to (A.49). This is valid for any τ, including $\tau = 0$. With $R_{xy}(0) = m_x m_y$ follows from (A.96)

$$\rho = 0 \tag{A.99}$$

and consequently from (A.94) with (A.97) and (A.98) follows

$$p_{xy}(x, y) = p_x(x)p_y(y) \tag{A.100}$$

Hence, we conclude that uncorrelated Gaussian processes $X(t)$ and $Y(t)$ are even statistically independent.

A.7.3 Complex Gaussian Random Process

Let $X(t)$ and $Y(t)$ be WSS real-valued Gaussian lowpass processes with properties given in Sect. A.7.1. Then they constitute a complex Gaussian random lowpass process

$$Z(t) = X(t) + jY(t) \tag{A.101}$$

A.7.4 Gaussian Bandpass Process

Any real-valued bandpass process in general is given by (A.76)

$$n(t) = x(t)\cos(2\pi f_0 t) - y(t)\sin(2\pi f_0 t)$$

If $X(t)$ and $Y(t)$ are stationary Gaussian lowpass processes with properties given in Sect. A.7.1, we denote $N(t)$ as a Gaussian bandpass process. We know from Sect. A.6 that $N(t)$, $X(t)$, and $Y(t)$ have zero mean and identical mean power. Furthermore, if the two Gaussian lowpass processes $X(t)$ and $Y(t)$ are uncorrelated then they are even statistically independent.

A.8   Sampling of a Stochastic Process

The receive signal in a digital communication system is a random process. Before detection the signal is sampled. In this section the main basics for sampling of a stochastic process are summarized.

A.8.1   Prerequisites

$X(t)$ shall be a WSS stochastic process. Then $X(t)$ and $X(t + a)$ $\forall a$ have the same density function, because $p_x(x)$ is independent of t. Consequently, after sampling with a sampling frequency $f_S = \frac{1}{T}$ the resulting samples $X(kT)$ exhibit the same probability density function $p_x(x)$ as $X(t)$. Furthermore, $X(kT) = X_S(k)$ can be considered as a sequence of equidistant random variables of the the process $X(t)$, which constitute a stationary discrete-time stochastic process, where $k \in \mathbb{Z}$ is the discrete-time.

A.8.2   Auto- and Cross-Correlation Function of a Discrete-Time Stochastic Process

$X(kT)$ for every fixed $k \in \mathbb{Z}$ can be considered as a random variable. Consequently, we apply (A.40) to get the autocorrelation function as $R_{xx}(\tau) = \mathbf{E}[X^*(kT)X(kT + \tau)]$. Obviously, $X(kT) = X_S(k)$ is only defined for discrete-time arguments. Therefore $R_{xx}(\tau)$ can take on defined values only for $\tau = lT$ with $l \in \mathbb{Z}$ and the autocorrelation function (autocorrelation sequence) of $X_S(k)$ will become a function of a discrete variable to be be written as

$$R_{x_S x_S}(l) = \mathbf{E}[X_S^*(k)X_S(k + l)] \qquad (A.102)$$

Interestingly

$$R_{x_S x_S}(l) = R_{xx}(lT) \qquad (A.103)$$

can also be considered as the sampled version of the continuous-time autocorrelation function $R_{xx}(\tau)$. It is straightforward to show that the properties of (A.41) hold similarly as

$$R_{x_S x_S}(-l) = R_{x_S x_S}^*(l) \; ; \; R_{x_S x_S}(0) \geq \left| R_{x_S x_S}(l) \right| \qquad (A.104)$$

With the same arguments as for (A.47) we can define the *cross-correlation function* of two discrete-time WSS stochastic processes $X_S(k)$ and $Y_S(k)$ as

$$R_{x_S y_S}(l) = \mathbf{E}[X_S(k + l)Y_S^*(k)] \qquad (A.105)$$

Symmetry properties and conditions for uncorrelated processes are similar as for the continuous time processes.

A.8.3   Power Spectral Density

According to the Wiener–Khintchine theorem we find the power density spectrum $S_{x_S x_S}(f)$ of the WSS stochastic process $X_S(k)$ by the Fourier transform of the sampled autocorrelation function. Applying ideal sampling on $R_{xx}(\tau)$ yields[2]

$$R_{xx,S}(\tau) = R_{xx}(\tau) \sum_{l=-\infty}^{\infty} T\delta(\tau - lT) \tag{A.106}$$

With the Fourier correspondence

$$\sum_{l=-\infty}^{\infty} T\delta(\tau - lT) \longmapsto \sum_{m=-\infty}^{\infty} \delta(f - m\frac{1}{T}) \tag{A.107}$$

and with $R_{xx}(\tau) \longmapsto S_{xx}(f)$ we obtain from (A.106)

$$R_{xx,S}(\tau) \longmapsto S_{x_S x_S}(f) = S_{xx}(f) * \sum_{m=-\infty}^{\infty} \delta(f - m\frac{1}{T})$$

which results after executing the convolution integral in

$$S_{x_S x_S}(f) = \sum_{m=-\infty}^{\infty} S_{xx}(f - m\frac{1}{T}) \tag{A.108}$$

We see that the spectrum is a periodic repetition of the baseband power spectral density $S_{xx}(f)$, where the period is given by the sampling frequency $\frac{1}{T}$.

---

[2]We multiply the Dirac impulses by T to ensure that the autocorrelation functions on both sides of the equation have the same physical dimension.

Appendix B
Some Fundamentals of Linear Algebra

B.1 Eigenvalue Decomposition

In this section we review some properties of the eigenvalue decomposition of a matrix $\mathbf{A}$, assuming for the moment that such a decomposition shall exist for the given matrix. The eigenvalue-eigenvector problem of linear algebra can be stated as follows: Given a $N \times N$ matrix $\mathbf{A} \epsilon \mathbb{C}^{N \times N}$ with in general complex entries a_{ij}, a column vector $\mathbf{v}_i \epsilon \mathbb{C}^{N \times 1}$, and a scalar factor λ_i. We are looking for the vector $\mathbf{A}\mathbf{v}_i$, which is equal to the vector $\lambda_i \mathbf{v}_i$

$$\mathbf{A}\mathbf{v}_i = \lambda_i \mathbf{v}_i \; ; \; i = 1, ..., N \tag{B.1}$$

with $\mathbf{v}_i \neq \mathbf{0}$, otherwise we would have the trivial solution, which is of no interest. A matrix can change the length (by its determinant) and the direction of the vector after multiplication. Thus, we are looking for the vector $\mathbf{A}\mathbf{v}_i$ with the same direction as $\mathbf{v}_i$ but with the length changed by λ_i. The non-trivial solutions λ_i and $\mathbf{v}_i$ are called eigenvalues and eigenvectors of the matrix, respectively. The set of all eigenvalues is denoted as spectrum and the absolute value of the largest eigenvalue is referred to as spectral radius. We can rewrite (B.1) with matrix notation as

$$\mathbf{A}\mathbf{V} = \mathbf{V}\Lambda \tag{B.2}$$

where

$$\mathbf{V} = \begin{pmatrix} \mathbf{v}_1 & \mathbf{v}_2 & ... & \mathbf{v}_N \end{pmatrix} \epsilon \mathbb{C}^{N \times N} \tag{B.3}$$

is the matrix of eigenvectors and

$$\Lambda = \text{diag}\,(\lambda_1, \lambda_2, ..., \lambda_N) \tag{B.4}$$

© Springer Nature Switzerland AG 2021
J. Speidel, *Introduction to Digital Communications*,
Signals and Communication Technology,
https://doi.org/10.1007/978-3-030-67357-4

is a diagonal matrix composed of the eigenvalues of $\mathbf{A}$. To solve the eigenvalue-eigenvector problem (B.1) can be written as

$$\mathbf{A}\mathbf{v}_i - \lambda_i \mathbf{v}_i = \mathbf{0} \iff (\mathbf{A} - \lambda_i \mathbf{I}_N)\,\mathbf{v}_i = \mathbf{0} \; ; \; i = 1, ..., N \qquad (B.5)$$

This system of homogeneous equations has a non-trivial solution only if

$$\det(\mathbf{A} - \lambda_i \mathbf{I}_N) = 0 \; ; \; i = 1, ..., N \qquad (B.6)$$

with the $N \times N$ identity matrix $\mathbf{I}_N$. Equation (B.6) is called characteristic equation for the matrix $\mathbf{A}$ and the left-hand side is the characteristic polynomial with degree N as a function of λ_i. We can conclude that the eigenvalues of the matrix $\mathbf{A}$ are the roots of the characteristic polynomial. After all λ_i are calculated from (B.6) we can insert each into (B.5) and find the corresponding eigenvectors $\mathbf{v}_i$. Note that the solution for each $\mathbf{v}_i$ contains at least one free parameter, because (B.5) is a homogeneous system of equation and thus the rank of the matrix $\mathbf{A} - \lambda_i \mathbf{I}_N$ is

$$\text{rank}\,(\mathbf{A} - \lambda_i \mathbf{I}_N) \leq N - 1 \qquad (B.7)$$

The free parameters have to be used to normalize all eigenvectors such that

$$\mathbf{v}_i^H \mathbf{v}_i = \|\mathbf{v}_i\|^2 = 1 \; ; \; i = 1, ..., N \qquad (B.8)$$

Please note that the resulting eigenvectors associated to different eigenvalues are non-orthogonal in general, i.e.

$$\mathbf{v}_i^H \mathbf{v}_j = 0 \; ; \; i = 1, ..., N \; ; \; i \neq j \qquad (B.9)$$

does not hold. Equation (B.9) is true for Hermiteian and symmetric matrices with real entries, see Sect. B.3. On the other hand we will see later that the singular value decomposition of a matrix yields pairwise orthogonal eigenvectors. The components of the eigenvectors are in general complex, where $\mathbf{v}_i^H = \left(\mathbf{v}_i^*\right)^T = \left(\mathbf{v}_i^T\right)^*$ is the conjugate transpose vector (Hermiteian vector) to $\mathbf{v}_i$. The inverse matrix $\mathbf{V}^{-1}$ is obtained from the relation

$$\mathbf{V}^{-1}\mathbf{V} = \mathbf{V}\mathbf{V}^{-1} = \mathbf{I}_N \qquad (B.10)$$

Taking $\mathbf{V}$ we can now transform $\mathbf{A}$ into diagonal form Λ. The procedure is also called principal axis transformation or eigenvalue decomposition of $\mathbf{A}$. For that purpose we multiply (B.2) from the left by $\mathbf{V}^{-1}$ and obtain with (B.10)

$$\mathbf{V}^{-1}\mathbf{A}\mathbf{V} = \Lambda \qquad (B.11)$$

$\mathbf{V}$ is therefore called transform matrix. We see that there is no need for orthogonal eigenvectors. If we multiply (B.11) in a first step from the left with $\mathbf{V}$ and secondly from the right with $\mathbf{V}^{-1}$ we get

$$\mathbf{A} = \mathbf{V}\Lambda\mathbf{V}^{-1} \tag{B.12}$$

which is another form of the eigenvalue decomposition or diagonalization of $\mathbf{A}$. Such an eigenvalue decomposition will not exist for all square matrices, in other words not all square matrices are diagonalizable. In any case the inverse matrix $\mathbf{V}^{-1}$ must exist. Alternative formulation are: $\mathbf{V}$ must be a non-singular matrix or all eigenvectors $\mathbf{v}_i$ are linearly independent of each other. In the following let us pinpoint some special $N\times N$ matrices, which are diagonalizable, and which are of importance for MIMO systems.

B.2   Normal Matrices

By definition $\mathbf{A}$ is a *normal matrix*, if and only if

$$\mathbf{A}^H\mathbf{A} = \mathbf{A}\mathbf{A}^H \tag{B.13}$$

$\mathbf{A}^H$ is called the *Hermiteian matrix*[3] or *conjugate transpose matrix* of $\mathbf{A}$. The Hermiteian operator $(...)^H$ is defined as

$$\mathbf{A}^H = \left(\mathbf{A}^*\right)^T = (\mathbf{A}^T)^* \tag{B.14}$$

It can be shown [4] that every normal matrix is unitarily diagonalizable, i.e., an eigenvalue decomposition exists, where the transform matrix $\mathbf{V}$ is a unitary matrix defined by

$$\mathbf{V}^H\mathbf{V} = \mathbf{V}\mathbf{V}^H = \mathbf{I}_N \iff \mathbf{V}^H = \mathbf{V}^{-1} \tag{B.15}$$

Then from (B.11) follows

$$\Lambda = \mathbf{V}^H\mathbf{A}\mathbf{V} \tag{B.16}$$

It should be noted that (B.13) is only a sufficient and not a necessary condition for diagonalizable matrices. That means there are diagonalizable matrices which are not normal. An example is the matrix $\begin{pmatrix} 0 & 1 \\ 4 & 0 \end{pmatrix}$. Please note that the Hermiteian operator can also be applied to non-square matrices $\mathbf{A}\epsilon\,\mathbb{C}^{M\times N}$ and $\mathbf{B}\epsilon\,\mathbb{C}^{N\times M}$ with the property similar to the transposition operation

$$(\mathbf{AB})^H = \mathbf{B}^H\mathbf{A}^H \tag{B.17}$$

---

[3]Charles Hermite, French mathematician.

B.3   Hermiteian Matrices

Definition of a Hermiteian Matrix

By definition, $\mathbf{A} \in \mathbb{C}^{N \times N}$ is called a Hermiteian matrix, if and only if

$$\mathbf{A}^H = \mathbf{A} \tag{B.18}$$

It is easy to show that a Hermiteian matrix is also a normal matrix. For the proof, we insert (B.18) into (B.13) and obtain

$$\mathbf{A}^H \mathbf{A} = \mathbf{A}\mathbf{A} = \mathbf{A}\mathbf{A}^H \tag{B.19}$$

As a consequence, Hermiteian matrices are also unitarily diagonalizable given by (B.16).

Quadratic Form

Let $\mathbf{A} \in \mathbb{C}^{N \times N}$ be a Hermiteian matrix with eigenvalues λ_i ; $i = 1, ..., N$. Then this matrix can be defined as

$$\mathbf{A} = \mathbf{a}\mathbf{a}^H \tag{B.20}$$

with the column vector $\mathbf{a} \in \mathbb{C}^{N \times 1}$. With the column vector $\mathbf{z} \in \mathbb{C}^{N \times 1}$ we define the quadratic form

$$\mathbf{z}^H \mathbf{A} \mathbf{z} \tag{B.21}$$

which has the property

$$\mathbf{z}^H \mathbf{A} \mathbf{z} \geq 0 \; \forall \mathbf{z} \neq 0 \tag{B.22}$$

For the proof we insert (B.20) into (B.22) and find

$$\mathbf{z}^H \mathbf{A} \mathbf{z} = \mathbf{z}^H \mathbf{a}\mathbf{a}^H \mathbf{z} = \mathbf{z}^H \mathbf{a} \left(\mathbf{z}^H \mathbf{a}\right)^H = \left|\mathbf{z}^H \mathbf{a}\right|^2 \geq 0 \; \forall \mathbf{z} \neq 0 \tag{B.23}$$

Eigenvalues of a Hermiteian Matrix

All eigenvalues λ_i of a Hermiteian matrix are positive, i.e.

$$\lambda_i \geq 0 \; ; \; i = 1, ..., N \tag{B.24}$$

This implies that all eigenvalues of a Hermiteian matrix are real. For the proof let $\mathbf{v}_i$ be the eigenvector associated with the eigenvalue λ_i. Then

$$\mathbf{A}\mathbf{v}_i = \lambda_i \mathbf{v}_i \tag{B.25}$$

The corresponding positive semi-definite quadratic form according to (B.23) then is

$$\mathbf{v}_i^H \mathbf{A} \mathbf{v}_i = \mathbf{v}_i^H \lambda_i \mathbf{v}_i = \lambda_i \|\mathbf{v}_i\|^2 \geq 0 \tag{B.26}$$

from which we conclude the proposition (B.24).

Eigenvectors of a Hermiteian Matrix

Lemma
The eigenvectors $\mathbf{v}_i \in \mathbb{C}^{N\times1}$ and $\mathbf{v}_j \in \mathbb{C}^{N\times1}$ of a Hermiteian matrix $\mathbf{A}$ associated with two different non-zero eigenvalues $\lambda_i \neq \lambda_j \neq 0$ are (pairwise) orthogonal, i.e.

$$\mathbf{v}_i^H \mathbf{v}_j = \mathbf{0} \; ; \; i = 1, ..., N \; ; \; i \neq j \tag{B.27}$$

For the proof we use the definition of the eigenvectors $\mathbf{A}\mathbf{v}_i = \lambda_i \mathbf{v}_i$ and $\mathbf{A}\mathbf{v}_j = \lambda_j \mathbf{v}_j$. Then we calculate

$$\mathbf{v}_i^H \mathbf{v}_j \lambda_i = (\lambda_i \mathbf{v}_i)^H \mathbf{v}_j = (\mathbf{A}\mathbf{v}_i)^H \mathbf{v}_j = \mathbf{v}_i^H \mathbf{A}^H \mathbf{v}_j = \mathbf{v}_i^H \mathbf{A}\mathbf{v}_j = \mathbf{v}_i^H \mathbf{v}_j \lambda_j \tag{B.28}$$

and the result is $\mathbf{v}_i^H \mathbf{v}_j \lambda_i = \mathbf{v}_i^H \mathbf{v}_j \lambda_j$. As the eigenvalues are unequal and unequal to zero, proposition (B.27) follows and the proof is finished.

Please note that (B.27) also holds, if the eigenvectors are not normalized, which can be easily proven by checking (B.27) with the vectors $\alpha_i \mathbf{v}_i$ and $\alpha_j \mathbf{v}_j$.

Orthogonal and normalized vectors are called orthonormal. For the matrix $\mathbf{V}$ of eigenvectors in (B.3) then follows

$$\mathbf{V}^H \mathbf{V} = \begin{pmatrix} \mathbf{v}_1^H \\ \mathbf{v}_2^H \\ \vdots \\ \mathbf{v}_N^H \end{pmatrix} \begin{pmatrix} \mathbf{v}_1 & \mathbf{v}_2 & ... & \mathbf{v}_N \end{pmatrix} = \mathbf{I}_N \tag{B.29}$$

and with (B.10) we conclude that $\mathbf{V}$ is unitary, $\mathbf{V}^H = \mathbf{V}^{-1}$.

B.4 Unitary Matrices

Definition
$\mathbf{V} \in \mathbb{C}^{N\times N}$ is called a unitary matrix, if and only if

$$\mathbf{V}^{-1} = \mathbf{V}^H \tag{B.30}$$

The inverse matrix $\mathbf{V}^{-1}$ is the solution of

$$\mathbf{V}^{-1}\mathbf{V} = \mathbf{V}\mathbf{V}^{-1} = \mathbf{I}_N \tag{B.31}$$

Consequently, with (B.30) follows

$$\mathbf{V}^H\mathbf{V} = \mathbf{V}\mathbf{V}^H = \mathbf{I}_N \tag{B.32}$$

$\mathbf{V}$ is composed of orthonormal column vectors.

$$\mathbf{V} = \begin{bmatrix} \mathbf{v}_1 & \mathbf{v}_2 & \cdots & \mathbf{v}_N \end{bmatrix} \tag{B.33}$$

satisfying (B.8) and (B.9), i.e.

$$\mathbf{v}_i^H \mathbf{v}_j = \begin{cases} 1 \; ; \; i = j = 1, ..., N \\ 0 \; ; \; i, \; j = 1, ..., N \; ; i \neq j \end{cases} \tag{B.34}$$

Properties

- An inverse matrix is defined only for a square matrix. Therefore, all unitary matrices are square matrices.
- (B.32) is also the property of a normal matrix. Consequently, unitary matrices are a subset of normal matrices and thus unitarily diagonalizable. With a unitary transform matrix $\mathbf{V}$ the eigenvalue decomposition of (B.12) can be written as

$$\mathbf{A} = \mathbf{V}\boldsymbol{\Lambda}\mathbf{V}^H \tag{B.35}$$

- All eigenvalues ϱ_i of a unitary matrix $\mathbf{V} \in \mathbb{C}^{N \times N}$ have absolute values equal to 1

$$|\varrho_i| = 1 \; ; \; i = 1, ..., N \tag{B.36}$$

Proof: We calculate the scalar product of the two vectors and obtain with (B.32)

$$(\mathbf{V}\mathbf{v}_i)^H \, \mathbf{V}\mathbf{v}_i = \mathbf{v}_i^H \mathbf{V}^H \mathbf{V}\mathbf{v}_i = \|\mathbf{v}_i\|^2 \tag{B.37}$$

On the other hand the eigenvalue-eigenvector condition

$$\mathbf{V}\mathbf{v}_i = \varrho_i \mathbf{v}_i \tag{B.38}$$

holds. The left-hand side of (B.37) yields with (B.38)

$$(\mathbf{V}\mathbf{v}_i)^H \, \mathbf{V}\mathbf{v}_i = \varrho_i^* \varrho_i \mathbf{v}_i^H \mathbf{v}_i = |\varrho_i|^2 \, \|\mathbf{v}_i\|^2 \tag{B.39}$$

The left-hand sides of (B.37) and (B.39) are identical. Consequently, this must also hold for the right-hand sides

$$\|\mathbf{v}_i\|^2 = |\varrho_i|^2 \, \|\mathbf{v}_i\|^2 \qquad (B.40)$$

from which the proposition (B.36) directly follows and the proof is finalized.

- The input signal $\mathbf{s}$ and output signal $\mathbf{y} = \mathbf{V}\mathbf{s}$ of a system described by a unitary matrix $\mathbf{V}$ have the same mean power

$$\mathbf{E}\left[\|\mathbf{y}\|^2\right] = \mathbf{E}\left[\|\mathbf{s}\|^2\right] \qquad (B.41)$$

Proof:

$$\mathbf{E}\left[\|\mathbf{y}\|^2\right] = \mathbf{E}\left[(\mathbf{V}\mathbf{s})^H \, \mathbf{V}\mathbf{s}\right] = \mathbf{E}\left[\mathbf{s}^H \mathbf{V}^H \mathbf{V}\mathbf{s}\right] = \mathbf{E}\left[\mathbf{s}^H \mathbf{I}_N \mathbf{s}\right] = \mathbf{E}\left[\|\mathbf{s}\|^2\right] \quad (B.42)$$

B.5   Norm of a Vector, Norm of a Matrix

The squared norm of a vector $\mathbf{v} = \begin{pmatrix} v_1 & v_2 & \cdots & v_N \end{pmatrix}^T$ with complex components is given by the sum of the squared absolute values of the components

$$\|\mathbf{v}\|^2 = \sum_{i=1}^{N} |v_i|^2 \qquad (B.43)$$

The squared norm (Frobenius norm) of a matrix $A \in \mathbb{C}^{M \times N}$ with complex entries a_{ij} is given by the sum of the squared absolute values of the entries

$$\|A\|_F^2 = \sum_{i=1}^{M}\sum_{j=1}^{N} |a_{ij}|^2 = \sum_{j=1}^{N} \|\mathbf{a}_j\|^2 \qquad (B.44)$$

Alternatively, the Frobenius norm can be calculated as the sum of the squared norms of the column vectors $\mathbf{a}_j$ or row vectors of $\mathbf{A}$, respectively.

B.6   Singular Value Decomposition

The Procedure

The Singular Value Decomposition (SVD) of a matrix $\mathbf{H} \in \mathbb{C}^{N \times M}$ is given by

$$\mathbf{H} = \mathbf{U}\mathbf{D}\mathbf{V}^H \qquad (B.45)$$

with

$$
\mathbf{D} =
\begin{pmatrix}
\sqrt{\lambda_1} & 0 & 0 \cdots & 0 & 0 \cdots 0 \\
0 & \sqrt{\lambda_2} & 0 \cdots & 0 & 0 \cdots 0 \\
& & \ddots & & \\
0 & 0 & 0 \cdots \sqrt{\lambda_P} & 0 \cdots 0 \\
0 & 0 & 0 \cdots & 0 & 0 \cdots 0 \\
& \vdots & \cdots & & \ddots \\
0 & 0 & 0 \cdots & 0 & 0 \cdots 0 \\
0 & 0 & 0 \cdots & 0 & 0 \cdots 0
\end{pmatrix}
=
\begin{pmatrix}
\Lambda_P^{\frac{1}{2}} & 0 \cdots 0 \\
0 & 0 \cdots 0 \\
\vdots & \vdots \cdots \vdots \\
0 & 0 \cdots 0
\end{pmatrix}
\in \mathbb{R}^{N \times M} \qquad (B.46)
$$

$\lambda_1 \geq \lambda_2 \geq \cdots \lambda_P > 0$, and $\lambda_{P+1} = \lambda_{P+2} = \cdots = \lambda_N = 0$ are the N eigenvalues of the Hermiteian matrix

$$
\mathbf{Q}_N = \mathbf{H}\mathbf{H}^H \in \mathbb{C}^{N \times N} \qquad (B.47)
$$

and

$$
P = \mathrm{rank}\,(\mathbf{Q}_N) \qquad (B.48)
$$

is the rank of the matrix $\mathbf{Q}_N$. In general the rank of a matrix $\mathbf{H} \in \mathbb{C}^{N \times M}$ is defined as the number of linearly independent rows or columns of the matrix, thus

$$
\mathrm{rank}\,(\mathbf{H}) \leq \min\{M, N\} \qquad (B.49)
$$

From this definition follows for (B.48)

$$
P \leq N \qquad (B.50)
$$

$\sqrt{\lambda_i}$; $i = 1, ..., P$ are called the singular values of the matrix $\mathbf{H}$. $\mathbf{U} \in \mathbb{C}^{N \times N}$ and $\mathbf{V} \in \mathbb{C}^{M \times M}$ are unitary matrices, thus

$$
\mathbf{U}^{-1} = \mathbf{U}^H \; ; \; \mathbf{V}^{-1} = \mathbf{V}^H \qquad (B.51)
$$

hold. Furthermore, $\mathbf{U}$ is the matrix of the normalized eigenvectors with respect to the eigenvalues $\lambda_1, \lambda_2, \ldots, \lambda_N$. Let

$$
\Lambda_N = \mathrm{diag}\,(\lambda_1, \lambda_2, ..., \lambda_P, 0, ..., 0) \; \in \mathbb{R}^{N \times N} \qquad (B.52)
$$

be a diagonal matrix composed of the eigenvalues of $\mathbf{Q}_N$. Then the eigenvalue decomposition of $\mathbf{Q}_N$ is

$$
\mathbf{U}^H \mathbf{Q}_N \mathbf{U} = \Lambda_N \qquad (B.53)
$$

One method to find the matrix $\mathbf{V}$ in (B.45) is the eigenvalue decomposition of the matrix

$$
\mathbf{Q}_M = \mathbf{H}^H \mathbf{H} \in \mathbb{C}^{M \times M} \qquad (B.54)
$$

which is

$$\mathbf{V}^H \mathbf{Q}_M \mathbf{V} = \mathbf{\Lambda}_M \qquad (\text{B.55})$$

with the diagonal matrix

$$\mathbf{\Lambda}_M = \text{diag}\ (\lambda_1, \lambda_2, ..., \lambda_P, 0, ..., 0)\ \in \mathbb{R}^{M \times M} \qquad (\text{B.56})$$

$\mathbf{V}$ is the matrix of eigenvectors of $\mathbf{Q}_M$ with respect to the eigenvalues $\lambda_1, \lambda_2, \ldots, \lambda_M$. Note that the eigenvalues $\lambda_1, \lambda_2, \ldots, \lambda_P$ are the same as for the matrix $\mathbf{Q}_N$. Furthermore

$$\text{rank}\ (\mathbf{Q}_M) = \text{rank}\ (\mathbf{Q}_N) = P \qquad (\text{B.57})$$

holds and $\mathbf{\Lambda}_M$ as well as $\mathbf{\Lambda}_N$ contain the same diagonal matrix

$$\mathbf{\Lambda}_P = \text{diag}\ (\lambda_1, \lambda_2, ..., \lambda_P)\ \in \mathbb{R}^{P \times P} \qquad (\text{B.58})$$

of the P eigenvalues, which are unequal to zero. Note that in (B.46)

$$\mathbf{\Lambda}_P^{\frac{1}{2}} = \text{diag}\ \left(\sqrt{\lambda_1}, \sqrt{\lambda_2}, ..., \sqrt{\lambda_P} \right) \qquad (\text{B.59})$$

holds.

Notes

- In contrast to the eigenvalue decomposition, which is only feasible for square matrices $(M = N)$, the SVD in (B.45) can be done for any matrix $\mathbf{H} \in \mathbb{C}^{N \times M}$ with arbitrary M and N.
- Exercise: Consider the SVD of a square matrix $\mathbf{H} \in \mathbb{C}^{M \times M}$.
- The matrix $\mathbf{D}$ in (B.46) contains the square matrix $\mathbf{\Lambda}_P^{\frac{1}{2}} \in \mathbb{R}^{P \times P}$ defined in (B.59). As the remaining elements in $\mathbf{D}$ are zero, the SVD can also be formulated with $\mathbf{D} = \mathbf{\Lambda}_P^{\frac{1}{2}}$, a non-square matrices $\mathbf{U} \in \mathbb{C}^{N \times P}$, and $\mathbf{V} \in \mathbb{C}^{M \times P}$.

Proof of SVD Lemma

Proof of (B.45) and (B.46). We prove that with the eigenvalue decomposition (B.53) of $\mathbf{Q}_N$ and with a unitary matrix $\mathbf{V} \in \mathbb{C}^{M \times M}$ the proposition (B.45) with (B.46) follows. First we easily see that $\mathbf{Q}_N$ in (B.47) is a Hermiteian matrix, i.e. $\mathbf{Q}_N = \mathbf{Q}_N^H$ holds. We know from Sects. B.2 and B.3 that for any Hermiteian matrix an eigenvalue decomposition exists according to (B.53). By inserting (B.47) we obtain

$$\mathbf{U}^H \mathbf{H} \mathbf{H}^H \mathbf{U} = \mathbf{\Lambda}_N \qquad (\text{B.60})$$

Now we introduce the following identity matrix $\mathbf{I} = \mathbf{V}\mathbf{V}^H$ making use of the prerequisite (B.51) that $\mathbf{V}$ is a unitary matrix, which yields

$$\mathbf{U}^H \mathbf{H} \mathbf{V} \mathbf{V}^H \mathbf{H}^H \mathbf{U} = \mathbf{U}^H \mathbf{H} \mathbf{V} \left(\mathbf{U}^H \mathbf{H} \mathbf{V} \right)^H = \mathbf{\Lambda}_N \qquad (\text{B.61})$$

Next we decompose the diagonal matrix on the right-hand side into the product of two matrices

$$\Lambda_N = DD^H \tag{B.62}$$

Inserting (B.62) into (B.61) results in

$$U^H HV \left(U^H HV\right)^H = DD^H \tag{B.63}$$

and by comparison of the left- and right-hand part we obtain

$$U^H HV = D \tag{B.64}$$

from which we conclude the proposition $H = UDV^H$ and the proof ends. The only condition we have imposed so far on $V \in \mathbb{C}^{M \times M}$ is the requirement that V is a unitary matrix. Moreover we see that the derived SVD holds for arbitrary matrices $H \in \mathbb{C}^{N \times M}$.

Proof of (B.55)

We now prove that $V \in \mathbb{C}^{M \times M}$ can be obtained by the eigenvalue decomposition of $Q_M = H^H H$. Assume that the singular value decomposition of H is given by (B.45). From this equation follows by applying the Hermiteian operation on both sides

$$H^H = VD^H U^H \tag{B.65}$$

Multiplying (B.65) from the right-hand side with (B.45) and knowing that $U^H U = I_N$ results in

$$H^H H = VD^H U^H UDV^H = VD^H DV^H \tag{B.66}$$

From (B.46) follows

$$D^H D = \Lambda_M \tag{B.67}$$

and we obtain from (B.66)

$$Q_M = H^H H = V \Lambda_M V^H \tag{B.68}$$

From (B.68) follows by multiplication with V^H and V directly the eigenvalue decomposition $V^H Q_M V = \Lambda_M$ of Q_M in (B.55). Consequently, V must be the matrix of eigenvectors associated to the eigenvalues given in Λ_M. As Q_M is a Hermiteian matrix, we know from Sect. B.3 that V is unitary. Furthermore, we see from (B.68) and (B.53) together with (B.52) and (B.56) that Q_M and Q_N have the same positive eigenvalues $\lambda_1, \lambda_2, ..., \lambda_P$ and that their remaining eigenvalues are zero. This finalizes the proof.

B.7 Some Lemmas of Determinants

The proofs can be found in [4].

In the following, we assume "compatible" matrices $\mathbf{A}$, $\mathbf{B}$, $\mathbf{C}$, and $\mathbf{I}$, which means that their dimensions allow multiplication and addition.

- Determinant of the product of two matrices

$$\det(\mathbf{AB}) = \det(\mathbf{A})\det(\mathbf{B}) \tag{B.69}$$

- Determinant of the sum of matrices

$$\det(\mathbf{A} + \mathbf{BC}) = \det(\mathbf{A} + \mathbf{CB}) \; ; \; \text{if } \mathbf{AB} = \mathbf{BA} \tag{B.70}$$

- Determinant of the sum of matrices with cyclic permutation

$$\det(\mathbf{I} + \mathbf{ABC}) = \det(\mathbf{I} + \mathbf{BCA}) = \det(\mathbf{I} + \mathbf{CAB}) \tag{B.71}$$

- Let $\lambda_1, \lambda_2, ..., \lambda_N$ be the eigenvalues of the matrix $\mathbf{A} \in \mathbb{C}^{N \times N}$. Then

$$\det(\mathbf{A}) = \lambda_1 \lambda_2, ..., \lambda_N \tag{B.72}$$

B.8 Trace of a Matrix

Definition of Trace

Given the square matrix

$$\mathbf{A} = (a_{ik}) \in \mathbb{C}^{N \times N} \tag{B.73}$$

The trace of $\mathbf{A}$ is defined as

$$\text{tr}(\mathbf{A}) = a_{11} + a_{22} + ... + a_{NN} \tag{B.74}$$

With a scalar factor α follows

$$\text{tr}(\alpha \mathbf{A}) = \alpha \, \text{tr}(\mathbf{A}) \tag{B.75}$$

Note, for a non-square matrix the trace does not exist, because there is no main diagonal. The proof of the following Lemmas is straightforward.

Cyclic Permutation

Let $\mathbf{A} \in \mathbb{C}^{N \times M}$; $\mathbf{B} \in \mathbb{C}^{M \times N}$ and $\mathbf{C} \in \mathbb{C}^{N \times N}$. Consequently, the product $\mathbf{ABC}$ is an $N \times N$ square matrix. Then

$$\text{tr}\,(\mathbf{ABC}) = \text{tr}\,(\mathbf{BCA}) = \text{tr}\,(\mathbf{CAB}) \neq \text{tr}\,(\mathbf{ACB}) \tag{B.76}$$

In particular

$$\text{tr}\,(\mathbf{AB}) = \text{tr}\,(\mathbf{BA}) \tag{B.77}$$

This also holds for $M = N$.

Trace of the Sum of Matrices

Let $\mathbf{A} \in \mathbb{C}^{N \times N}$ and $\mathbf{B} \in \mathbb{C}^{N \times N}$ be square matrices of the same dimension. Then

$$\text{tr}\,(\mathbf{A} + \mathbf{B}) = \text{tr}\,(\mathbf{A}) + \text{tr}\,(\mathbf{B}) \tag{B.78}$$

With (B.77) follows

$$\text{tr}\,(\mathbf{AB} - \mathbf{BA}) = 0 \tag{B.79}$$

Trace and Eigenvalues

Let $\lambda_1, \lambda_2, ..., \lambda_N$ be the eigenvalues of the matrix $\mathbf{A} \in \mathbb{C}^{N \times N}$. Then

$$\text{tr}\,(\mathbf{A}) = \sum_{i=1}^{N} \lambda_i \tag{B.80}$$

and

$$\text{tr}\,(\mathbf{A}^{-1}) = \sum_{i=1}^{N} \lambda_i^{-1} \tag{B.81}$$

For the latter, the eigenvalues must be unequal to zero.

B.9 Differentiation of a Scalar Function $f(\mathbf{Z})$ with Respect to a Matrix $\mathbf{Z}$

For the proof see [5].

Definition

Differentiation of a scalar function with respect to a matrix $\mathbf{Z}$ is a short-hand notation meaning that the scalar function is partially differentiated with respect to all matrix elements of $\mathbf{Z}$ and arranged in a matrix. Example:

$$\mathbf{Z} = \begin{pmatrix} z_{11} & z_{12} & z_{13} \\ z_{21} & z_{22} & z_{23} \end{pmatrix} \; ; \; f(\mathbf{Z}) = f(z_{11}, z_{12}, z_{13}, z_{21}, z_{22}, z_{23}) \; ; \; z_{ik} \in \mathbb{C} \tag{B.82}$$

Obviously, $f(\mathbf{Z})$ is a multi-variate scalar function of $z_{11}, ..., z_{23}$. Then we define

$$\frac{\partial f}{\partial \mathbf{Z}^*} = \begin{pmatrix} \frac{\partial f}{\partial z_{11}^*} & \frac{\partial f}{\partial z_{12}^*} & \frac{\partial f}{\partial z_{13}^*} \\ \frac{\partial f}{\partial z_{21}^*} & \frac{\partial f}{\partial z_{22}^*} & \frac{\partial f}{\partial z_{23}^*} \end{pmatrix} \tag{B.83}$$

For complex variables z_{ik} we define

$$\frac{\partial f}{\partial z_{ik}^*} = \frac{1}{2}\left(\frac{\partial f}{\partial \mathrm{Re}\,(z_{ik})} + j\frac{\partial f}{\partial \mathrm{Im}\,(z_{ik})}\right) \tag{B.84}$$

Differentiation of the Trace of a Matrix with Respect to a Matrix

We start with the differentiation of a constant α

$$\frac{\partial \alpha}{\partial \mathbf{Z}^*} = 0 \ ; \ \mathbf{Z} \in \mathbb{C}^{N \times M} \ ; \ \alpha \in \mathbb{C} \tag{B.85}$$

Please note in the following that the argument of tr (...) has to be a square matrix.

$$\frac{\partial \mathrm{tr}\,(\mathbf{Z}^H)}{\partial \mathbf{Z}^*} = \mathbf{I}_N \ ; \ \mathbf{Z} \in C^{N \times N} \tag{B.86}$$

$$\frac{\partial \mathrm{tr}\,(\mathbf{Z})}{\partial \mathbf{Z}^*} = 0 \ ; \ \mathbf{Z} \in \mathbb{C}^{N \times N} \tag{B.87}$$

$$\frac{\partial \mathrm{tr}\,(\mathbf{AZ}^H)}{\partial \mathbf{Z}^*} = \mathbf{A} \ ; \ \mathbf{A} \in \mathbb{C}^{N \times M} \ ; \ \mathbf{Z} \in \mathbb{C}^{N \times M} \ ; \ \mathbf{AZ}^H \in \mathbb{C}^{N \times N} \tag{B.88}$$

$$\frac{\partial \mathrm{tr}\,(\mathbf{AZ})}{\partial \mathbf{Z}^*} = 0 \ ; \ \mathbf{A} \in \mathbb{C}^{M \times N} \ ; \ \mathbf{Z} \in \mathbb{C}^{N \times M} \ ; \ \mathbf{AZ} \in \mathbb{C}^{M \times M} \tag{B.89}$$

$$\frac{\partial \mathrm{tr}\,(\mathbf{ZZ}^H)}{\partial \mathbf{Z}^*} = \mathbf{Z} \ ; \ \mathbf{Z} \in \mathbb{C}^{N \times M} \ ; \ \mathbf{ZZ}^H \in \mathbb{C}^{N \times N} \tag{B.90}$$

$$\frac{\partial \mathrm{tr}\,(\mathbf{AZZ}^H)}{\partial \mathbf{Z}^*} = \mathbf{AZ} \ ; \ \mathbf{A} \in \mathbb{C}^{N \times N} \ ; \ \mathbf{Z} \in \mathbb{C}^{N \times M} \ ; \ \mathbf{AZZ}^H \in \mathbb{C}^{N \times N} \tag{B.91}$$

With cyclic permutation Lemma (B.76) we obtain from (B.91)

$$\frac{\partial \mathrm{tr}\,(\mathbf{ZZ}^H\mathbf{A})}{\partial \mathbf{Z}^*} = \frac{\partial \mathrm{tr}\,(\mathbf{AZZ}^H)}{\partial \mathbf{Z}^*} = \mathbf{AZ} \ ; \ \mathbf{A} \in \mathbb{C}^{N \times N} \ ; \ \mathbf{Z} \in \mathbb{C}^{N \times M} \ ; \ \mathbf{ZZ}^H\mathbf{A} \in \mathbb{C}^{N \times N}$$
$$\tag{B.92}$$

$$\frac{\partial \mathrm{tr}\,(\mathbf{Z}\mathbf{A}\mathbf{Z}^H\mathbf{B})}{\partial \mathbf{Z}^*} = \mathbf{B}\mathbf{Z}\mathbf{A}\ ;\ \ \mathbf{A} \in \mathbb{C}^{M \times M}\ ;\ \ \mathbf{Z} \in \mathbb{C}^{N \times M}\ ;\ \ \mathbf{B} \in \mathbb{C}^{N \times N}\ ;\ \ \mathbf{Z}\mathbf{A}\mathbf{Z}^H\mathbf{B} \in \mathbb{C}^{N \times N}$$

$$(B.93)$$

References

1. Papoulis, A., Pillai, S.U.: Probability, Random Variables, and Stochastic Processes, 4th edn. McGraw-Hill, Boston (2002)
2. Papoulis, A.: Probability, Random Variables, and Stochastic Processes. McGraw-Hill, Boston (1965)
3. Proakis, J.G., Salehi, M.: Digital Communications. McGraw-Hill, Boston (2007)
4. Horn, R.A., Johnson, C.R.: Matrix Analysis. Cambridge University Press, Cambridge (2013)
5. Hjorungnes, A.: Complex-Valued Matrix Derivatives with Applications in Signal Processing and Communications. Cambridge University Press, Cambridge (2011)

Index

© Springer Nature Switzerland AG 2021
J. Speidel, *Introduction to Digital Communications*,
Signals and Communication Technology,
https://doi.org/10.1007/978-3-030-67357-4

Printed in the United States
by Baker & Taylor Publisher Services